AF522270

Biotechnology in India

Biotechnology in India

Initiatives and Accomplishments

Niladri Bag
Chief Editor

R. Murugan
Editor

Arundhati Bag
Editor

Sikkim University (A Central University)
Gangtok, Sikkim

NEW INDIA PUBLISHING AGENCY
New Delhi – 110 034

NEW INDIA PUBLISHING AGENCY
101, Vikas Surya Plaza, CU Block, LSC Market
Pitam Pura, New Delhi 110 034, India
Phone: + 91 (11)27 34 17 17 Fax: + 91(11) 27 34 16 16
Email: info@nipabooks.com
Web: www.nipabooks.com

Feedback at feedbacks@nipabooks.com

ISBN : 978-93-85516-25-2

Composed, Designed and Printed in India.

Preface

Biotechnology is a key technology for the new millennium. It has wide range of applications in plant science, agriculture, medicine and medical science, food processing, environmental sciences, mining, nanotechnology and nanoelectronics etc. The potential for altering the genetic structure and characteristics of living organisms, including humans, plants and animals, has opened a huge opportunity for development and also raised several concerns about safety and ethical implications of the new technologies. Industrial biotechnology involves the use of enzymes and microorganisms to produce value-added chemicals from renewable sources and it is being used in big ways by chemical, pharmaceutical, food, and agricultural industries. The areas which may be greatly impacted by such applications/ innovations and research on biotechnology are very wide ranging and may include diverse fields such as agriculture, floriculture and horticulture, medicinal & aromatic plants and forestry sector, soil fertility management, microbial products, traditional fermented foods and beverages, vaccine production, animal health including reproduction & nutrition, aquaculture, human health and as well as gene pool conservation and utilization, etc.

Plant cell and tissue culture has been an important tool for several decades for improving the understanding of cell biology, particularly in the areas of plant physiology and biochemistry. Micropropagation or plant tissue culture is one of the methods of clonal plant propagation. It is a specialized "high-tech" technique of producing a large number of genetically identical (true to type to mother plant or clonal) plants by culturing plant cells, tissues or organs in an artificial nutrient medium under germ-free or aseptic conditions. Maintenance of aseptic conditions in the laboratory, establishment of aseptic culture, standardization or development of the culture medium, culture conditions – *i.e.*, temperature, light intensity in the culture room, photoperiod *etc.* are most important considerations for success of micropropagation. Now a day, tissue culture is one of the vital components of plant biotechnology and permits the regeneration of plants as clones and transgenics. It has gained popularity not only as a method for 'micropropagation', but also for 'conservation' of rare or endangered species that are on the verge of extinction.

As civilisation progressed the demand for energy is increasing exponentially all over the world, specifically the demand for petroleum derived fuels. Increasing demand of petroleum based fuels has been serious concerns and is one of the major factors leading to climate change and global warming. Biofuel, an organic, non-toxic and biodegradable fuel made from renewable resources, cuts down CO_2 emissions, and is going to be one of the effective alternatives towards energy sufficiency in the coming days.

In this backdrop this is an attempt to bring together work of some young talented researchers in a volume entitled "Biotechnology in India: Initiatives and Accomplishments". The book provides information in its various chapters on some of the latest aspects of micropropagation, microbiology, biofertilizer, nanotechnology, RNA interference, metabolomics, biosurfactants, secondary metabolites, cancer therapy, biofuel and biostatistics etc. All the chapters have been well compiled with great dedication and sincerity and authors own the authenticity of information available in the book. We hope the book will be useful to large sections of people of diverse occupations including students, researchers, teachers, scientists, planners and policymakers, general readers and intellectuals etc. Authors have expressed their experiences in the respective chapters and have taken the liberty to access to the research findings and other required information from national and international literatures, which have been duly acknowledged and referred.

Gangtok, Sikkim
May 2015

Niladri Bag, *Chief Editor*
R. Murugan, *Editor*
Arundhati Bag, *Editor*

List of Contributors

Anirudha Chattopadhyay	Department of Plant Pathology, C.P. College of Agriculture, S. D. Agricultural University, S. K. Nagar, Gujarat
A.K. Kathireshan	Department of Microbiology, School of Life Sciences, VELS University, Velan Nagar, Pallavaram Chennai-600117, Tamilnadu, India
Akhilesh Kumar Singh	Department of Biotechnology, Amity University, Lucknow - 226028, Uttar Pradesh, India
Allen John Henry	Department of Microbiology, School of Life Sciences, VELS University, Velan Nagar, Pallavaram Chennai-600117, Tamilnadu, India
Arundhati Bag	Department of Medical Biotechnology, Sikkim Manipal University, 5th Mile, Tadong, Gangtok – 737102, India
Aseesh Pandey	G.B. Pant Institute of Himalayan Environment & Development, Kosi - Katarmal, Almora - 263 643, Uttarakhand, India
B. Abitha	School of Biosciences and Technology, VIT University, Vellore-632014, Tamilnadu, India
B. Agieshkumar	Central Inter-Disciplinary Research Facility (CIDRF), A Unit of Sri Balaji Educational and Charitable Public Trust, Puducherry -607403, India

D. Shalini	Department of Microbiology, School of Life Sciences, VELS University, Velan Nagar, Pallavaram, Chennai-600117, Tamilnadu, India
Dharmendra Pratap	Department of Horticulture, Sikkim University, 6th Mile, Gangtok -737102, India
Diwakar Aggarwal	Department of Biotechnology, Multani Mal Modi College Patiala-147001, Punjab, India
Ishfaq Nabi Najar	Department of Microbiology, School of Life Sciences, Sikkim University, Gangtok 737 102, India
Jagjit Kaur	Department of Biotechnology, Multani Mal Modi College, Patiala-147001, Punjab, India
Joe M. Melvin	Department of Environmental and Biological Chemistry, Chungbuk National University, Cheongju, Chungbuk, 361-763, Republic of Korea
K.P. Arun	Centre for Advanced Research in Indian System, of Medicine (CARISM), SASTRA University, Thanjavur Tamilnadu – 613 401, India
K.V. Prabhanjana	School of Chemical and Biotechnology, SASTRA University, Thanjavur – 613 401, India
Kulbhushan Chaudhary	Advanced Centre of Plant Virology, IARI, New Delhi – 110 012, India
Kuldeep Kumar	Department of Biotechnology, Multani Mal Modi College Patiala-147001, Punjab, India
Laxuman Sharma	Department of Horticulture, Sikkim University, 6th Mile, Gangtok - 737102, India

M. Gomathi	Department of Microbiology, School of Life Sciences, VELS University, Velan Nagar, Pallavaram Chennai-600117, Tamilnadu, India
M. Gopalakrishnan	Department of Botany, Pachaiyappa's College, Chennai – 600 030, India
M. Ayyanar	Department of Botany & Microbiology, A.V.V.M., Sri Pushpam College, Poondi – 613 503, Thanjavur, India
Mingma Thundu Sherpa	Department of Microbiology, School of Life Sciences, Sikkim University, Gangtok 737 102, India
Mithilesh Singh	G.B. Pant Institute of Himalayan Environment and Development, Sikkim Unit, Pangthang, Gangtok, Sikkim - 737101, India
Nagendra Thakur	Department of Microbiology, School of Life Sciences, Sikkim University, Gangtok 737 102, India
Niladri Bag	Department of Horticulture, 6th Mile, Sikkim University Gangtok -737102, India
P.S. Manoharan	Indira Gandhi Institute of Dental Sciences, Puducherry - 605 006, India
Pankaj Kumar	Department of Microbiology, Dolphin (PG), College of Science & Agriculture, Chunnikalan, Fatehgarh Sahib-140307, Punjab, India
R. C. Dubey	Department of Botany and Microbiology, Gurukul Kangri Vishwavidyalaya, Haridwar - 249404, Uttarakhand, India

R. Parthasarathy	Department of Microbiology, Faculty of Agriculture, Annamalai University, Annamalainagar-608002, Tamilnadu, India
Tongmin Sa	Department of Environmental and Biological Chemistry, Chungbuk National University, Cheongju, Chungbuk, 361-763, Republic of Korea
Sayak Das	Department of Microbiology, School of Life Sciences, Sikkim University, Gangtok 737 102, India
Sekar K. Chandra	G.B. Pant Institute of Himalayan Environment & Development, Kosi - Katarmal, Almora - 263 643 Uttarakhand, India
Shailendra Pradhan	Department of Dravyaguna, Uttarakhand Ayurved University, Gurukul Campus, Haridwar, Uttarakhand 249404, India
Soundravally R.	Department of Biochemistry, JIPMER, Puducherry 605 006, India
T. Sekar	Department of Botany, Pachaiyappa's College, Chennai 600 030, India

Contents

1

Metabolomics – A New Approach in Plant Biotechnology

Arun KP
Centre for Advanced Research in Indian System of Medicine (CARISM)
SASTRA University, Thanjavur, Tamilnadu – 613 401, India
Email: arun@carism.sastra.edu

Abstract

Dietary phytochemicals and herbal medicines are under intensive investigation for their potential as chemopreventive agents against various diseases due to the wide variety of microbial and plant functionally relevant secondary metabolites. The chemical diversity of phytochemicals, as well as phytochemicals and complex metabolic interactions between the biological systems may exceed the capacity of the platforms of traditional analysis, and therefore pose major challenges in the study of chemopreventive phytochemical compounds. In addition, the development of powerful genomics-based analysis tools, proteomics, metabolomics, bioinformatics and other 21st century technologies are greatly accelerates the identification and characterization of these natural products. Here we discuss reciprocal and synergistic benefits of linking these "omics technologies of ethnobotanical studies and robust ethnomedical traditional medicines provide medicines and critical treatments that are inexpensive, accessible, safe and reliable enhanced needed. A review of the current literature reveals a wide range of workflows that are currently used in metabolomics, plant and new methods for the analysis and presentation of data mass spectrometry (MS) in particular, liquid chromatography / mass spectrometry (LC-MS/MS) can determine the presence, quantity, and sometimes plant metabolites structures in complex herbal mixtures, with significant advantages compared to techniques such as nuclear magnetic resonance (NMR) spectroscopy and gas chromatography / mass spectrometry (GC-MS). Here, we describe the factors that must be in place to optimize the use of metabolomics in predictive biology. A key to achieving this vision is a time resolved collection of accurate data and abundance of spatially defined metabolites and associated metadata.

Introduction

Despite the great diversity derived synthetic development of combinatorial chemistry methods and high throughput screening in the last fifty years, natural products and related structures remain important elements of the pharmacopoeia. Looking ahead, natural and related structures products may become even more important for the development of new and better drugs, due to the variety of functionally relevant secondary metabolites of microbial and plant whose chemical species and genetic diversity are being revealed by sequencing the ultra fast and related genomics and bioinformatics tools DNA. Metabolomics, the analysis of a broad range of small molecule compounds in a biological system, has been used to provide a systems wide overview of plant metabolism associated with defense responses. Metabolomics provides a comprehensive assessment of traditional Chinese medicine or complex remedies and classification of samples of different biological states, origins or qualities in the samples, using chemometrics. Identification, treatment and the pharmaceutical preparation are the main large scale production procedures of traditional medicine preparations. Through comprehensive analysis, metabolomics plants address some of the shortcomings of the individual analysis and have great potential to become a sharp tool for assessing the quality of traditional medicine. The discovery of the molecular components and their functionality is the key to development of hypotheses about the organization and the regulation of metabolic networks. The iterative experimental test of these hypotheses is the path that eventually may allow accurate computational modeling and prediction of metabolic outcomes. This information can be particularly important for understanding the biology of natural products, whose metabolism is often poorly defined.

Metabolomics is an emerging issue that, along with genomics, transcriptomics, proteomics together constitute the "systems biology" (German *et al.* 2005). It is the branch of science that studies the quantitative understanding of the components of metabolites of integrated living systems and their dynamics to changes in both endogenous (such as those related to the physiology and development) and exogenous factors (such as environmental factors replies and xenobiotics). The emergence of new "omics" technologies, such as genomics, proteomics and metabolomics, is allowing researchers to identify the mechanism of action of traditional medicinal plants against various pathogenic diseases. In recent years, a wide range of metabolomic analytical techniques have been applied in research over traditional medicines whose qualitative analysis is difficult due to the complexity and diversity of its components. In general, one or two biomarkers are used for identification and authentication of herbal products. Plant Metabolomics is a relatively new technique (Fiehn *et al.* 2000), but has not been widely used because of the great challenges in detecting and analyzing chemical data. For example, while the chemical structure of genes, transcripts and proteins is well defined, there is wide variation in the structure and abundance of small molecules in plants (Dixon 2001). This diversity is an important global analytical challenge in metabolomics. However, the application of modern technologies negligence may challenge traditional knowledge and biodiversity that are the basis of traditional medicine to meet these challenges, while meeting the need to improve drugs encourage the development of regional centers omics technologies functionally related to regional centers of genetic resources, particularly in areas where the use of traditional medicines is common and essential to health. The curative effects of traditional medicine is based primarily on the synergistic effect the preparation of various ingredients multitarget, unlike modern pharmacology and drug development often focus on a single chemical entity. Therefore, the method using pharmacologically active constituents or markers to assess the quality and authenticity of the complex preparations have a number of serious challenges. Metabolomics can provide an effective platform for the analysis of complex samples. It is also reported that applies to the analysis of the quality of traditional medicines. Present review discusses the use of metabolomics in natural product research.

Metabolomics

Metabolomics is complete, unbiased, high throughput analysis of complex mixtures of metabolites ideally permits identification and quantification of each individual metabolite. Metabolomics has emerged as a functional genomics methodology that contributes to our understanding of the complex molecular interactions in biological systems (Nicholson and Wilson 2003). As such, metabolomics is the logical progression of large-scale analysis systems (Weckwerth 2003; Bino *et al.* 2004), but unlike the transcripts and proteins, metabolites diverse molecular identity is based on sophisticated instruments such as mass spectrometry (MS) and nuclear magnetic resonance (NMR) (Lei *et al.* 2011). The metabolites are the end products of the processes of cellular regulation, and their levels may be regarded as the maximum response of biological systems to genetic or environmental changes (Fiehn 2002). The metabolites are also dynamic in time and space and exhibit a wide range of structures that increase challenges in their measurement analytical procedures (Stitt and Fernie 2003). Metabolomics enables an overall assessment of a cell state within the context of the immediate environment, given the expression of genes, gene regulation, and the modified regulation in kinetic activity of enzymes, and changes in metabolic reactions (Griffin and Shockcor 2004; Mendes *et al.* 1992; Mendes *et al.* 1996). Therefore, metabolomics in comparison with genomic or proteomic reflect changes in the phenotype and function of a particular tissue, or organism. Therefore, the omics sciences are used in a complementary manner to carry out studies to characterize the phenotype. The variation in the expression of genes and proteins likely to produce changes in the metabolic profile of a cell, tissue or organism (Spratlin *et al.* 2009). In addition, determinations of metabolomics offer the same benefits as proteomics and transcriptomics techniques as all have the ability to analyze biological fluids and are relatively inexpensive, fast and automated, but the costs of such techniques are to be considered. Metabolomics has been first applied to the study of toxicology and pharmacology, inborn errors of metabolism and nutrition (Spratlin *et al.* 2009). There is currently a great interest in the application of metabolomics to characterize various pathological conditions of human diseases such as cancer, diabetes, autoimmune diseases, and coronary heart disease. In addition, metabolomics may provide valuable tools in a wide range of applications including microbial biotechnology, food technology, pharmacology, toxicology, the discovery of the enzyme, systems biology and plant biotechnology. Current research in plants makes extensive use because of the advantage of metabolomics product on human health of products of plant origin, including food, pharmaceuticals and industrial raw materials, and their use in plant breeding and evaluation of the nutrition. Furthermore, the great chemical diversity of the plants with respect to animals and microorganisms also contributes to the importance of plant metabolomics research (Akiyama *et al.* 2008). Recent technological advances in spectroscopic methods NMR and MS accepted two most used in measuring metabolites, improved sensitivity and spectral resolution of the analyzes carried out on samples of metabolomics in attempts to achieve full biochemical measure. Although all the strategies used in metabolomics are universally accepted, including the metabolite target analysis, metabolite profiling, metabolomics analysis metabolite flow and metabolic impressions.

Safety and effective regulation of herbal products need methods that can provide convincing evidence of clinical efficacy (in some cases by clinical trials) and tools powerful analytics that can determine the composition of herbal products and identify the active ingredients. Metabolomics appears ideal for this purpose. According to the home page Metabolomics Society, metabolomics can be defined as the "complete characterization of small molecule metabolites in a biological system" (http://www.metabolomics society.org/). In turn, these small metabolites represent the result of gene expression and define the biochemical phenotype of a cell, tissue, organ and body. The small metabolites include intermediate and end products of metabolism which comprise primary metabolites (e.g., sugars, amino acids, fatty acids and organic acids) and secondary metabolites (e. g., alkaloids) and phenylpropanoids. Metabolomics can also characterize the metabolome dynamic, reflecting changes in the abundance of small molecules during development and in response to external constraints.

There is a wide range of technologies that can be used to analyze metabolites in solvent plant extracts. The choice of the analysis will depend on the availability of equipment, but more importantly, it will depend on the question or hypothesis tested. The two main approaches are generally divided between metabolite fingerprinting and profiling metabolite. These approaches are distinct: fingerprints metabolite is a semi-quantitative technique with identification of putative metabolites in large part while the chemical metabolite profiling involves the identification and accurate quantification (Dunn and Ellis 2005). The approach fingerprint metabolite was a powerful technique to identify general trends in the metabolic response to stress or population studies (Field and Lake 2011). These fingerprinting techniques are primarily to analyze metabolites in a complex mixture by direct injection of the raw extract in a mass spectrometer a technique known as the direct injection mass spectrometry; Dunn and Ellis 2005 or by obtaining the nuclear magnetic resonance (NMR) spectra of the entire mixture (Forseth and Schroeder 2011). The advantage of NMR fingerprints on MS is that there is less bias in the number and classes of metabolites can be detected in the NMR systems. This is because detection does not depend on the ability to ionize metabolites, and structural information can be obtained for the unknown metabolites (Forseth and Schroeder 2011). Fingerprints resulting metabolites must always be checked by quantitative metabolic profiling techniques (Devaux *et al.* 1971). These profiling techniques separate the metabolites prior identification either by GC, GC × GC techniques where several GC columns together to improve the separation resolution and sample data acquisition time or LC. Dunn *et al.* (2011) describes the main tests for the sample preparation, derivation, and analysis by HPLC and GC-MS. Although Dunn *et al.* (2011) emphasize the human samples; these techniques can be applied directly to samples of plants. Whatever separation approach is used, the metabolites in the sample will be identified either by their molecular weight, fragmentation pattern using MS or NMR chemical shifts (Forseth and Schroeder 2011) provide advice on how these identification techniques can be used to assign function to the metabolites in the context of environmental studies.

Application of metabolomics in plant biotechnology

During 1970, GC-MS were used to analyze steroids, acids and drug metabolites in urine neutral and acidic (Devaux *et al.* 1971; Horning and Horning 1970; Horning and Horning 1971). Shortly thereafter, the concept of profiles using metabolic screen, diagnose and evaluate health began to spread (Cunnick *et al.* 1972; Mroczek 1972). However, it was not until used in the early 1990s that metabolite profiling to study plant organisms (Roessner *et al.* 2000; Maloney 2004). Plant metabolites are involved in many responses of resistance and stress and also contribute to the color, flavor, aroma and fragrance of fruits and flowers. The challenge of metabolomics is to find changes in the biochemical pathways and metabolic networks that can be correlated with the physiological and development of a cell, tissue or organism phenotype. One of the greatest achievements of plant biology is the realization of the full genome sequences of model plants like *Arabidopsis thaliana* and rice. ~ 27,000 genes in Arabidopsis is predicted based on information of the nucleotide sequence; however, only half of these genes were functionally annotated based on sequence similarity to known genes, including the function and only ~ 11% is confirmed by direct experimental evidence (MASC 2006). Therefore, the elucidation of the function of unknown genes is currently an important plant research challenge. Because there is very little information on the number of genes in a family of genes including a non model plant, it is necessary to know the profile of expression of these genes in different conditions and integration of metabolomics stimuli.

Plants produce over 200,000 metabolites, many of which perform specific functions that can be adapted to specific ecological niches (Oksman-Caldentey *et al.* 2004). Therefore, the main problems encountered when characterizing plant metabolome deal with the fact that compared to the transcriptome or proteome, the metabolome is highly complex in nature, due to the enormous diversity of chemical compounds. There is a wide range of concentrations of metabolites, which may vary over nine orders of magnitude (pM to mM). These large variations in the nature and concentration of the analytes to be studied pose challenges for all analytical technologies used in metabolomics strategies (Dunn and Bailey 2005). There are several examples in which the modification of certain metabolic pathways leading to the production of plants with a higher nutritional value. This is the case of Golden Rice (GR), which is rice that accumulates β-carotene in the GM endosperm (Ye *et al.* 2000). The productions of this variety of rice allowed alleviate vitamin A deficiency, a major nutritional problem worldwide. The nutritional value of GR later improved by the over expression of a gene for *phytoene synthase* that leads to obtaining the variety GR2, where higher amounts of carotenoids (84% of total is β-carotene) is accumulated (Paine *et al.* 2005). Mehta *et al.* (2002) were able to express *S-adenosylmethionine decarboxylase* gene under inducible tomato E8 promoter. The transgenic variety shows the highest levels of different polyamines during fruit ripening, including spermidine and spermine, leading to an increase in the metabolite lycopene, which prolongs the life of the vine and increased fruit juice and quality nutrients (Mehta *et al.* 2002). Other examples include plant engineering to improve the content of anthocyanin. Anthocyanins are flavonoid, a class of pigments that contribute to antioxidant qualities and color of plants. Moreover, these metabolites have been associated with protection against several human diseases.

Metabolomics in natural product research

Plants have provided humans with drugs since the dawn of civilization, and up to 80% of the population in many countries in Asia and Africa still depend on traditional herbal medicine for primary health care (WHO 2008). In Western countries, herbal remedies were supplanted by modern drugs with defined pharmaceutical ingredients, but their popularity is increasing again, especially for the relief of symptoms of mild disease. Approximately one third of the adult population in the UK now use herbal products (Gilbert 2011) and the market for herbal medicines in the US was $ US 5 billion in

2010 (Ouedraogo *et al.* 2012). Herbal medicines pose a number of security issues, such as the identification of plant materials in herbal products, active ingredients, method of preparation, dosing regimens, the potential to interact with other herbal remedies and conventional drugs, and assurance that herbal products are genuine and contain no toxins or contaminants. The lack of clear guidelines for safety of herbal medicinal plants has resulted in many reports of toxic effects in many herbal drugs. In a well known case of liver toxicity was probably caused by high levels of epigallocatechin gallate (EGCG) (Licata *et al.* 2012). Although low doses of EGCG appear to be safe and even beneficial, it is impossible to control the dosage of herbal products as effectively as formulated pharmaceuticals (Yang *et al.* 2007). Recent advances in genomics, informatics and related technologies omic 21st century are dramatically accelerating the pace of discovery and analysis. Important synergies and benefits for the development of better drugs and new drug can come from linking these powerful scientific tools for robust traditional medicines. However, ethical and socio-economic challenges also need to be addressed for developing better drugs and new drugs to be achieved, while also benefit those who are currently based on traditional medicines and accompanying natural products for health and wellness.

The inclusion of traditional medicines in developing models of 21st century treatments can help ensure your convenience, acceptability and accessibility. The use of potent omics technologies' makes unravel this complexity: metabolomics analyzes allow profiles of major and minor metabolites and bioactive components that contribute to the synergy; and computational methods to analyze multiple activities networks have become powerful tools for the definition of the principal components of mixtures with synergistic action modes for the prediction of drug metabolism and toxicity, high performance and prioritizing combinations of agents. Data mining approaches to identify active compounds in mixtures of natural products are being developed and will be essential for the development of multiple drugs effective agent of traditional medicine.

The regulation of herbal products marketing has been attempted through initiatives such as the Traditional Herbal Medicinal Products Directive (THMPD), which entered into force across the European Union (EU) in April 2011. Under this directive, a vegetable product is eligible for registration as a traditional herbal medicine only if it has been used as treatment for minor ailments specific for at least 30 years, including a minimum of 15 years in Europe. The efficacy and safety of this type of plant products only if a long history of use is demonstrated is permitted. However, clinical trials are required which provides traditional herbal medicines, apart from products based on herbs and defined newest pharmaceutical ingredients. In the US, herbal products are regulated as dietary supplements, and no claims that prevent diseases are allowed, so that the FDA registration and chemical analysis is not required. However, if specific health claims are made, the products must be registered as drugs and must undergo rigorous FDA approval.

The basic instrumentation useful for evaluating plant metabolome includes direct spectroscopy, chromatography (fingerprints) or high performance chromatography (GC or HPLC) combined with spectroscopy (IR, NIR, MS). Coupling MS chromatographic methods can substantially increase the depth coverage of metabolome add an extra dimension for metabolite identification and improve biological context through the more rigorous identification of a greater number of metabolites. GC-MS is one of the most widely used analytical metabolomics techniques in plant research. It is used to qualitatively and quantitatively analyze a wide range of non-volatile and / or derivatized with high thermal stability volatile metabolites. After separation, the eluted metabolites are identified by mass spectrophotometer. This analytical technique has high reproducibility and lower costs compared to other techniques scripts, such as LC-MS or LC-NMR (Bedair and Sumner 2008). In recent years, the use of LC-MS in plant metabolomics has been increasing. LC-MS is more suitable than GC-MS for labile compounds as well as those that are difficult to derivatize. Another technique used for quantitative analysis of metabolites of plants is the MS coupled to capillary electrophoresis (CE-MS) (Soga 2007). CE-MS was used to characterize the profile of amino acids in plant cell cultures (Williams *et al.* 2007), to characterize the metabolome of leaves orange-bacterial infection (Cevallos-Cevallos *et al.* 2009), and identify the metabolic profile in Illicium anisatum (Urakami *et al.* 2010). Fourier transform Infrared spectroscopy (FT-IR), near infrared resonance spectroscopy (NIR) and more recently, nuclear magnetic resonance (NMR) are in constant development of rapid, non-destructive techniques for a wide high throughput range of sample types. FT-IR has also been introduced as a technique of metabolic fingerprints within the plant sciences (Baranska *et al.* 2004, Kaderbhai *et al.* 2003). Raman spectroscopy with microscopy has recently been used with great success, enabling the identification and quantification of phytochemicals and their distribution directly from plant tissues (Baranska *et al.* 2005; Schulz *et al.* 2005; Schulz *et al.* 2003). NMR techniques can uniquely identify and quantify a wide range of organic compounds in the micromolar range (Kim *et al.* 2011; Schripsema 2010) simultaneously. NMR-based methods can be broadly classified into the solution and insoluble solid state NMR In one dimensional NMR, proton (1H) are usually observed (1H NMR) due to the sensitivity and the common occurrence of this magnetic core. More detailed analyzes, and identification of metabolites or test flow, can be obtained with other nuclei, particularly with 13C and 15N which engage 1H nuclei in two-dimensional or multidimensional analysis NMR.

Recently, NMR has been used to identify secondary metabolites involved in the host plant resistance (49). NMR also has other applications in investigating the quality of food and the standardization and control of different phytomedical preparations (Ward *et al.* 2007). In a workflow metabolomics type metabolites were determined by extraction of plant tissue of compounds in a solvent, and then the detection and quantification of the compounds using various chemical detection platforms (for example, MS, magnetic resonance nuclear, ultraviolet absorbance). MS metabolite detection is a powerful tool for metabolic investigations of the plant due to its sensitivity for low abundance molecules and flexibility for the detection of molecular several chemical classes. Therefore, MS is well suited to study signaling, physiological and other chemical events associated with plant defense. However, while many compounds related to plant defense are well defined, there is currently no extraction protocol metabolite or MS detection platform that can probe the entire defense metabolome in a single experiment. Metabolites extraction procedures are inherently biased towards solubility of the molecule in the choice of solvent, in particular differences in polarity or pH. In addition, the MS detection platforms are biased in their compatibility of a particular molecule for a user or ionization detection. For example, electro spray ionization (ESI) MS may be applied differently positive or negative charges to molecules, and some of the compounds vary in their propensity to form positively charged ions and negatively charged. The ability to generally very complex mixtures profile plant extracts is enhanced by coupling chromatography with MS detection. Thus, a "platform metabolomics" refers to the combination of chromatography and MS. Both platforms metabolomics most commonly used include mass spectrometry, liquid chromatography (LC-MS) and gas chromatography mass spectrometry (GC-MS) (Kopka *et al.* 2004). Following the acquisition and processing of data, MS metabolomics data is often expressed as a matrix of molecular characteristics defined by (i) time elution chromatography, (ii) mass (mass / charge ratio) in mass error of the instrument, and (iii) the intensity of a quantitative mass unit signal. While platforms GC-MS and LC offer superior sensitivity in the detection of a variety of compounds, annotate molecular function detected as a metabolite is the major bottleneck in MS metabolomics workflow (Wishart *et al.* 2011), and is even more complicated due to the great diversity of plant compounds.

LC-MS untargeted metabolomics allows complete analysis of complex plant tissues. The ability to analyze different types of metabolites by LC-MS strongly depends on the ionization source and the chromatographic method used for the elution and fractionation analytes. Untargeted LC-MS-based metabolomics combines high sensitivity and non-targeted approach to provide an ideal procedure for the analysis of a wide range of non-volatile metabolites. Several recent studies have examined the data processing challenge in untargeted metabolomics, which has benefited enormously from the development of automated procedures (Eliasson *et al.* 2012).

Recent IT tools such as MetAlign (http://www.metalign.wur.nl) MZmine (http://mzmine.sourceforge.net/) and XCMS (http://metlin.scripps.edu/download/) (Theodoridis *et al.* 2012) are designed to automatically extract relevant information from raw chromatographic data sets, enabling rapid processing of thousands of data points, which turns the concept of non-targeted metabolomics in practical reality. The main challenges to overcome these tools should include background subtraction, signal recognition, signal quantization and alignment of the signal, so that the output of an LC-MS chromatograms lot is usually a table listing the amount of each signal in all samples.

Both targeted and untargeted metabolomics reveal the expected behavior of known metabolites but only untargeted metabolomics allows detection of synergistic effects between the variables that cannot be observed individually. Untargeted Metabolomics is usually a hypothesis-free approach. The most widely used multivariate techniques are based on the projection. Projection methods combine the variables measured in the latent variables calls that can solve the problem under investigation. In many cases, the latent variables produced by projection can be interpreted in terms of variables individual measures, and simple models can be constructed using a subset of the original data. In particular, when considering individual classification problems putative markers can often be highlighted by exploring the latent structure of the projective model.

Several databases of LC-MS metabolite features have been met, as MASSBANK (http://www.massbank.jp/index.html) (Horai *et al.* 2010), METLIN (http://metlin.scripps.edu/) (Smith *et al.* 2005) and MS2T (http://prime.psc.riken.jp/lcms/ms2tview/ms2tview.html) (Matsuda *et al.* 2009), but because of the great diversity of plant metabolites the database remains inadequate. However, the determination patterns and fragmentation exact mass obtained by mass spectrometry in tandem (MS / MS), especially when combined with the techniques of soft ionization as ESI, may allow the structure elucidation of metabolites de novo. For example, high performance liquid chromatography diode array detection (HPLC-DAD), electrospray ionization spectrometry time of flight mass spectrometry (HPLC-ESI-TOF-MS) and electrospray ionization quadrupole ion mass spectrometry trap (HPLC-ESI-MS / MS) were used for the isolation, identification and structural analysis of soluble phenolic in water and constituents diterpenoid nonpolar in roots Dan Shen (Salvia miltiorrhiza) (Zhu *et al.* 2007) and flavonoid compounds in *Oldham Dendrocalamopsis* leaf extracts (Lv *et al.* 2011). In the latter study, the authors were able to determine the structures of 11 compounds, and identify four mono-C-glycosylflavones three O, C-diglycosylflavones and three together glycosylflavones, including three types of aglycone (luteolin, apigenin and Tricin). In another example, ESI-MS / MS was combined with reverse phase HPLC to identify proanthocyanidins in Saskatoon berry (Amelanchier alnifolia), confirming the presence of compounds of multimeric proanthocyanidin ranging from dimers through heptamers and polymers even higher (Hellstöm *et al.* 2007).

LC-MS untargeted metabolomics can unravel environment dependent plant cell composition. Plants are sessile organisms and must adapt to changes in their environment by modulating the responses of development, physiology and biochemistry. The plants benefit from the ability to synthesize a wide and diverse range of secondary metabolites, because such compounds allow them to adapt and survive when exposed to biotic and abiotic stress (Obata and Fernie, 2012; Moura *et al.*, 2010). The metabolic profile of plants therefore be changed according to the environment, which means that the active ingredients in herbal remedies qualitatively and quantitatively vary depending on growth conditions. From a quality control perspective, the active ingredients of medicinal herbs should be standardized, but this can be difficult in plants subjected to variables environments.

Liquid chromatography mass spectrometry in metabolomics

Liquid chromatography mass spectrometry (LC-MS / MS) based metabolomics is a stimulating activity in the characterization of detectable metabolites. The description of the metabolites accumulation models is desirable in many areas of research; *e.g.*, fuel, lead chemicals for the pharmaceutical, safety assessment, and playback. Using LC-MS / MS, a few hundreds of thousands of metabolites can be detected in the body. The plants are the largest producer of phytochemicals- metabolites which play an essential role in the interactions between plants and other organisms. In addition, when ingested, phytochemicals are metabolized in the body of the predator and suffers changes (Yuji and Masami 2013). Recent advances in hardware and software allow a full analysis of biological samples.

In general, isolation and complete identification of a detectable metabolite is the first step in the investigation of metabolic systems. However, classical instrumental analysis (MS, MS / MS, UV and NMR) can be used for very short survey metabolites in model organisms (Yuji and Masami 2013, Lv *et al.* 2011) that have established research infrastructure (e. g., genomics, transcriptomics, proteomics). Metabolomics can be integrated with other omics approaches metabolic pathways (Yuji and Masami 2013; Lv *et al.* 2011), but the links between metabolites and gene functions also have large gaps in data interpretation have led the UN (Eliasson *et al.* 2012, Leiss *et al.* 2011). Indeed, the levels of metabolites in an organism are modified dynamically by the effects of different levels (*e.g.*, genome, transcription, proteins and metabolites).

Detection without MS metabolite specific objective is an innovative methodology that does not depend on the following classical identification process: prepared biological samples, the acquisition of samples, data alignment and generating a data matrix (samples against metabolites), multivariate analysis and characterization of the metabolite structure changed significantly. The description of all candidates detectable metabolites in an organism is likely to include new biological data, and the goal is to clarify the complex metabolic systems structure-activity relationships and metabolite biosynthesis. To identify the structure of the detectable metabolite, methodology without specific target should be integrated with conventional validation.

The validation of the isolated metabolites using multiple instruments is the most reliable evidence, but this process requires a long time is an important factor in the interpretation of metabolomics to UN-led obstacle. LC-MS, combined with solid phase extraction-NMR, will be an effective tool novo structural narrowing down. Using the m / z of accurate high resolution, detectable metabolites values can be annotated as elemental compositions by using a database of metabolites. However, the monoisotopic mass of the metabolite cannot be assigned their elemental composition even in less than 1 mass ppm error FT-MS data. Thus, accurate mass data combined with natural abundance isotopic ion are used for analysis of elemental composition, and stable isotope labeling is also effective for the analysis of elemental composition.

MS / MS data from the literature are a valuable resource that data is not dependent on authentic compounds available, and a few hundred metabolites can be collected from the literature of each year. Using a hybrid / MS MS data resource, our data acquisition authentic compounds and literature data curated manually, we have successfully established a Web database of MS / MS and a new search algorithm. Confidence levels can be managed by MS / MS fragmentation -a association rules algorithm to discover common data fragmentation in MS / MS (Respect: http://spectra.psc.riken.jp/) (Yuji and Masami 2013). ReSpect is the first fully downloadable resource MS / MS data under a Creative Commons Attribution - Noncommercial 2.1 Japan License. However, other public databases are presented as a showcase, which cannot be reused for the development of the new methodology. We expect users of bioinformatics and metabolomics will develop algorithms and new methodologies by using our reusable data resources. More than a million tags MS / MS no specific target (MS2T) can be collected by scanning high-speed LC-Q-TOF-MS, but we have only a few thousand reference data. Given the sensitivity and extent of the detectable metabolites in the advanced MS instruments, the gap between the reference and information detected from spreading further. Due to the large difference between the query and the reference MS / MS, over 90% of detectable metabolites are unknown. Unfortunately, no methodology novo identification of unknown metabolites MS based metabolomics, as described above. Therefore, integrated omics approaches, such as genomics and metabolomics is believed to have great potential to reduce effectively by the structure of unknown candidate (Yuji and Masami 2013) metabolites. As a genetic resource for integrated omics approach, recombinant inbred lines (RILs) were established in model species and no plant model. In this review, we show recent methodologies for quantitative trait locus metabolite (QTL) analysis of non-white linkage mapping.

Liquid chromatography mass spectrometry and untargeted metabolomics

Untargeted metabolomics has recently been used to study the composition of specific herbs, which can be composed of one or more plants or parts of them. LC-DAD / ESI-MS / MS is used to determine the metabolic footprint of many medicinal plants. By comparing the metabolic profiles herbs acquired at different places, one can find important regional differences in the abundance of the major and minor metabolites. It was reported that the different tissues of the same plant (leaves, fruits, stem and root) were qualitatively similar in terms of metabolic profile, but the individual metabolites differ in abundance, suggesting that metabolomic analysis of different tissues plants could help determine who is likely to have more effective drugs. (Zhao *et al.* 2009) Although the untargeted LC-MS analysis was applied to specific herbs, many traditional preparations include many herbs, so the abundance of specific bioactive compounds can vary depending on the mixing ratio and the effect of different environments based on individual plant components (Steinmann and Ganzera 2011). The high selectivity, sensitivity and versatility of the LC-MS analysis, it is ideal for those based on complex herbal medicines. For example, Chinese licorice (GăN co) is one of the oldest and most popular medicinal plant in the world, from the root of Glycyrrhiza root and present in about 60% of all prescriptions of traditional Chinese medicine (Wang and Yang 2007; Zhang and Ye 2009). LC-ESI-MS / MS was used to identify the licorice flavonoids and saponins Si-Jun-Zi decoction, which includes four Chinese herbs, including gân căo. (Liu *et al.* 2006). LC-DAD / ESI-MS was used to analyze the preparation of Chinese medicine Gan-Lu-Yin, revealing 14 key compounds including liquiritigenin, liquiritin and glycyrrhizic acid căo gân. (Liu *et al.* 2006). LC-MS was also used to analyze PHY906, a pharmaceutical preparation modified Huangqin Tang, which is used in traditional Chinese medicine to treat diarrhea, nausea and abdominal cramps (Tilton *et al.* 2010). PHY906 four herbs - Scutellaria Georgi (S), Paeonia lactiflora Pall. (P), Glycyrrhiza uralensis Fisch. (G) and the plant Ziziphus. (Z) - in a ratio of 3: 2: 2: 2. LC-MS identified or characterized provisionally 64 peaks, including flavonoids, triterpene saponins and glycosides monoterpene, all of which may be assigned to four individual herbs, and whose some were reported for the first time (Ye *et al.* 2007). In the future, the untargeted metabolomics could be used not only to characterize single herbs and blends, but also to integrate data from different experimental platforms. The sensitivity and LC-MS versatility led to the development of Phytomics QC concept, which combines chemical analysis, an analysis of bioresponse animal pharmacology and determine the batch to batch reproducibility. This platform integrates molecular fingerprints, quantitative analysis and comparisons of statistical models for the analysis of natural remedies. The Herbalome project (Zhang *et al.* 2012) aims to globalize traditional medicines using LC-MS to determine the composition of different products and the structure and function of their components, and to establish a standard Resource Library to unravel the synergistic effects more of these components on multiple targets.

Conclusion

Recently, natural herbal remedies have become very popular in Western countries, but, unlike the modern drug, its chemical composition is still partially unknown and difficult to determine. Moreover, the chemical constituents of herbal products may vary from batch to batch, as they are greatly affected by the environmental conditions of growth of the original plants and manufacturing practices. In order to ensure the safeness of these products, it is necessary to have a deeper understanding of these natural remedies. Untargeted metabolomics based on LC-MS is particularly suitable for the analysis of natural products. The presence, amount and structure of plant metabolites can be unraveled by using this approach. Metabolomic studies require expertise in many areas such as biochemistry, biology, physiology and bioinformatics. Metabolomics has the potential to make a big impact in different areas of biology, such as human health, plant biotechnology, toxicology and pharmacology, among others. Using the reference data derived from large-scale data acquisition for authentic compounds and reports in the literature, coverage of annotations on data from the UN aimed improve dramatically in the activities of the future metabolomics. Also effective Web-based integration between databases, and can be achieved by reusing resources fully downloadable data or establishing an application programming interface in each database. The analysis and understanding of the data require the use of bioinformatics, as well as integration with data from genomic and proteomic analysis. With the development of more sensitive technologies and computational tools for statistical analysis and interpretation of data, metabolomics has the potential to help us better understand the molecular mechanisms of disease. Moreover, the identification and characterization of novel biomarkers allow the diagnosis and prevention of many diseases and the discovery of new drugs. However, in most disease analysis also other biomarkers, protein markers as well as evaluation of physiological state for the diagnosis and treatment are required. In the area of plant biotechnology, metabolomics, along with studies of genomics and proteomics, has identified new genes or genes with new functions. In this area, metabolomics has gained importance in the assessment of transgenic plants, food quality, resistance to disease, and herbicides or increase tolerance to salinity. In addition, metabolomics applied to the study of plants becomes extremely important in efforts being made to improve human health. As there are several examples in which the manipulation of multiple pathways lead to an increase in the production of a particular metabolite, modification of metabolic pathways of plants could lead to the production of new drugs that could be used for treatment of many diseases.

References

Akiyama K., Chikayama E., Yuasa H. *et al.*, (2008) "PRIMe: a web site that assembles tools for metabolomics and transcriptomics," In Silico Biology. 8(3-4): 339–345.

Baranska M., Schulz H., R¨osch P., Strehle M. A. and Popp J. (2004) "Identification of secondary metabolites inmedicinal and spice plants by NIR-FT-Raman microspectroscopic mapping," Analyst. 129(10): 926–930.

Baranska M., Schulz H., Kr¨uger H. and Quilitzsch R. (2005) "Chemotaxonomy of aromatic plants of the genus Origanum via vibrational spectroscopy," Analytical and Bioanalytical Chemistry. 381(6): 1241–1247.

Bedair M. and Sumner L.W. (2008) "Current and emerging massspectrometry technologies for metabolomics," Trends in Analytical Chemistry. 27(3): 238–250.

Bino R.J., Hall R.D., Fiehn O., *et al.*, (2004) "Potential of metabolomics as a functional genomics tool," Trends in Plant Science. 9(9): 418–425.

Cevallos-Cevallos J.M., Rouseff R. and Reyes-De- Corcuera J.I. (2009) "Untargeted metabolite analysis of healthy and Huanglongbing-infected orange leaves by CE-DAD," Electrophoresis. 30(7): 1240–1247.

Cunnick W.R., Cromie J.B., Cortell R. *et al.*, (1972) "Value of biochemical profiling in a periodic health examination program: analysis of 1,000 cases," Bulletin of the New York Academy of Medicine. 18(1): 5–22.

Davey M.P., Burrell M.M., Woodward. F.I. and Quick W.P. (2008) Population-specific metabolic phenotypes of Arabidopsis lyrata ssp petraea. New Phytologist. 177: 380–388.

Devaux P.G., Homing M.G. and Homing B.C. (1971) "Benyzloxime derivative of steroids, a new metabolic profile procedure for human urinary steroids," Analytical Letters. 4: 151.

Dixon R.A. (2001). Natural products and plant disease resistance. Nature 411: 843–847.

Dunn W.B. and Ellis D.I. (2005) Metabolomics: current analytical platforms and methodologies. Trends in Analytical Chemistry. 24: 285–294.

Dunn W.B., Bailey N.J.C. and Johnson H.E. (2005) "Measuring the metabolome: current analytical technologies," Analyst. 130(5): 606–625.

Eliasson M., Rännar S., Madsen R., Donten M.A., Marsden-Edwards E., *et al.* (2012) Strategy for optimizing LC-MS data processing in metabolomics: a design of experiments approach. Anal Chem 84: 6869–6876

Fiehn O. (2002) "Metabolomics—the link between genotypes and phenotypes," Plant Molecular Biology. 48(1-2): 155– 171.

Fiehn O., Kopka J., Dormann P., Altmann T., Trethewey R. N., Willmitzer L. (2000) Metabolite profiling for plant functional genomics. Nat. Biotechnol. 18: 1157–1161.

Field K. and Lake J. (2011) Environmental metabolomics links genotype to phenotype and predicts genotype abundance in wild plant populations. Physiologia Plantarum. 142: 352–360.

Forseth R.R. and Schroeder F.C. (2011) NMR-spectroscopic analysis of mixtures: from structure to function. Current Opinion in Chemical Biology. 15: 38–47.

German J.B., Hammock B.D., Watkins S.M. (2005) Metabolomics: building on a century of biochemistry to guide human health. Metabolomics. 1: 3-9.

Gilbert N. (2011). Regulation: Herbal medicine rule book. Nature 480: 598-599.

Griffin J.L. and Shockcor J.P. (2004) "Metabolic profiles of cancer cells," Nature Reviews Cancer. 4(7): 551–561.

Hellstöm J., Sinkkonen J., Karonen M., Mattila P. (2007) Isolation and structure elucidation of procyanidin oligomers from saskatoon berries (Amelanchier alnifolia). J Agric Food Chem. 55: 157–164.

Horai H., Arita M., Kanaya S., Nihei Y., Ikeda T., *et al.* (2010) MassBank: a public repository for sharing mass spectral data for life sciences. J Mass Spectrom. 45: 703–714

Horning B.C. and Horning M.G., (1970) Metabolic Profiles: Chromatographic Methods for Isolation and Characterization of a Variety of Metabolites in Man, Yearbook Medical Publishers, Chicago, Ill, USA.

Horning B.C. and Horning M.G. (1971) "Human metabolic profiles obtained by GC and GC/MS," Journal of Chromatographic Science. 9: 129–140.

Kaderbhai N.N., Broadhurst D.I., Ellis D.I., Goodacre R. and Kell D.B. (2003) "Functional genomics via metabolic footprinting: Monitoring metabolite secretion by Escherichia coli tryptophan metabolism mutants using FT-IR and direct injection electrospray mass spectrometry," Comparative and Functional Genomics. 4(4): 376–391.

Kim H.K., Choi Y.H., and Verpoorte R. (2011) "NMR-based plant metabolomics: where do we stand, where do we go?" Trends in Biotechnology. 29(6): 267–275.

Kopka J., Fernie A., Weckwerth W., Gibon Y. and Stitt M. (2004). Metabolite profiling in plant biology: platforms and destinations. Genome Biol. 5(6): 109.

Leiss K.A., Choi Y.H., Verpoorte R. and Klinkhamer P.G.L. (2011) "An overview of NMR-based metabolomics to identify secondary plant compounds involved in host plant resistance," Phytochemistry Reviews. 10(2): 205–216.

Lei Z., Huhman D.V. and Sumner L.W., (2011) "Mass spectrometry strategies in metabolomics," Journal of Biological Chemistry. 286(29): 25435–25442.

Licata A., Macaluso F.E., Craxi A. (2012) Herbal hepatoxicity: a hidden epidemic. Intern Emerg Med.

Liu Y., Yang J., Cai Z. (2006) Chemical investigation on Sijunzi decoction and its two major herbs Panax ginseng and Glycyrrhiza uralensis by LC/MS/MS. J Pharmaceut Biomed. 41: 1642–1647.

Lin I.H., Lee M.C., Chuang W.C., (2006) Application of LC/MS and ICP/MS for establishing the fingerprint spectrum of the traditional Chinese medicinal preparation Gan-Lu-Yin. J Sep Sci. 29: 172–179.

Lv Z., Dong J. and Zhang B. (2011) Rapid identification and detection of flavonoid compounds from bamboo leaves by LC-(ESI)-IT-TOF/MS. BioResources. 7: 1405–1418.

Maloney V. (2004) "Plant metabolomics," Biotechnology Journal. 2: 92–99.

MASC, The Multinational Coordinated Arabidopsis thaliana Functional Genomics Project—Annual Report 2007, 2007.

Matsuda F., Yonekura-Sakakibara K., Niida R., Kuromori T., Shinozaki K., *et al.* (2009) MS/MS spectral tag-based annotation of non-targeted profile of plant secondary metabolites. Plant J. 57: 555–577

Mehta R.A., Cassol T., Li N., Ali N., Handa A.K., and Mattoo A.K., (2002) "Engineered polyamine accumulation in tomato enhances phytonutrient content, juice quality, and vine life," Nature Biotechnology. 20(6): 613–618.

Mendes P., Kell D.B. and Westerhoff H.V. (1992) "Channelling can decrease pool size," European Journal of Biochemistry. 204(1): 257–266.

Menedes P., Kell D.B. and Westerhoff H.V. (1996) "Why and when channelling can decrease pool size at constant net flux in a simple dynamic channel," Biochimica et Biophysica Acta. 1289(2): 175–186.

Moura J.C.M.S., Bonine C.A.V., De Oliveira Fernandes Viana J., Dornelas M.C., Mazzafera P. (2010) Abiotic and biotic stresses and changes in the lignin content and composition in plants. J Integr Plant Biol. 52: 360–376.

Mroczek W.J. (1972) "Biochemical profiling and the natural history of hypertensive diseases," Circulation. 45(6): 1332–1333.

Nicholson J.K. and Wilson I.D. (2003) Opinion: understanding ¼global¼ systems biology: metabonomics and the continuum of metabolism. Nat Rev Drug Discov. 2: 668–676.

Obata T., Fernie A.R., (2012) The use of metabolomics to dissect plant responses to abiotic stresses. Cell Mol Life Sci. 69: 3225–3243.

Oksman-Caldentey K.M., Inz´e D., and Ore¡si¡c M., (2004) "Connecting genes to metabolites by a systems biology approach," Proceedings of the National Academy of Sciences of the United States of America. 101(27): 9949–9950.

Ouedraogo M., Baudoux T., Stevigny C., Nortieer J., Colet J.M., *et al.*, (2012) Review of current and 'omics' methods for assessing the toxicity (genotoxicity, teratogenicity and nephrotoxicity) of herbal medicines and mushrooms. J. Ethnopharmacol. 140: 492-512.

Paine J.A., Shipton C.A., Chaggar S. *et al.*, (2005) "Improving the nutritional value ofGolden Rice through increased pro-vitamin A content," Nature Biotechnology. 23(4): 482–487.

Roessner U., Wagner C., Kopka J., Trethewey R. N. and Willmitzer L. (2000) "Simultaneous analysis of metabolites in potato tuber by gas chromatography-mass spectrometry," Plant Journal. 23(1): 131–142.

Schulz H., Baranska M., Quilitzsch R., Sch¨utze W. and L¨osing G. (2005) "Characterization of pepper corn, pepper oil, and pepper oleoresin by vibrational spectroscopy methods," Journal of Agricultural and Food Chemistry. 53(9): 3358–3363.

Schulz H., Schrader B., Quilitzsch R., Pfeffer S. and Kr¨uger H. (2003) "Rapid classification of basil chemotypes by various vibrational spectroscopy methods," Journal of Agricultural and Food Chemistry. 51(9): 2475–2481.

Schripsema J. (2010) "Application of NMR in plant metabolomics: techniques, problems and prospects," Phytochemical Analysis. 21(1): 14–21.

Smith C.A., O'Maille G., Want E.J., Qin C., Trauger S.A., *et al.* (2005) METLIN: a metabolite mass spectral database. Ther Drug Monit. 27: 747–751

Soga T. (2007) "Capillary electrophoresis-mass spectrometry for metabolomics,"Methods inMolecular Biology. 358: 129– 137.

Spratlin J. L., Serkova N. J. and Eckhardt S. G. (2009) "Clinical applications of metabolomics in oncology: a review," Clinical Cancer Research. 15(2): 431–440.

Steinmann D., Ganzera M. (2011) Recent advances on HPLC/MS in medicinal plant analysis. J Pharm Biomed Anal. 55: 744–757.

Stitt M. and Fernie A.R. (2003) "From measurements of metabolites to metabolomics: an "on the fly" perspective illustrated by recent studies of carbon-nitrogen interactions," Current Opinion in Biotechnology. 14(2): 136–144.

Theodoridis G.A., Gika H.G., Want E.J. and Wilson I.D. (2012) Liquid chromatography–mass spectrometry based global metabolite profiling: a review. Anal Chim Acta. 711: 7–16.

Tilton R., Paiva A., Guan J.Q., Marathe R., Jiang Z., *et al.* (2010) A comprehensive platform for quality control of botanical drugs (Phytomics QC): a case study of Huangqin Tang (HQT) and PHY906. Chinese Med. 5: 30.

Urakami K., Zangiacomi V., Yamaguchi K. and Kusuhara M. (2010) "Quantitative metabolome profiling of Illicium anisatum by capillary electro-phoresis time-of-flight mass spectrometry," Biomedical Research. 31(2): 161–163.

Wang Y.C., Yang Y.S. (2007) Simultaneous quantification of flavonoids and triterpenoids in licorice using HPLC. J Chromatogr. B850: 392–399.

Ward J.L., Baker J.M. and Beale M.H. (2007) "Recent applications of NMR spectroscopy in plant metabolomics," FEBS Journal. 274(5): 1126–1131.

Weckwerth W. (2003) "Metabolomics in systems biology," Annual Review of Plant Biology. 54: 669–689.

Williams B.J., Cameron C.J., Workman R., Broeckling C.D., Sumner L.W. and Smith J. T. (2007) "Amino acid profiling in plant cell cultures: an inter-laboratory comparison of CE-MS and GCMS," Electrophoresis. 28(9): 1371–1379.

Wishart D.S. (2011). Advances in metabolite identification. Bioanalysis. 3: 1769–1782.

World Health Organization (2008). Traditional medicine: fact sheet No. 34. Accessed 2012 Oct 19.

Yang C.S., Lambert J.D., Ju J., Lu G. and Sang S. (2007) Tea and cancer prevention: molecular mechanisms and human relevance. Toxicol Appl Pharmacol. 224: 265-273.

Ye M., Liu S.H., Jiang Z., Lee Y., Tilton R., *et al.* (2007) Liquid chromatography/mass spectrometry analysis of PHY906, a Chinese medicine formulation for cancer therapy. Rapid Commun Mass Spectrom. 21: 3593–3607.Yuji S., Masami Y.H., (2013) Integrated LC-MS/MS system for plant metabolomics. Comput Struct Biotechnol J; 4.

Ye X., Al-Babili S., Kloti A. *et al.*, (2000) "Engineering the provitamin A (β-carotene) biosynthetic pathway into (carotenoid-free) rice endosperm," Science. 287(5451): 303–305.

Zhao Z.Y., Dong L.L. and Lin F. (2009) Fingerprint analysis of *Euonymus alatus* (Thuhb) siebold by LC-DAD and LC–ESI-MS.Chromatographia. 69: 429–436.

Zhang Q., Ye M. (2009) Chemical analysis of the Chinese herbal medicine Gan-Cao (licorice), J Chromatogr A. 1216: 1954–1969.

Zhang X., Liu Y., Guo Z., Feng J., Dong J., *et al.* (2012) The herbalome-an attempt to globalize Chinese herbal medicine. Anal Bioanal Chem. 402: 573–581.

Zhu Z.Y., Zhang H., Zhao L., Dong X., Li X., *et al.* (2007) Rapid separation and identification of phenolic and diterpenoid constituents from Radix Salvia miltiorrhizae by high-performance liquid chromatography diode-array detection, electrospray ionization time-of-flight mass spectrometry and electrospray ionization quadrupole ion trap mass spectrometry. Rapid Commun Mass Spectrom. 21: 1855–1865.

2

Biotechnological Approaches for Biosynthesis of Camptothecin, an Anticancer Alkaloid from *Nothapodytes nimmoniana* (Grah.) Mabb.

M. Ayyanar[1*], M. Gopalakrishnan[2] and T. Sekar[2]
[1]Department of Botany and Microbiology, A.V.V.M. Sri Pushpam College, Poondi – 613 503, Thanjavur, India.
[2]Department of Botany, Pachaiyappa's College, Chennai – 600 030, India
**E-mail: asmayyanar@yahoo.com*

Abstract

Biotechnology offers an opportunity to exploit the cell, tissue, organ or entire organism by growing them *in vitro* and to genetically manipulate them to obtain desired compounds. It plays a pivotal role in search for alternatives to production of desirable biomedical compounds. There are several approaches have been used to increase the yield of metabolites from cell cultures such as manipulation of nutrient medium, application of exogenous growth stimulators, chemical treatment, precursor feeding, elicitation, transformed cultures, metabolic engineering, micropropagation, bioreactor cultures and *etc*. Camptothecin (CPT) is a potential anticancer alkaloid isolated from *Camptotheca acuminata* and *Nothapodytes nimmoniana* is very expensive and supply is inadequate on a global scale. *N. nimmoniana* is globally used by the pharmaceutical industries due to the presence of CPT with anticancer therapeutics. Because of CPT's unique structure, its artificial synthesis is difficult and one has to depend on the natural sources for raw material. In general, while the demand for medicinal plants is increasing, their survival in natural habitats becomes threat and a number of medicinal plants have been assessed as threatened due to over harvesting in the wild. Simultaneously, medicinal plants are collected extensively by the traditional practitioners, plant collectors, traders and herbalists for their role in medicinal preparation. The studies on biosynthesis of CPT showed higher yield of biomass and bioactive molecules by large-scale cultivation of cell cultures. From this review, it may be inferred that among the different culture strategies employed to enhance the CPT biosynthesis in cell cultures, variation in concentration of plant growth regulators resulted in maximum product accumulation and it might reduce the pressure on natural plant population when properly scale up thorough plant tissue culture medium requirements were established.

Keywords: Anticancer compound, Camptothecin, *Camptotheca acuminata*, Cell suspension culture, *Nothapodytes nimmoniana*, 9-methoxy Camptothecin.

Introduction

Cancer continues to represent the largest cause of mortality and claims over 6 million lives each year all over the world and an extremely promising strategy for cancer prevention today is chemoprevention, which is defined as the use of synthetic or natural agents to block the development of cancer in human beings. A large number of plants, marine and microbial sources have been tested as leads and many compounds were identified as the potential leads. Herbal medicines used in traditional medicine have been accepted as one of the main sources of cancer chemopreventive drug discovery and development (Abdullaev 2001). For last few decades, a greater emphasis has been given towards the researches on complementary and alternative medicine that deals with cancer management.

Early examples of anticancer agents developed from higher plants are antileukemic alkaloids vinblastine and vincristine from *Catharanthus roseus*. Since the early 1950s, identification and development of new lead compounds for anticancer chemotherapy has been partially driven by broad plant screening programs and new antineoplastic agents like 'taxol' (*Taxus brevifolia*) and an alkaloid 'camptothecin' (*Camptotheca acuminata*) were identified (Voss *et al.* 2006). Etoposide is an epipodophyllotoxin, highly active molecule isolated from *Podophyllum peltatum* and *P. emodi* which has produced high cure rates in testicular cancer when used in combination with bleomycin, another plant derived molecule and cisplatin (Williams *et al.* 1987; Stahelin 1973). Kelland (2000) reported that flavopiridol, a synthetic flavone derived from an alkaloid rohitukine which was isolated from the leaf and stem of *Amoora rohituka* and later from *Dysoxylum binectariferum*.

Discovery of Camptothecin

The extensive effort by many groups towards the synthesis of plant antitumor agent has culminated in some successful achievements from the plant *Camptotheca acuminata* (Meyars *et al.* 1973). It is a tree belonging to the family Nyssaceae native to south China. Wall (1998) revealed that, of the samples of 1000 ethanolic plant extracts testing for antitumor activity, only one extract from the plant *C. acuminata* showed some astonishing result with high activity then used as one of the standard test systems. Camptothecin (CPT, a monoterpenoid indole alkaloid) is now recognized as 3rd most promising alkaloid of the 21st century (Lorence and Nessler 2004).

Camptothecin having unique pentacyclic structure and was first isolated in 1966 by Wall and Wani from the wood of the *C. acuminata*. It is yellow and is highly insoluble in organic and aqueous solvents. This insolubility has impeded clinical development of CPT and its analogs (Ciddi and Shuker 2000). CPT contains a pentacyclic ring system that includes a pyrroll (3,4-b) quinoline moiety (ring A, B and C). It has gained attention of many researchers and pharmaceutical industries all over the world because of its potent clinical applications in anticancer therapeutics. CPT was isolated from *C. acuminata* with 0.005 % (Wall *et al.* 1966) however its bark accumulates a lower concentration of CPT, but young leaves produce 50 and 250 % higher CPT (up to 0.4 % dry weight) than fruits and bark, respectively (Lopez-Meyer *et al.* 1994). It was also reported that CPT content in different parts of *C. acuminata* are 0.012 % in bark, 0.02 % in roots), 0.05 % in wood and 0.02 % in fruits (Weidenfeld *et al.* 1997). Soil fertility and coppicing are found to be affecting CPT levels in *C. acuminata* seedlings (Liu *et al.* 1998). Since CPT is produced by the plant as a defence strategy against herbivory and CPT being a secondary metabolite which is generally synthesized under stress conditions, it will be important and interesting to study what kind of stress conditions are responsible for high levels of CPT in the plant. For this, multi-location field trials at different agro climatic zones are a prerequisite. So studies on effect of various fertilizers on CPT levels, impact of stresses like water stress, difference in CPT levels across plantations in rainy versus dry areas could be future course of action (Patwardhan 2010).

Pharmacological activities of Camptothecin

Camptothecin and its derivatives are unique in their ability to inhibit DNA topoisomerase I by stabilizing a covalent reaction termed the cleavable complexes which ultimately cause tumor cell death. Irinotecan and topotecan are two water-soluble derivatives of CPT gained approval by the Food and Drug Administration of the United States of America for treating colorectal and ovarian cancer (Sriram *et al.* 2005). Other camptothecins such as 9-aminocamptothecin (9-ACPT), 9-nitrocamptothecin (9-NCPT), and 7-(4-methyl piperazinomethylene)-10, 11-ethylene dioxy camptothecin have also shown remarkable potential in the treatment of carcinoma which are under clinical trials and have promising chemotherapeutic efficacy (Wall and Mani 1986).

CPT and its analogues have a unique mechanism of action; they produce DNA damage in the presence of topoisomerase-I by binding to and stabilizing a covalent DNA-topoisomerase-I complex in which one strand of DNA gets broken (Puri *et al.* 2005). Clinically CPT also exhibit remarkable anti-tumor and anti-leukemia activity (Premalakshmi *et al.* 2012). In 1970s, CPT was first proven to be effective in curing digestive system tumors and melanomas, further clinical trials with CPT were abandoned during 1970s due to its severe side effects (Chang *et al.* 2014). Since the early 1990s, 2 semisynthetic CPT derivatives, *i.e.*, irinotecan and topothecan with decreased cytotoxicity, better anticancer activities, and water solubility were approved and are now used in cancer chemotherapy including for treating colorectal, ovarian, lung, stomach, and small cell lung cancers, as well as in inhibiting HIV replication (Priel *et al.* 1991).

CPT has also been reported to have a number of biological activities. CPT and four of its analogues were studied against pathogenic protozoan *Leishmania donovani in vitro* and substantial enhance in the antileishmanial activity was reported (Werbovetz *et al.* 2000). African trypanosomes (*Trypanosoma brucei*) are parasitic protozoa that cause lethal disease in humans and cattle are treated by CPT which was cytotoxic to *T. brucei* and related pathogenic hemoflagellates (Bodley *et al.* 1995a). A series of CPT analogues tested against *T. Brucei* showed that modifications to the pentacyclic nucleus of CPT ablated antiparasitic activity (Bodley *et al.* 1995b). They also reported that (1998) in *Plasmodium falciparum*, CPT trapped protein–DNA complexes inhibited nucleic acid biosynthesis and have cytotoxic effect against on erythrocytic malaria parasite *in vitro*.

Sources of Camptothecin

CPT has been extracted from root, root bark and fruits of *C. acuminata*, but recent results have indicated that CPT exists in high levels in juvenile leaves than fruits (Zhang and Yang, 1997). Apart from *Camptotheca acuminata*, the plants, *Nothapodytes nimmoniana*, *Merrilliodendron megacarpum* and *Pyrenacantha klaineana* of Icacinaceae, *Chonemorpha fragrans* and *Ervatamia heyneana* of Apocynaceae, *Ohpiorrhiza pumila* and *O. mungos* of Rubiaceae and *Mostuea brunonis* of Loganiaceae has also been reported to synthesize varied levels of CPT. Among these, *Ervataemia heyneana* (Gunasekara *et al.* 1979) *Ophiorrhiza mungos* (Tafur *et al.* 1976) and *Ophiorrhiza pumilla* (Saito *et al.* 2001) has been reported to produce a very low concentration of CPT and 9-methoxycamptothecin, whereas and *Nothapodytes nimmoniana* is reported with maximum concentration of CPT (0.01%) and 9-methoxycamptothecin (0.001%) from the root cultures (Fulzele *et al.* 2001). CPT content varies among species and varieties but even more significantly within the various plant parts with different tissues, tissue age and between seasons. This implies that analyzing various tissues, at different ages, in different seasons and in different forest habitats for CPT content may come out with better options for sustainable extraction.

Biotechnological approaches for biosynthesis of biomolecules

The production of alkaloids in plant cell cultures is a result of enormously complex set of interactions between cellular and extra cellular compartments. The plant cell culture is a dynamic system in which smallest change can have large and sometimes fatal consequences and hence provides many opportunities for manipulation to improve product yield. Biotechnology offers an opportunity to exploit the cell, tissue, organ or entire organism by growing them *in vitro* and to genetically manipulate them to obtain desired compounds and it plays a vital role in search for alternatives to production of desirable biomedical compounds (Kehie *et al.* 2014). Hence, plant tissue culture is used as a major tool in propagating the elite species and screening of high yielding lines in tissue culture studies result in large number of samples for quantification (Puri *et al.* 1999; Fulzele *et al.* 2002). *In vitro* processes for secondary metabolite production have fallen far short of expectations and never approached commercial status and still, the arena of secondary product formation in cell cultures remains as an industrial pursuit. Biochemical scientists are currently joining forces on a global scale to develop new strategies for streamlining the critical bioprocesses (Karuppusamy 2009).

The strong and growing demand in today's marketplace for natural and renewable products has refocused attention on *in vitro* plant materials as potential factories for secondary phytochemical products has paved the way for new research exploring secondary product expression in *in vitro*. When plant cell cultures are developed *in vitro* as suspension cultures, the yield of metabolites is generally low, as secondary metabolite production is often associated with tissue differentiation (Karwasara and Dixit 2013). The development of favourable culture conditions is of prime significance to the growth and biosynthetic capacity of *in vitro* cultured cells. Slight changes in culture conditions including carbon source, nitrogen source, nitrate to ammonia ratio, phosphorous, micronutrients, plant growth regulators, precursors, etc. causes great variations in biological and biosynthetic behavior (Merillon *et al.* 1984).

The discovery of cell cultures capable of producing specific medicinal compounds at a rate similar or superior to that of intact plants have accelerated in the last decade only. Karwasara and Dixit (2013) revealed that biosynthetic potential of cultured cells has been exploited for production of a number of metabolites with limited commercial success. The poor understanding of controlling factors of underlying biosynthetic pathway and genetic instability of cell lines contributed to low product yield. Sakato and Misawa (1974) reported that tissue culture of *C. acuminata* produces 2.5μg CPT per g dry wt, whereas the intact plant contains 120 μg per g dry wt basis of CPT. Van-Hengel *et al.* (1992) found 4 μg per g dry wet of alkaloid in suspension culture using MS medium with NAA (4 mg l^{-1}). The callus cultures of *C. acuminata* produces CPT at 2.04 - 2.36 mg per g dry wt and 10-hydroxy-CPT at 80-100 μg per g dry wt (Wiedenfeld *et al.* 1997).

To establish a sustainable CPT production system, there are several attempts have been made to produce CPT by callus and cell cultures mostly with *C. acuminata* and *N. nimmoniana*. Hairy roots and crown galls induced by transformation with *Agrobacterium rhizogenes* were found to be a potentially favourable system of producing high value pharmaceuticals because of their stable and high productivity in phytohormone free culture conditions with growth rates comparable to those of the fastest growing cell suspension culture (Hu and Du 2006). It was reported that hairy roots of *Ophiorrhiza pumila* was successfully cultivated on large scale in a bioreactor and maintain CPT production even after 5 years (Sudo *et al.* 2002). The crown gall culture of *C. acuminata* is able to produce 0~0.039 % of CPT, large cell aggregates occur in the liquid culture and limit the development of scaling-up to bioreactor culture (Chang *et al.* 2007).

The formation in tissue culture of several types of indole alkaloids was inseparably connected with morphological differentiation of cells. De Luca *et al.* (1986) demonstrated that synthesis of vinblastine was connected to morphological differentiation of seedlings of *Catharanthus roseus*. Similarly in *Cinchona* seedlings alkaloid formation is abundantly expressed (Aerts *et al. 1990) but* in suspension cultures productivity is very low. Also, the formation of indole alkaloids in *Tabernaemontana* tissue cultures proved to be dependent on morphological differentiation (Sierra *et al.* 1991). Cell cultures are capable of producing alkaloids, which had not yet been detected in the plant. Nabila *et al.* (2003) reported the presence of rosmarinic acid in cell suspension cultures of *Salvia fruticosa* (5.1 mg/ 100mg) which was higher than the amount present in the root cultures of the same plant (2.62mg/ 100mg). Similarly, anthraquinone production (2.9 g/l) was obtained from cell suspension cultures of *Morinda elliptica* (Abdullah *et al.* 1998).

Authentication of medicinal plants for standardization and quality control

The phytochemical composition of plant material and resulting quality can vary due to several factors such as geographical location, soil quality, climatic factors (temperature, rainfall, etc.) time of collection, method of collection, cultivation, harvesting, drying and storage conditions (Sapna *et al.* 2004). According to WHO general guidelines for methodologies on research and evaluation of medicinal plants, first step is assuring quality, safety and efficacy for correct identification (Khan *et al.* 2011).

A chemical method for evaluation covers the isolation, identification and purification of drug is done to assess the potency of source material in terms of their active principles, identification and characterization of crude drug with respect to phytochemical constituent as marker. The bioactive extract should be standardized on the basis of active principles or major compounds along with the chromatographic fingerprints like TLC, HPTLC and HPLC which will provide qualitative and semi-quantitative data about main active constituents present in the medicinal plant as chemical markers for evaluation. Pharmacological activity of certain drugs has been applied to evaluate and standardize them with assays on living animal, antibacterial, antifungal and *in vitro* models (cell line studies) can strengthen the effect of medicinal plants on human delivery system. Phytochemical standardization for identification of plant material can be carried out by obtaining chemical fingerprint through chromatographic technique or bioassay or DNA fingerprinting.

DNA-based techniques have been widely used for authentication of plant species of medicinal importance and this is highly useful in case of frequently substituted or adulterated with other species or varieties that are morphologically or phytochemically indistinguishable. The main components of such analysis include genetic identity (molecular fingerprinting), relatedness (molecular diversity), structure (genotypic) and gene location (DNA sequence / chromosomal). Literature on genetic variations of medicinal plants is scanty despite the fact that such variations hold a greater promise. It is a well known fact that the loss of genetic (intraspecific) diversity is much more serious and occurs much earlier than total extinction of the particular species itself.

Nothapodytes nimmoniana: distribution of the plant

Nothapodytes nimmoniana (Graham) Mabberley (synonyms: *Nothapodytes foetida* (Slemure) Wight; *Mappia foetida* Miers) is a rare medicinal plant distributed in India, Sri Lanka, Myanmar, Thailand, Malaysia and China. In India the plant is distributed in moist and shola forests of Western Ghats in Karnataka, Kerala and Tamil Nadu and in some patches of north-east India. It is a small, lower canopy tree belonging to family Icacinaceae. The ripened fruits resemble Jamun (*Syzygium cumini*) fruits. The distribution is highly patchy and found in different types of forests ranging from secondary forests, moist deciduous forests, semi-evergreen forests and evergreen forests. It prefers forest edges as micro-habitat and commonly referred to as 'stinking tree' because of the foetid smell it emits while flowering. Locally the plant is known as *Araali, Perumpulagi* and *Kalkurinji* in Tamil, *Durvasane mara* in Kannada. It extends up to 30-40 individuals/ha at some localities such as shoal forests of Nilgiris, where populations of *N. nimmoniana* are surviving against the severe threat of destruction by the indigenous traders. However the World Bank study warned the rate of exploitation of *N. nimmoniana* has increased rapidly and will soon become threatened.

Taxonomic description of *N. nimmoniana*

N. nimmoniana is a small tree grows upto 3-8 m height. Bark is smooth, grey, wrinkled, 5mm thick, branchlets slightly angled with prominent leaf scars. Leaves simple, alternate, slightly leathery, spiral, clustered at twig ends; petiole 1.2-6 cm long, flat above, puberulous; lamina 9-30 x 5-14 cm, broadly ovate-oblong to elliptic-oblong, apex acuminate, base acute to rounded, attenuate or asymmetric, margin entire with strong nerves, dark green above, pale beneath, glabrous, midrib flat. Flowers are in terminal corymbose cymes, bisexual, creamy yellow, petals hairy inside, emitting an unpleasant odour at the time of flowering. Fruit is a drupe, purplish red, black when ripe, smooth, oblong with single seed. The plant flowers during April to June and fruits between August to October.

Market value of Camptothecin

CPT is extracted from the wood chips of *N. nimmoniana* and the herbal extract carrying the trade name "Ghanera". The annual demand from Japan for dried stem of *N. nimmoniana* was 200-300 tonnes in 1994 and since then there has been increasing trend. The trade volume increased to 1600 tons in the year 2002. In 2006-08, the reported trade is exceeding 1000 tonnes, whereas unreported trade is thought to be at least twice as reported. The ever increasing worldwide market of Irinotecan and Topotecan has also reached $1000 million which represents approximately 1 ton of CPT in terms of raw material (Watase *et al.* 2004). To meet this ever increasing demand for CPT related drugs from worldwide, more plants are cut, dried and exported and this export business is completely managed by private sector. The collectors of the plant have trained local tribal and rural labourers in cutting and drying processes and they are paid Rs. 10-15/ kg of dried stems and exported by the traders at the price of $1500 to $2000/kg (Padmanabha *et al.* 006).

In Indian market, the demand for CPT's biomass is 500-700 metric tonnes (Maslekar 1994) and total loss recorded from certain areas. As CPT accumulates in stem and root bark of *N. nimmoniana*, whole tree needs to be cut for generating good biomass for extraction. In Maharashtra, overexploitation and habitat destruction for raw material of *N. nimmoniana* has led to population decline by 50-80 % in last decade (CAMP 2001). Currently the species population density is as low as 1-2 individuals/ha in some areas of the state and on the other hand it extends up to 30-40 individuals/ha at some localities such as forest of Mahabaleshwar region where populations of *N. nimmoniana* are surviving against the severe threat of destruction (Patwardhan 2010).

Biosynthesis of Camptothecin from *N. nimmoniana*

Camptothecin (Fig.1) accumulates in stem and root bark of *N. nimmoniana* and therefore the whole tree is cut for extraction. CPT content in *N. nimmoniana* is higher than in other CPT containing plants and thus this plant represents most convenient source for large-scale isolation of this alkaloid. It has been reported to contain 0.06–0.1 % of CPT and 0.001 – 0.02 % of 9-Met-CPT (Govindachari and Viswanathan 1972). Fulzele and Satdive (2005) examined various morphological parts of *N. foetida* and noted maximum concentration in bark (0.27 % CPT and 0.11 % 9-Met-CPT). Puri *et al.* (1999) demonstrated that *N. foetida* cultivated in the north-western area of Jammu accumulated 0.1 % DW CPT in roots and seeds, whereas the bark produced lower concentrations of CPT. Among the various tissues of *N. nimmoniana* analyzed, the root bark (0.33 % DW) yielded the highest CPT content followed by stem bark (0.24 % DW). The root and stem contain 0.18 and 0.14 % of CPT respectively. The leaves yielded the lowest CPT content as 0.08 % (Padmanabha *et al.* 2006).

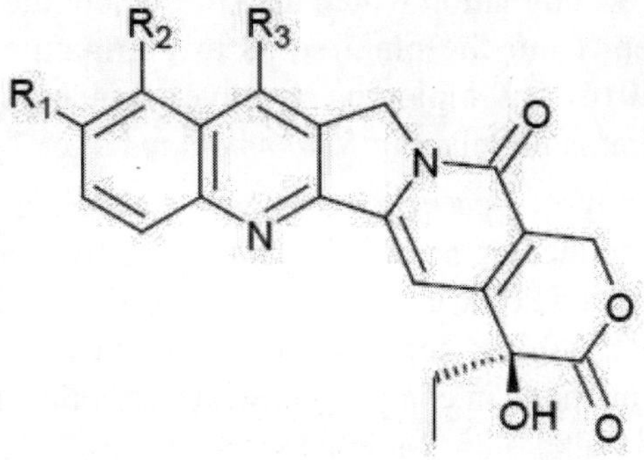

	R_1	R_2	R_3
Camptothecin	H	H	H
9-Aminocamptothecin	H	NH_2	H
Topotecan	OH	$CH_2N(CH_3)_2$	H
Irinotecin	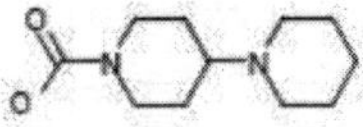	H	CH_2CH_3
SN38	OH	H	CH_2CH_3

Fig. 1 : Major chemical constituents of *Nothapodytes nimmoniana*

Countries like Japan, USA and Spain are in commercialization of CPT as drug and these countries import dried raw material from India. Research on clinical trials of CPT in India is only at laboratory scale. Since there is no synthetic prominent source of CPT and the dependence for the raw material from natural population, *Camptotheca acuminata* (0.1% per dry wt.) and *N. nimmoniana* (0.14-0.24% per dry wt.) is increasing (Puri *et al.* 1999; Padmanabha *et al.* 2006).

The coppicing strategy can be used as a successful harvest practice when it comes to commercial plantation for the production of CPT. According to biosynthesis pathway, CPT is synthesized in leaf and mainly accumulated in stem and root bark of *N. nimmoniana*. As the plant matures, CPT accumulation will go on increasing till some concentration after which concentration will remain in more or less with bark thickness. Studies on this pattern of accumulation would help to find out the appropriate age of harvest of stem from the plant so as to get maximum concentration (Patwardhan 2010). By employing effective coppicing strategy one can get 2-3 kg dry biomass per plant in *N. nimmoniana* after 7 yrs of plantation.

An endophytic *Neurospora* sp. was isolated from the seed of *N. foetida* and examined as potential source of anticancer compound, when grown in Sabouraud liquid culture media under shake flask conditions. The presence of CPT in this endophyte was confirmed by chromatographic and spectroscopic methods in comparison with authentic CPT (Rehman *et al.*, 2008). Two new alkaloids, 9-methoxy-18,19-dehydrocamptothecin and 5-hydroxyl-mappicine-20-O-beta-glucopyranoside together with nine known compounds were isolated from the immature seeds of *N. foetida*, in which one compound has exhibited significant cytotoxicity against cancer cell lines (Wu *et al.* 2008). HPLC profiles were developed for qualitative and quantitative analysis of these active constituents in different parts of the plant including mature and immature seeds, leaves, stems and roots. Puri *et al.* (2007) reported that, immune-modulatory activity of an extract of the novel fungal endophyte *Entrophospora infrequens* (found to synthesize camptothecin) was isolated from *Nothapodytes foetida*.

Comparative studies on CPT content from *N. foetida* and *Ophiorrhiza rugosa* by inhibition of the enzyme DNA topoisomerase I was carried out by Roja (2006) and results indicated highest yields of CPT and 9-Met.CPT in *N. foetida*. Hsiao *et al.* (2008) studied the CPT content in DMSO extracts of *N. foetida* by direct injection capillary electrophoresis and this method was proven to be very suitable for monitoring the amount of CPT. Fulzele and Satdive (2006) investigated the distribution of anticancer drug CPT in *N. foetida* and results showed that, bark contained 0.27% of CPT and 0.11% of 9-Met.CPT followed by root stem, and leaf; immature seeds has higher concentrations of CPT (0.32%) and 9-Met. CPT (0.16%) compared to mature seeds; highest concentration of CPT (0.42%) and 9-Met. CPT (0.18%) was accumulated in the cotyledons of immature seeds.

Separation of CPT and 9-Met.CPT from aerial parts of *N. foetida* by HPLC was done by Puri *et al.* (2005) and purity of isolates was determined by physicochemical data and liquid chromatography-mass spectrometry. Quantitative analysis using ^{1}H-NMR method has been developed for the determination of CPT derivatives and trigonelline in *N. foetida* root, stem and leaf and quantity of compounds were calculated by the ratio of intensity of each compound to known amount of internal standard 3,4,5-trimethoxybenzaldehyde (Li *et al.* 2005). Fulzele and Satdive (2005) reported that, extraction methods using stirring extraction, soxhlet extraction, ultrasonic extraction and microwave-assisted extraction were evaluated for percentage extraction of CPT and 9-Met.CPT from *N. foetida* and the extracts were analyzed by HPLC. The results showed that the percentage extraction of CPT and 9-Met.CPT from *N. foetida* by microwave-assisted extraction (2.67% CPT) was more efficient followed by soxhlet extraction, ultrasonic and stirring extraction methods.

Biotechnological approaches for biosynthesis of CPT from *N. nimmoniana*

The use of chromatographic techniques and marker compounds to standardize botanical preparations has limitations because of their variable sources and chemical complexity. Thin layer chromatography (TLC), high performance liquid chromatography (HPLC) and high pressure thin layer chromatography (HPTLC) methods were witnessed their use in quantification of CPT (Zhao *et al.* 2006). However published literature indicates the need for identification and optimization of simple and cost effective methods for quantification of CPT when there are many samples. CPT is very commonly quantified by HPLC (Sakato *et al.* 1994) however use of HPLC method for quantification is very expensive and demands a simple and cost effective method.

N. nimmoniana is prolonged juvenile period and restricted cultivation areas and areas under cultivation take years to yield commercially viable quantity of CPT and this resulted in threat to the wild resources. In Western Ghats, overexploitation and habitat destruction for raw material led to population decline by 50-80 % in last decade (CAMP 2001). Since there is no convenient synthetic source for CPT, one has to depend on raw material from natural populations. Recently, several researchers were addressed the need to conserve the species, to chemically characterize populations, to identify populations or individuals with high CPT yields and to explore the possibilities of identifying high yielding individuals for the development of *in vitro* production systems (Ciddi and Shuker 2000; Thengane *et al.* 2003; Fulzele and Satdive 2003; Padmanabha *et al.* 2006). Therefore, there is an urgent need to conserve this plant through *in vitro* methods and also to maintain the genetic uniformity of species. It was reported that, plant tissue culture emerged as a viable route for production of taxol and provides a model for production of anticancer agents from woody plants (Cragg *et al.* 1993) and this might be applicable to CPT too (Ciddi and Shuker 2000).

Culture condition provides environment for optimum production of alkaloid in cells and optimization of culture condition requires knowledge of biosynthesis regulation which influences productivity of cultured cells (Isah and Mujib 2014). There are several approaches in use to increase the yield of metabolites from cell cultures such as manipulation of nutrient medium, application of exogenous growth stimulators, chemical treatment, precursor feeding, elicitation, transformed culture, metabolic engineering, micropropagation and bioreactor cultures (Karwasara and Dixit 2013). The CPT production ability of *N. nimmoniana*, has generally not been acceptable, as the contents of CPT was only 1/100~1/1000 of those from soil grown plants (Roja and Heble 1994; Ciddi and Shuler 2000; Fulzele *et al.* 2003).

Induction of callus from different explants of *N. nimmoniana* at variable frequency on Murashige and Skoog medium amended with auxin alone or with cytokinin and cultured explants were developed into callus after 8–9 weeks on medium amended with various PGRs (Table 1). Similar to other alkaloids, CPT accumulates well in cultured cells during stationary phase of growth and differentiated tissues than undifferentiated ones. Besides their influence on cyto-differentiation, PGRs affect CPT biosynthesis in concentration dependent manner and for callus induction, 2,4-D is more efficient than others but CPT yield is higher with NAA induction (Isah and Mujib 2014).

Biosynthesis of CPT from regenerated plants

Fulzele and Satdive (2003) evaluated the CPT content in two year old plants grown in the greenhouse after transfer from the tissue culture laboratory and the results observed that regenerated plant roots yielded 0.20 % of CPT with 0.097 % of 9-Met-CPT, regenerated plant stems yielded 0.11 % of CPT with 0.06 % of 9-Met-CPT and regenerated plant leaves yielded 0.045 % of CPT with 0.013 % of 9-Met-CPT. These results indicated that alkaloid synthesis gradually increased with the age of plant.

When adventitious shoots induced from the callus (derived from nodal explants) of *N. nimmoniana* on semisolid and liquid MS medium supplemented with various concentrations of BA, Kn and 2-iP, highest number was developed from liquid medium with BAP at 2.0 μM

Table 1 : Tissue culture studies in production of Camptothecin from the various parts of Nothapodytes nimmoniana

Explant/ culture type used	Media and PGRs used	CPT source	CPT yield (dry weight)	Reference
Immature embryo	MS with BA + 2,4-D	Callus	Traces of CPT + 0.0001% 9-Met-CPT	Roja and Beble
		Well developed plantlets	Traces of CPT + 0.0007% 9-Met-CPT	(1994)
Embryo	MS with Picloram (8.28 μM)	Callus	0.00095% with traces of 9-Met-CPT	Ciddi and Shuler (2000)
Embryo		Callus	0.95 % + traces of 9-Met-CPT	Veeresham and Shuler (2000)
Somatic embryo	NAA (10.74 μM) + BA (2.22 μM)	Cell suspension culture	0.035% + 0.026% 9-Met-CPT	Fulzele *et al.* (2001)
Immature zygotic embryo	MS with NAA (71.36 μM) and BA (8.87 μM) MS with NAA (71.36 μM) and Kn (9.29 μM)	Untransformed root culture Proliferated callus with root 9-Met-CPT	0.01% + 0.0016% 9-Met-CPT 0.00017% + 0.000058%	Fulzele *et al.* (2002)
Immature	MS with 2,4-D (9.05 μM) + BA	Callus and globular embryos	Small traces of CPT and 9-Met-CPT	
zygotic embryo	(4.44 μM), + Kn (2.32 μM)	Somatic embryos	0.011% + 0.0028% 9-Met-CPT	
		In vitro plantlets	0.061% + 0.0053% 9-Met-CPT	Fulzeleand
		Plantlets in paper cups	0.094% + 0.0088% 9-Met-CPT	Satdive
		Regenerated plant roots	0.20% + 0.097% 9-Met-CPT	(2003)
		Regenerated plant stems	0.11% + 0.06% 9-Met-CPT	
		Regenerated plant leaves	0.045% + 0.013% 9-Met-CPT	
Cotyledon	MS with 2,4-D (4.52 μM) + BA (2.22 μM)	Callus	1.30% with traces of 9-Met-CPT	Thengane *et al.* (2003)

Contd...

Leaf	MS with Picloram (8.28 μM) + BA (4.44 μM) + GA3 (5.20 μM)	Callus	2.9% + 0.4% 9-Met-CPT	Sundaravelan *et al.* (2004)
Regenerated leaf and stem (Nodal explant)	MS (liquid and semi-solid) with BAP (2.0 μM)	Leaf in semi-solid medium	0.083 %	Dandin and Murthy (2012)
		Stem in semi-solid medium	0.037 %	
		Leaf in liquid medium	0.028 %	
		Stem in liquid medium	0.021 %	
Leaf	MS (modified) with NAA (2.0 mg l^{-1})+Kn (0.5 mg l^{-1})	Callus Callus	0.00298% DW	Karwasara and Dixit (2013)
Cell suspension culture	MS (modified) with NAA (10.74 μM) + Kn (0.93 μM) + 2 mg/l ascorbic acid		0.0051.7% DW	
Immature fruit	MS with BAP (4.0 μM) + GA_3 (8.0 μM)	Callus Regenerated plants	1.18% + 0.47% 9-Met-CPT 2.86% + 1.14% 9-Met-CPT	Sharma *et al.* (2014)
Hairy root culture (leaf and stem)	MS without PGRs	Hairy roots in liquid medium	0.021 % DW	Chang *et al.* (2014)

(Dandin and Murthy 2012). They reported that CPT content in regenerated leaf explants in semi-solid and liquid medium are 0.083 and 0.028% respectively; whereas, CPT content in regenerated stem explants in semi-solid and liquid medium are 0.037 and 0.021 % respectively. Sharma *et al.* (2014) initiated callus culture from immature fruits of *N. foetida* on MS medium supplemented with different combinations of PGRs (BAP at 4.0 μM with GA_3 at 8.0 μM) and the maximum CPT (2.86%) and 9-Met-CPT (1.14%) content was recorded in plants transferred in earthen pots where as minimum CPT (1.18%) and 9-Met-CPT (0.47%) content was found in callus biomass.

Biosynthesis of CPT from callus cultures

The callus induced from immature cotyledons of *N. nimmoniana* in MS medium with different concentrations of PGRs such as 2,4-D and BA was studied by Thengane *et al.* (2003). Among these, one month old callus fortified with 2,4-D (4.52 μM) + BA (2.22 μM) yielded 1.30% of CPT along with trace amount of 9-Met-CPT. Induction of very high amount of 9-Met-CPT was observed when concentration of BA was increased and increase in the concentration of NAA along with BA inhibited the yield of CPT production. Their study suggested that different media combinations with 2,4-D and BA is most suitable for alkaloid production especially CPT.

Callus culture was established from excised embryos of *N. nimmoniana* on MS medium supplemented with picloram (8.28 μM) and embryos developed into callus (Ciddi and Shuler 2000). After 4 weeks of culture, CPT content in callus was 0.00095% with traces of 9-Met-CPT and this may be due to genetic differences resulting from geographical differences in seeds or effect of a different culture conditions like PGRs. In contrary, callus cultures produced from immature embryos of *N. foetida* produced small amount of CPT and 0.0001% 9-Met-CPT in Ms medium supplemented with BA and 2,4-D (Roja and Heble 1994).

Karwasara and Dixit (2013) studied the effect of culture medium nutrients on growth and alkaloid production by plant cell cultures of *N. nimmoniana* with a view to increasing the production of the alkaloid CPT and among the various sugars (glucose, fructose, maltose and sucrose) tested with MS medium, maximum yield of CPT was recorded with the addition of sucrose. Also, enhanced amount of nitrate in media supports the increase in biomass, while high ammonium stimulates the amount of CPT content.

Sundaravelan *et al.* (2004) stated that MS medium supplemented with Picloram (8.28 μM), BA (4.44 μM) and GA_3 (5.20 μM) was greatly suitable for the development of callus from leaf explants of *N. nimmoniana* and highest amount of callus biomass was recorded in young leaves followed by immature embryos. The study also showed that CPT content in callus cultures yielded 2.9% and 0.4% of CPT and 9-Met-CPT respectively.

Immature zygotic embryos of *N. foetida* on MS medium supplemented with different concentrations of NAA and BA developed untransformed root cultures and it was evaluated after eight weeks (Fulzele *et al.* 2002) and the presence of NAA with BA was shown to influence the growth and production of alkaloids in the culture. Culture medium containing NAA (71.36 μM) and Kn (9.29 μM) proliferated callus interspersed with roots which synthesised CPT (0.00017 %) and 9-Met-CPT (0.000058 %).

Biosynthesis of CPT from cell suspension cultures

The optimized cell suspension culture derived from leaf callus of *N. nimmoniana* showed 0.0051 % CPT content in modified MS medium fortified with NAA (10.74μM), Kn (0.93μM) and ascorbic acid (2 mg/l) and the callus initiated from the leaf explant reported to have 0.022 % CPT content (Karwasara and Dixit 2013). Fulzele *et al.* (2001) reported as cell suspension culture established from the somatic embryos in MS medium with NAA (10.74 μM) and BA (2.22 μM) yields 0.035 % of CPT and 0.026 % of 9-Met-CPT.

Biosynthesis of CPT from hairy root cultures

Chang *et al.* (2014) cultured ten clones of *N. nimmoniana* seeds on MS medium without adding any PGRs. Leaf and stem of *in vitro* grown seedlings were used as explants to induce infection by 3 wild-type *Agrobacterium rhizogenes* strains and hairy roots were developed. After 30 days of cultivation in MS liquid medium, the weight of hairy roots were increased with 5.3 fold growth rate and the CPT content in hairy roots of liquid medium was recorded as 0.021% of dry weight. The study revealed that, with continual harvesting of CPT from the liquid medium and hairy root culture system will benefit the development of large-scale bioreactor culture for the better yield of CPT in the future. Immature zygotic embryos of *N. foetida* yielded maximum concentration of CPT (0.01 %) and 9-Met-CPT (0.0016 %) by untransformed root cultures incubated on MS medium supplemented with NAA (71.36 μM) and BA (8.87 μM) (Fulzele *et al.* 2002).

Biosynthesis of CPT from *in vitro* grown plantlets

Well developed plants produced from immature embryos of *N. foetida* produced very small amount of CPT and 0.0007 % of 9-Met-CPT in MS medium supplemented with BA and 2,4-D (Roja and Heble 1994). When somatic embryos developed from immature zygotic embryos of *N. foetida* in MS medium fortified with 2,4-D (9.05 μM), BA (4.44 μM) and Kn (2.32 μM) callus cultures and white globular embryos were developed and callus biomass were synthesized trace amou.nts of alkaloids (Fulzele and Satdive 2003). The somatic embryos in the culture showed the presence of CPT at 0.011 % and 9- Met-CPT at 0.0028 %. Then well developed *in vitro* plantlets grown in MS medium with BA recorded with the production of 0.061 % of CPT and 0.0053 % of 9-Met-CPT; and *in vitro* grown plants in paper cups synthesized CPT at 0.094% and 9-Met-CPT at 0.0088%.

Precursor feeding to increase the production in CPT

We investigated the addition of appropriate precursors L-tryptophan an intermediate in the biosynthesis of Indole alkaloids to stimulate CPT compounds to the culture media fortified with MS. After establishing the callus biomass and differentiation cultures (multiple shoots) and were further subjected to the medium containing the L-tryptophan with varying concentrations of 10mg/l to 40mg/l for callus culture and 10 mg/l to 30mg/l for shoot cultures. Callus biomass was induced with BA (8.87 μM), NAA (10.74 μM), 2,4-D (9.05 μM) and addition of L-tryptophan with a range of 10mg/l to 40mg/l to the medium to stimulate the production of CPT as effective precursor feeding method. The effect of feeding precursor on CPT production resulted in increase in the concentration of L-tryptophan to the optimum level of 20mg/l and produce 0.05 % of CPT. Further increasing the concentration of L-tryptophan, the yield of CPT level was decreased in the callus biomass (Fig. 2 C and D).

After the elongation of multiple shoot proliferation, the cultures were imperiled to the same medium with the addition of L-tryptophan (10 mg/l to 30mg/l) as a feeding precursor to enhance the production of CPT level. The highest accumulation of 0.0041 % of CPT was observed with 20mg/l of L-tryptophan, while increasing the optimal level of feeding precursor decreased in the yield of CPT.

Fig. 2 : (A) Flowering twig of *Nothapodytes nimmoniana* collected from Nilgiri hills of Tamilnadu, (B) Multiple shoot proliferation from leaf explants of *N. nimmoniana*, (C) Callus culture from leaf explants, (D) Callus biomass from leaf explants

Conclusion

Camptothecin is a potential anticancer and anti-AIDS alkaloid which is extremely expensive and its supply is inadequate on a global scale. The previous studies on biosynthesis of CPT showed that higher yield of biomass and bioactive molecules can be achieved by large-scale cultivation of cell cultures in bioreactors. Therefore, large scale cultivation of *N. nimmoniana* cells and organ cultures for production of CPT in bioreactors might be an alternative source to accomplish the global demands for this high value bioactive anticancer drug. CPT production in cells, tissues and organ cultures was attempted by several researchers in the last two decades to increase the CPT accumulation. Manipulation of culture strategies like immobilization of cells, precursor feeding, cell selection, elicitors treatments, hairy root culture, influence of various elements, nutrient stress, pH stress have been adopted to enhance *in vitro* biosynthesis of CPT in *N. nimmoniana*. It may be inferred that among the different culture strategies used for the higher production of CPT employed in cell cultures as shown in table 1, variation in concentration of plant growth regulators resulted maximum product accumulation.

Due to the high amount of CPT content in wood chips of *N. nimmoniana*, bark and roots, it is harvested to fulfill the commercial requirement and for getting 1 kg of CPT around 1,000–1,500 kg of *N. nimmoniana* wood chips are required and this leads to reduction in its population from forest habitats. The bark of *N. nimmoniana* is unfeasible for initiation of *in vitro* cell lines however the leaf, stem and young seeds of the plant were mostly used to initiate *in vitro* cell cultures (Karwasara and Dixit 2013). As it is a very rare species in wild, plant tissue culture will be a better option for enhancing CPT production and conserving this significant species and maintaining its genetic uniformity. So *in vitro* cell culture scheme gives promising results for CPT yield and it might reduce the load on its natural population when properly scale up and thorough culture medium requirements were established. Also, further critical analysis on detailed study of alternative species for the production of CPT is necessary.

References

Abdullaev F.I. (2001) Plant-derived agents against cancer. In: Gupta SK, Ed. Pharmacology and Therapeutics in the New Millennium. New Delhi: Narosa Publishing House, pp. 345–354.

Abdullah M.A., Ali A.M., Marziah M., Lajis N.H. and Ariff A.B. (1998) Establishment of cell suspension cultures of *Morinda elliptica* for the production of anthraquinones. Plant Cell Tissue Org. Cult. 54: 173-182.

Aerts R.J., Van der Leer T., Van der Heijden R. and Verpoorte R. (1990) Developmental regulation of alkaloid production in *Cinchona* seedlings. J. Plant Physiol. 13: 86–91.

Bodley A.F., Cumming J.N. and Shapiro T.A. (1998) Effect of camptothecin, a topoisomerase I inhibitor, on *Plasmodium falciparum*. Biochem Pharma. 55: 709–711.

Bodley A.L., Wani M.C., Wall M.E. and Shapiro T.A. (1995) Antitrypanosomal activity of camptothecin analogs structure activity correlations. Biochem. Pharmacol. 50: 937-942.

Bodley A.L. Shapiro T.A. (1995) Molecular and cytotoxic effects of Camptothecin, a topoisomerase I inhibitor, on trypanosomes and Leishmania. Proc. Natl. Acad. Sci. USA 92: 3726-3730.

CAMP. (2001) Conservation Assessment and Management Plan for medicinal plants in of Maharashtra State, FRLHT.

Chang S., Chen F., Tsay J., Chen J., Huang C., Lu W. and Hol C (2014) Establishment of Hairy Root Cultures of *Nothapodytes nimmoniana* to Produce Camptothecin. Taiwan J. For. Sci. 29: 193-204.

Chang S.H., Ho C.K., Tsay J.Y., Chen K.F. and Huang C.Y. (2007) Crown gall culture and Camptothecin production of *Camptotheca acuminata*. Taiwan J. For. Sci. 22: 395-404.

Ciddi V. and Shuler M.L. (2000) Camptothecin from callus cultures of *Nothapodytes foetida*. Biotech. Lett. 22: 129-132.

Cragg G.M., Schepartz S.A., Suffness M., Grever M.R. (1993) The taxol supply crisis. New NCI policies for handling the large-scale production of novel natural product anticancer and anti-HIV agents. J. Nat. Product 56: 1657–1668.

Dandin V.S. and Murthy H.N. (2012) Enhanced *in vitro* multiplication of *Nothapodytes nimmoniana* Graham using semisolid and liquid cultures and estimationof camptothecin in the regenerated plants. Acta Physiol. Plant 34: 1381–1386.

De Luca V., Balsevich J., Tyler R.T., Eilert U., Panchuk B.D., and Kurz W. G. W. (1986) Biosynthesis of indole alkaloid: developmental regulation of the biosynthesis pathway from tabersonine to vindoline in *Catharanthus roseus*. J. Plant Physiol. 125: 147 – 156.

Fulzele D.P. and Satdive R.K. (2003) Somatic embryogenesis, plant regeneration and the evaluation of the camptothecin content in *Nothapodytes foetida. Soc In vitro Biol* 39: 212-216.

Fulzele D.P. and Satdive R.K. (2005) Comparison of techniques for the extraction of the anti-cancer drug camptothecin from *Nothapodytes foetida*. J. Chromatogr. A. 1063: 9-13.

Fulzele D.P. and Satdive R.K. (2006) Distribution of anticancer drug camptothecin in *Nothapodytes foetida*. Fitoterapia 76: 643-648.

Fulzele D.P., Satdive R.K. and Pol B.B. (2001) Growth and production of camptothecin by cell suspension cultures of *N. foetida*. Planta Med. 67: 150–152.

Fulzele D.P., Satdive R.K. and Pol B.B. (2002) Untransformed root cultures of *Nothapodytes foetida* and production of camptothecin. Plant Cell Tissue Organ Cult. 69: 285-288.

Govindachari T.R. and Viswanathan N. (1972) Alkaloids of *Mappia foetida*. Phytochemistry 11: 3529–3531.

Gunasekera S.P., Badawi M.M., Cordell G.A., Farnsworth N.R. and Chitnis M. (1979) Plant anticancer agents X. Isolation of camptothecin and 9-methoxycamptothecin from *Ervatamia heyneana*. J. Nat. Prod. 142: 475-477.

Hsiao H.Y., Cheng T.J., Yang G.M., Huang I.J. and Chen R.L. (2008) Determination of camptothecins in DMSO extracts of *Nothapodytes foetida* by direct injection capillary electrophoresis. Phytochem Anal. 19: 136-140.

Hu Z.B. and Du M. (2006) Hairy root and its application in plant genetic engineering. J. Integ. Plant Biol. 48:121-7.

Isah T. and Mujib A. (2014) *In vitro* propagation and camptothecin production in *Nothapodytes nimmoniana* Plant Cell Tissue Organ Cult. DOI 10.1007/s11240-014-0683-1.

Karuppusamy S. (2009) A review on trends in production of secondary metabolites from higher plants by in vitro tissue, organ and cell cultures. J. Med. Plants Res. 3: 1222–1239.

Karwasara VS, Dixit VK, (2013) Culture medium optimization for camptothecin production in cell suspension cultures of *Nothapodytes nimmoniana* (J. Grah.) Mabberley. Plant Biotechnol Rep 7: 357–369.

Kehie M., Kumaria S., Tandon P. and Ramchiary N. (2014) Biotechnological advances on *in vitro* capsaicinoids biosynthesis in capsicum: a review. Phytochem Rev. DOI 10.1007/s11101-014-9344-6.

Kelland L.R. (2000) Flavopiridol, the first cyclin-dependent kinase inhibitor to enter the clinic: current status. Expert Opin. Invest Drugs 9: 2903–2911.

Khan S., Mirza K.J., Tyagi M.R. and Abdin M.Z. (2011) Development of RAPD markers for authentication of *Ruta graveolens* (L.) and its adulterant. MAPSB 5: 58-61.

Khan N., Tamboli E.T., Sharma V.K. and Kumar S. (2013) Phytochemical and pharmacological aspects of *Nothapodytes nimmoniana*, an overview. Herva Pelon. 59: 53-56.

Li C.Y., Lin C.H. and Wu T.S. (2005) Quantitative analysis of camptothecin derivatives in *Nothapodytes foetida* using ^{1}H-NMR method. Chem. Pharm. Bull. (Tokyo). 53: 347-349.

Liu Z., Carpenter S., Bourgeois W., Yu Y., Constantin R., Falcon M. and Adams J. (1998) Variations in secondary metabolite camptothecin in relation to tissue age and season in Camptotheca acuminata. Tree Physiol. 18: 265- 270.

Lopez-Meyer M., Nessler C.L. and McKnight T.D. (1994) Sites of accumulation of the antitumor alkaloid camptothecin in *Camptotheca acuminata*. Planta Med. 60: 558–560.

Merillon J.M., Rideau M. and Chenieux J.C. (1984) Influence of sucrose on levels of ajmalicine, serpentine and tryptamine in *Catharanthus roseus* cells *in vitro*. Planta Med. 50: 561–567.

Meyers A.I., Nolene R.L., Collington R.W., Narwid T.A. and Strickland R.C. (1973) A total synthesis of camptothecin and deethyldeoxycamptothecinl. J. Org. Chem. 38: 1974 – 1982.

Nabila S.K., Fawzia M., Jawad N.A. and Rida A.S. (2003) Growth and rosmarinic acid accumulation in callus, cell suspension, root cultures of wild *Salvia fruticosa*. Plant Cell Tissue Organ Cult. 73: 117–121.

Padmanabha B.V.M., Chandrashekar B.T., Ramesha G.H.C., Hombe R.P., Gunaga S., Suhas R., Vasudeva Ganeshaiah K.N. and Umashaanker R. (2006) Patterns of accumulation of camptothecin, an anti-cancer alkaloid in *Nothapodytes nimmoniana*, in the Western Ghats, India: Implications for identifying high yielding sources of the alkaloid. Curr. Sci. 90: 1-10.

Patwardhan A. (2010) Domestication of *Nothapodytes nimmoniana* (Grah.) Mabb. : an endangered medicinal tree from Western Ghats of India. Project report submitted to The Rufford Small Grants Foundation. Research and Action in Natural Wealth Administration (RANWA), C-26/1 Ketan heights, Kothrud, Pune, India.

Premalakshmi V., Devi Chitra S. and Theradimani M. (2012) Effect of camptothecin a potent anti cancer drug from *Mappia foetida* on breast cancer cell line MDA-MB-231. J. Med. Plants Res. 6: 5522-5526.

Puri S.C., Amna T., Khajuria A., Gupta A., Arora R., Spiteller M. and Qazi G.N. (2007) Immunomodulatory activity of an extract of the novel fungal endophyte *Entrophospora infrequens* isolated from *Nothapodytes foetida* (Wight) Sleumer. Acta Microbiol. Immunol. Hung. 54: 237-260.

Puri S.C., Handa G., Bhat B.A., Gupta V.K., Amna T., Verma N., Anand R., Dhar K.L. and Qazi G.N. (2005) Separation of 9-methoxycamptothecin and camptothecin from *Nothapodytes foetida* by semipreparative HPLC. J. Chromatogr Sci. 43: 348-350.

Puri S.C., Handa G., Gupta R.K., Srivastava T.N., Somal P. and Sharma S.N. (1999) Quantitation of camptothecin in Nothapodytes foetida. J. Indian Chem. Soc. 76: 370–371.

Rehman S., Shawl A.S., Verma V., Kour A., Athar M., Andrabi R., Sultan P. and Qazi G.N. (2008) An endophytic *Neurospora* sp. from *Nothapodytes foetida* producing camptothecin. Prikl. Biokhim. Mikrobiol. 44: 225-231.

Roja G. (2006) Comparative studies on the camptothecin content from Nothapodytes foetida and *Ophiorrhiza* species. Nat. Prod. Res. 20: 85-88.

Roja G. and Heble M.R. (1994) The quinoline alkaloids camptothecin and 9-methoxy camptothecin from tissue cultures and mature trees of *Nothapodytes foetida*. Phytochemistry 36: 65–66.

Saito K., Sudo H., Yamazaki M., Koseki-Nakamura M., Kitajima M., Takayama H. and Aimi N. (2001) Feasible production of camptothecin by hairy root culture of *Ophiorrhiza pumila*. Plant Cell Rep. 20: 267-271

Sakato K. and Misawa M (1974) Effects of chemical and physical conditions on growth of *Camptotheca acuminata* cell cultures. Agric. Biol. Chem. 38: 491–497.

Sakato K., Tanaka H., Mukai N. and Misawa M. (1994) Isolation and identification of camptothecin from *Camptotheca acuminata* suspension cultures and mature trees of Nothapodytes foetida. Phytochemistry 36: 65–66.

Sapna S.M., Maheswari A.U., Suganthi T.K. and Ravi. (2004) WHO Guidelines for Herbal Drug Standardization, http://www.Pharmainfo.net

Sharma P., Gurav A.M. and Sharma A. (2014) Micropropagation and estimation of camptothecin and 9-methoxycamptothecin variation in regenerated plants of *Nothapodytes foetida*. Nat. Prod. Chem. Res. 2: 154.

Sriram D., Yogeeswari P., Thirumurugan R. and Bal T.R. (2005) Camptothecin and its analogues: a review on their chemotherapeutic potential, Nat. Prod. Res. 19: 393-412.

Stahelin H. (1973) Actvity of a new glycosidic lignan derivative (VP-16-213) related to podophyllotoxin in experimental tumors. Eur. J. Cancer 9: 215-221.

Sudo H., Yamakawa T., Yamazaki M., Aimi N. and Saito K. (2002) Bioreactor production of Camptothecin by hairy root cultures of *Ophiorrhiza pumila*. Biotechnol. Lett. 24: 359-363.

Sundravelan R., Desireddy B. and Ciddi V. (2004) Production of Camptothecin from callus culture of *N. foetida* (Wright) Sleumer sian. J. Biotechnol. 3: 452–453.

Tafur S., Nelson J.D., Delong D.C. and Svoboda G.H. (1976) Antiviral components of *Ophiorrhiza mungos* isolation of camptothecin and 10–methoxycamptothecin. Lloydia 39: 261-262.

Thengane S.R., Kulkarni D.K., Shrikhande V.A., Joshi S.P., Sonawane K.B. and Krishnamurthy K.V. (2003) Influence of medium composition on callus induction and camptothecin (s) accumulation in *Nothapodytes foetida.* Plant Cell Tissue Org. Cult. 72: 247-251.

Van-Hengel A.J., Harkes M.P., Wichers H.J., Hessleink P.G.M. and Buitelaar R.M. (1992) Characterization of callus formation and Camptothecin production by cell lines of *Camptotheca acuminata.* Plant Cell. Tissue Org. Cult. 28: 11–18.

Voss C., Eyol E. and Berger M.R. (2006) Identification of potent anticancer activity in Ximenia americana aqueous extracts used by African traditional medicine. Toxicol. Appl. Pharmacol. 211: 177 – 187

Wall M.E. (1998) Camptothecin and taxol: discovery to clinic. Med. Res. Rev. 18: 299-314.

Wall M.E and Wani M.C. (1986) In The Camptothecins from Discovery to the Patient, P. Patazis, B.C. Giovanella, M.L. Rothinberg (Eds), pp. 1–12, The New York Academy of Sciences, New York.

Wall M.E., Wani M.C., Cook C.E., Palmer K.H., McPhail A.T. and Sim G.A. (1966) Plant antitumour agents I. The isolation and structure of camptothecin, a novel alkaloidal leukaemia and tumour inhibitor from *Camptotheca acuminata*. J. Am. Chem. Soc. 88: 3888–3889.

Weidenfeld H., Furmanowa M., Roeder E., Guzewska J. and Gustowski W. (1997) Camptothecin and 10-hydroxycamptothecin in callus and plantlets of *Camptotheca acuminata*. Plant Cell Tissue Organ Cult. 49:213–218.

Werbovetz K.A., Bhattacharjee A.K., Brendle J.J. and Scovill J.P. (2000) Stereoelectronic Features Modulating the Biological Activity of Camptothecin Analogs. Bioorg. Med. Chem. 8: 1741-1747.

Williams S.D., Birch R., Einhorn L.H., Irwin L., Greco F.A. and Loehrer P.J. (1987) Treatment of disseminated germ-cell tumors with cisplatin, bleomycin, and either vinblastine or etoposide. New Engl. J. Med. 316: 1435-1440.

Wu S.F., Hsieh P.W., Wu C.C., Lee C.L., Chen S.L., Lu C.Y., Wu T.S., Chang F.R. and Wu Y.C. (2008) Camptothecinoids from the seeds of Taiwanese *Nothapodytes foetida.* Molecules 16: 1361-1371.

Zhao Y.X., Gao J.Q., Qiao H.L., Chen H.L. and Liang W.Q. (2006) Development and validation of a sensitive reversed-phase HPLC method to determine intracellular accumulation of hydroxycamptothecin. J. Pharm. Biomed. Anal. 41: 1007–1010.

3

In vitro Propagation of Oak (*Quercus* sp.) of Kumaun Himalaya

Aseesh Pandey and K. Chandra Sekar*
G.B. Pant Institute of Himalayan Environment & Development
Kosi - Katarmal, Almora - 263 643, Uttarakhand, India
**E-mail: kcsekar1312@rediffmail.com*

Abstract

Quercus species of family Fagaceae are very important from both economic and ecological perspectives. These exhibit irregular fruiting, and produce seeds recalcitrant to storage, thereby difficult to propagate through conventional procedures. Again because of their very difficult to root nature, extremely limited success has been observed in vegetative propagation. Application of new technological approaches like *in vitro* propagation may be of interest for accelerating tree improvement programs to these hardwood species. This review provides a summary of the advancement in the application of *in vitro* propagation to oak species of Kumaun West Himalaya India.

Keywords: Quercus; *in vitro* propagation, Kumaun, Himalaya.

Introduction

Forest plays a dynamic role in maintaining and protecting the complex, diverse and fragile mountain ecosystems. The major component for the formation of forest communities is the woody species (Khumbongmayum *et al.* 2005) and trees are the largest and among the most complex organisms in the world with millions of diverse life forms (Zobel and Talbert 1984). *Quercus,* an angiospermic genus of family Fagaceae is anemophilous (wind pollinated) with trees or shrubs, either deciduous or evergreen found in the northern temperate zone, subtropical and tropical Asia, and the Andes of South America (Shrestha 2003). This genus is widely distributed in the northern hemisphere with 500 species (Yong-Wook 2000). Nearly 400 species distributed throughout the temperate regions of the world (Johnson *et al.* 2002). There are more than 35 species of *Quercus* reported from Himalayan region (Negi and Naithani 1995). Out of these, five evergreen species, namely *Q. leucotrichophora* A. Camus. (banj oak), *Q. semecarpifolia* Sm. (kharsu; brown oak), *Q. glauca* Thunb. (phaniyat oak), *Q. lanuginosa* Don. (rianj oak) and *Q. floribunda* Lindl. (tilonj oak) have been found growing in the central Himalaya between 1000 and 3600 m asl (Champion and Seth 1968) and represent the climax vegetation for the region (Singh and Singh 1992). *Quercus* species in general are late successional dominants in drier forests and occur as mid successional on the more mesic sites (Spurr and Barnes 1973).

In the present scenario natural forests of oaks are degrading because of over exploitation, irregular fruiting and low seedling recruitment rate (Pandey 2013). Major factors responsible for over exploitation are-unplanned urbanization, large population, increasing demand for fuel, fodder and timber other than these climate change is also a contributing factor. Besides, oak decline syndrome affects oaks worldwide and is of great concern. It is an episodic and gradual phenomenon characterized by a general loss of vigor caused by complex interactions between different biotic and abiotic factors and which is not limited to any one species or group of species (Thomas *et al.* 2002; Repo *et al.* 2008). Some major disease identified throughout the world are depicted in Table-1.

These threats highlight a need to take conservation action, however, *ex situ* conservation of oak has challenges (Toppila 2012). Recalcitrant acorns lose viability when dried, making their long-term storage in seed bank facilities futile (Connor 2004). Tissue culture and cryopreservation are challenging for most oaks due to the presence of tannins (Kramer and Pence 2012), though this has been carried out successfully in some species. Several propagation approaches like, stem cutting, air-layering, somatic embryogenesis, micropropagation *etc.* instead of natural means have been employed to protect, conserve and sustainably use these important species throughout the word. *In vitro* propagation techniques *viz.* micropropagation provides a means of large scale plant propagation without transmitting disease in to the progeny within a short time span. Enormous progress has been made in the application of biotechnological tools to the most important species of *Quercus* during the last decade. This study comprises current status of oak propagation in India and concentrates on the present status of six species growing in central Himalaya.

Table 1: Major diseases identified in genus *Quercus*

S. No.	Pathogen	Type	Effected part	Disease	References
1.	*Armillaria* sp.	Fungal	Root system	Root colonizing	Marçais and Breda 2006
2.	*Biscoyniauxia mediterranea*		Bole bark	Bole colonizing	Marçais and Breda 2006
3.	*Agrilus* sp.	Insect	Bark	Opportunistic parasite	Marçais and Breda 2006
4.	*Tortrix viridiana*	Moth	Leaf	Green oak leaf roller	Schroeder and Degen 2008
5.	*Erysiphe alphitoides*	Fungus	Leaf	Oak powdery mildew	Hajji *et al.* 2009
6.	*Phytophthora* sp.	Fungus	Root	Lesions at root	Vettraino *et al.* 2002
7.	*Phytophthora ramorum*	Fungus	Whole tree	Sudden oak death	Brown and Allen Diaz 2009

Source : Vieitez *et al.* (2012)

In vitro propagation

There are two most studied methods for *in vitro* propagation of *Quercus* species ‘i’ indirect *in vitro* propagation or propagation through somatic embryogenesis and ‘ii’ direct *in vitro* propagation or direct shoot initiation from nodes/meristematic parts. *Somatic embryogenesis* has been regarded as the *in vitro* system of choice with the potential for eventual mass propagation of superior and genetically engineered forest tree genotypes in both coniferous and hardwood species (Gupta *et al.* 1991). In somatic embryogenesis embryo or whole plant is derived from a single somatic cell or somatic tissue (leaf, protoplast *etc.*) instead of genetic cells and organogenesis from explants occur either *(direct)* without formation of callus and/or organogenesis takes place following to callus induction *(indirect)*

(Fig. 1d). Somatic embryogenesis is also perceived the best system for propagation of superior genotypes, mostly because both root and shoot meristems are present simultaneously (Kim 2000). Somatic embryogenesis has been reported for some *Quercus* species by Wilhelm (2000), but as far as Indian oak species are concern only *Q. floribunda* (Purohit *et al.* 2002c) and *Q. semecarpifolia* (Tamta *et al.* 2009) has been propagated using somatic embryogenesis method (Table 2). In the case of *Q. serrata* somatic embryos were produced but further organogenesis was not observed even after six month of culturing embryonic callus and became black (Pandey 2013).

Direct in vitro propagation is a method in which meristematic parts (nodes) used as explant, either derived from *in vitro* grown seedlings i.e. axillary shoot *Q. serrata* (Ide and Yamamoto 1987), cotyledonary node *Q. floribunda* (Purohit *et al.* 2002a); *Q. leucotrichophora* and *Q. glauca* (Purohit *et al.* 2002b), *Q. leucotrichophora* (Pandey and Tamta 2012) (Fig. 1a), *Q. serrata* (Pandey and Tamta 2014) (Fig. 1g-j), petiolar tube *Q. semecarpifolia* (Tamta *et al.* 2008) etc. or from natural/ *ex vitro* conditions i.e. nodal explants derived from mature tree *Q. serrata* (Pandey 2013), stump sprouts/ forced-flushed cuttings *Q. alba, Q. bicolor* and *Q. rubra* (Vieitez *et al.* 2009) etc. (Table 2).

Fig. 1: *In vitro* propagation studies in Kumaun West Himalayan *Quercus spp.* a-d. *Quercus leucotrichophora* (a. shoot multiplication in *in vitro* derived nodal segments; b. rhizogenesis; c. acclimatization of *in vitro* raised micro shoots; d. indirect somatic embryogenesis), e-f. *Q. floribunda* (e. Inoculation of nodal explants derived from mature tree; f. shoot bud induction), g-j. *Q. serrata* (g. *in vitro* seed germination; h. *in vitro* shoot multiplication; i. *in vitro* rooted plantlets j. well-developed *in vitro* grown plant of *Q. serrata* facing direct sunlight). Bar indicating 1 cm scale.

Table 2 : Studies on micropropagation through biotechnological approaches in Indian oak species

S.N.	Quercus species	Explant	Medium	Response	PGRs	References
1.	*Q. floribunda*	Cotyledonary node	MS, WP	Multiple shoots, plantlets	BA, BA+GA3	Purohit *et al.* 2002a
2.	*Q. glauca*	Intact embryos and cotyledonary nodes	MS, WP	Multiple shoots, plantlets	BA, BA+GA3, BA+IBA	Purohit *et al.* 2002b
3.	*Q. leucotrichophora*	Intact embryos and cotyledonary nodes	MS, WP	Multiple shoots, plantlets	BA, BA+GA3, BA+IBA	Purohit *et al.* 2002b
4.	*Q. floribunda*	Cotyledonary node	MS, WP	Somatic embryos, plantlets	BA, BA+NAA	Purohit *et al.* 2002c
5.	*Q. semecarpifolia*	Petiolar tube	MS, WP,	Multiple shoots, plantlets	BA	Tamta *et al.* 2008, Pandey 2013
6.	*Q. leucotrichophora*	Nodal segments derived from *in vitro* raisedseedlings	MS, WP	Multiple shoots,plantlets	BA, 2-ip, Kinetin	Pandey and Tamta 2012
7.	*Q. serrata*	Cotyledonary nodes	WP	Multiple shoots, plantlets	BA, BA+GA3, BA+IAA	Pandey and Tamta 2014
8.	*Q. serrata*	Intact embryos	MS, WA,	Plantlets WP	Different light conditions	Pandey and Tamta 2013a
9.	*Q. leucotrichophora*	Intact embryos	WA, WP	Multiple shoots,plantlets	BA	Pandey and Tamta 2013b
10.	*Q. semecarpifolia*	Nodal segments derived from mature tree	WA, WP	Multiple shoots	BA,Kinetin BA+GA_3	Pandey and Tamta 2013b
11.	*Q. semecarpifolia*	Cotyledonary node	WP, GD	Somatic embryos, plantlets	BA, GA_3	Tamta *et al.* 2009

MS: Murashige and Skoog 1962; WP: Woody Plant (Lloyd and McCown 1980); GD: Gresshoff and Doy, 1972; WA: Water Agar medium; 2iP: 2-isopentyladenine; BAP: 6-benzylaminopurine; IBA: Indole-3-butyric acid; IAA: Indole-3- acetic acid; NAA: α-Naphthalene acetic acid; GA3 :Gibberellic acid.

Shoot Multiplication

The most widely used plant regulator at the shoot multiplication stage is 6-benzylaminopurine (BA) (used at different concentrations either alone or in combination of others -GA_3, NAA IBA, IAA (Table 2), although the effect of thidiazuron in stimulating shoot proliferation was also evaluated by Chalupa (1988), Pandey and Tamta (2014).

Shoot multiplication in oak cultures does not occur profusely but by the production of only a few, elongated shoots (Fig. 1f, h). Cultures establishment is also difficult due to basal callus formation and necrosis. Various growth adjuvants were studied with species specific responces, casein hydrolysate (CH, 500 mg L^{-1}) promoted shoot multiplication rate significantly in comparison to silver nitrate and also eliminated the basal callus formation in *Q. serrata* (Pandey and Tamta 2014), ascorbic acid (AA, 0.02%, w/v) promoted shoot initiation by reducing phenolic excretion in *Q. semecarpifolia* (Tamta *et al.* 2008), citric acid (CA, 20mg/l) was added to control leaching of phenolic compounds and avoiding basal callus formation in *Q. leucotrichophora* (Pandey and Tamta 2012) (Fig. 1a). Sometime elongation of shoots might be required prior to *in vitro* rhizogenesis. Elongation medium differs from multiplication medium only in that the concentration of BA was either reduced (San-José *et al.* 1988) or addition of GA_3 along with BA in multiplication medium. Incorporation of GA_3 to the multiplication medium improved the proliferation of brown oak (Pandey 2013). Vieitez *et al.* (2009), observed that in American oak, replacement of BA with Zeatin in multiplication medium improved shoot cultures. During shoot multiplication of *Q. semecarpifolia*, it was observed that shoot multiplication was increase after successive subcultures but there was a concomitant decrease in shoot height (Tamta *et al.* 2008).

Rooting and Hardening

Microshoots exhibited a good rooting response in WPM medium especially half strength WPM was preferred for rhizogenesis. IBA and NAA are the most tested auxins in genus *Quercus*. Continuous exposure of auxins to microshoot resulted basal callus formation in *Q. leucotrichophora* and *Q. glauca* (Purohit *et al.* 2002b); *Q. semecarpifolia* (Tamta *et al.* 2008); *Q. leucotrichophora* (Pandey and Tamta 2012) (Fig. 1b); *Q. serrata* (Pandey and Tamta 2014) (Fig.1i); *Q. rubra* (Vengadesan and Pijut 2009).Therefore, two-step rooting procedure is quite efficient and frequently used now a days. In two step rooting procedure, the excised microshoots were inoculated in higher concentrations of IBA (50-100μM) for 24, 48, 72h or 1week in WPM and then transferred in plant growth regulator (PGR) free half strength WPM supplemented with (3%, w/v) sucrose, (0.24%, w/v) claregel but devoid of PGRs. Root induction in microshoots starts appearing visually within 10 days of culture. This method of rooting was also found suitable in *Q. suber* (Manzanera and Pardos, 1990); temprate bamboo (Bag *et al.* 2000); *Q. glauca, Q. floribunda and Q. leucotrichophora* (Purohit *et al.* 2002a,b); *Q. semecarpifolia* (Tamta *et al.* 2008), *Q. rubra* (Vengadesan and Pijut 2009); *Berberis chitria* (Pandey *et al.* 2013) *Q. serrata* (Pandey and Tamta 2014). In high phenolic secreting species addition of activated charcoal to the medium is important for the reduction of light intensity at the base of the microshoots and also for the phenolics absorption.

A disadvantage of micropropagation is the high mortality rate during the transfer of *in vitro* plantlets to *ex vitro* (Lovato *et al.* 1999). It may be a result of a lack of vascular connection as in *Q. rubra* (Vengadesan and Pijut 2009). Since *in vitro* raised plantlets are very delicate and sensitive to *ex vitro* conditions so their acclimatization takes place gradually. Rooted plantlets were transplanted into the mixture of soil and farmyard (3:1) (Fig. 1c) and immediately covered under a plastic bags for maintaining high air humidity and maintained in culture room conditions (25±2°C under 16/8h (light/ darkness) photoperiod with an irradiance of 60 μmol $m^{"2}$ $s^{"1}$). After 3 weeks plastic bags were removed to reduce humidity gradually to normal. The rooted plants were successfully acclimatized under greenhouse conditions and subsequently transferred to the garden soil into the bigger containers and placed outside. A detailed schematic representation for the micropropagation procedure of *Quercus serrata* Thunb is depicted in Fig. 2.

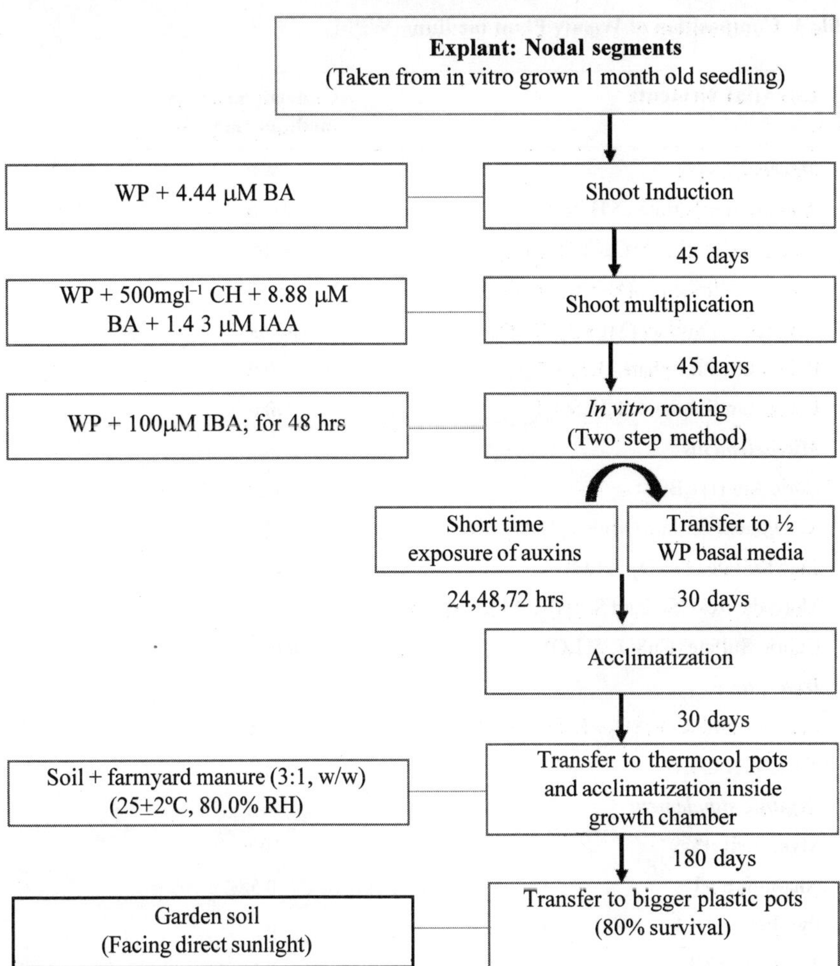

Fig. 2 : Schematic representation for the micropropagation procedure of *Quercus serrata* Thunb. (*Source:* Pandey and Tamta, 2014)

Table 3: Composition of Woody Plant medium (WPM)

S. No.	Essential elements	Concentration in medium (mg l"[1])
	Macroelements	WP
1.	Ammonium Nitrate (NH_4NO_3)	400.0
2.	Calcium Chloride * ($CaCl_2 \cdot 2H_2O$)	96.00
3.	Calcium Nitrate ($Ca(NO_3)_2$. $4H_2O$	386.34
4.	Magnesium Sulfate ($MgSO_4 \cdot 7H_2O$)	180.69
5.	Potassium Phosphate (KH_2PO_4)	170.0
6.	Potassium Sulphate (KH_2SO_4)	990.0
	Microelements	
2.	Boric Acid (H_3BO_3)	6.20
3.	Manganese Sulfate ($MnSO_4 \cdot H_2O$)	22.3
4.	Zinc Sulfate ($ZnSO_4 \cdot 7H_2O$)	8.60
5.	Molybdic Acid (Sodium Salt) ($Na_2MoO_4 \cdot 2H_2O$)	0.25
6.	Cupric Sulfate ($CuSO_4 \cdot 5H_2O$)	0.025
	Iron source	
1.	Ferrous Sulfate ($FeSO_4 \cdot 7H_2O$)	27.8
2.	$Na_2EDTA \cdot 2H_2O$	37.3
	Organic supplement	
1.	Myoinositol*	100
2.	Nicotinic acid	0.5
3.	Pyridoxine-HCl	0.5
4.	Thiamine-HCl	0.5
5.	Glycine	2.0
	Carbon source*	
1.	Sucrose	30 000
	Jelling agent*	
1.	Agar-agar	0.8%
2.	Phytagel/ clarigel	0.24%

*directly added per liter of medium. *Source* : Lloyd and McCown (1981)

Conclusion

Conventional propagation is very tidies task and comprises of several steps. Species of the genus *Quercus*, such as *Q. semecarpifolia*, *Q. robur* and *Q. petraea*, usually reach seed-bearing age at between 35, 40 and 100 years old, and the mast seed crops vary according to the individual tree and year (Ducousso and Bordacs 2004). *Quercus* are likely to produce good seed crops once every 3-5 years. In addition, vegetative propagation is of extremely limited success because it is very difficult to root cuttings of *Quercus* species if they are taken from selected mature trees. Grafting is another method of vegetative propagation for oaks, but delayed compatibility for grafting has been observed in them. Based on these limitations in conventional propagation approaches biotechnological techniques like *in vitro* propagation has been proved to be an alternative and suitable means of effective propagation. For *in vitro* propagation nutrient media plays an important role, since it is required for growing plants in controlled conditions. It is a composition of macro nutrients, micro nutrients, vitamins, iron source and carbon source, either jelled with agar-agar/ phytagel and/or liquefied, containing plant growth regulators (PGRs) and/or PGR free (Table 3). The composition of the nutrient media is not decisive in initiating cultures, various media are available for the *Quercus* propagation worldwide. Woody Plant Medium (WPM) (Lloyd and McCown 1980) and Gresshoff and Doy (1972) medium (GD) were superior to other nutrient media for shoot multiplication cultures viz. Murashige and Skoog (1962) (MS) (Schenk and Hilderbrandt 1972) (SH). For Indian species WPM was found to be the best medium for *in vitro* propagation (Table 2). Composition of WPM is depicted in table-3. Besides propagation, through *in vitro* propagation techniques transgenic trees may constitute, which could be an important tool for speeding up breeding cycles (Flachowski *et al.* 2009), developing disease resistant varieties (Fenning 2006), improving wood quality (Haggman *et al.* 2006).

In addition, efforts should be involve for the isolation and characterization of genes related to specific traits *i.e.* disease resistance, wood quality, tree form, which would help for improving these species. Further taxonomic classification of *Quercus* genus has ambiguities and it is represented by 200 (Neger and Münch 1950), 320 (Krahl-Urban 1959), 400 (Johnson *et al.* 2002), 450 (Krussmann 1978) or 500 species (Yong-Wook 2000). The differences in these numbers are partly explained by the definition of interspecific hybrids as separate species and the sub-division of ecological forms into species (Kleinschmit 1993). Molecular characterization of these species is need of the hour since identification of suitable genotype is prerequisite for any propagation and breeding programme.

References

Bag N., Chandra S., Palni L.M.S. and Nandi S.K. (2000) Micropropagation of Devringal (*Thamnocalamus spathiflorus*) A temperate bamboo, and comparison of *in vitro* propagated plants and seedlings. Plant Sci. 156: 125-135.

Chalupa V. (1988) Large scale micropropagation of *Quercus robur* L. using adenine-type cytokinins and thidiazuron to stimulate shoot proliferation. Biol. Plant. 30:414-421.

Champion H.G. and Seth S.K. (1968) A revised survey of the forest types of India, Manager of Publications, Delhi.

Connor K. (2004) Storing acorns. Native Plants J. 52: 161–166.

Ducousso A. and Bordacs S. (2004) EUFORGEN Technical Guidelines for genetic conservation and use for pedunculate and sessile oaks (*Quercus robur* and *Q. petraea*). International Plant Genetic Resources Institute, Rome, Italy pp.6

Fenning T.M. (2006) The use of genetic transformation procedures to study defence and disease resistance traits of trees. In: Fladung M, Ewald D (eds) Tree transgenesis. Springer, Heidelberg, pp 201-234.

Flachowski H., Hanke M.V., Peil A., Strauss S.H., Fladung M. (2009) A review on transgenic approaches to accelerate breeding of woody plants. Plant Breed. 128:217-226.

Gresshoff P.M. and Doy C.H. (1972) Development and differentiation of haploid *Lycopersicon esculentum* (tomato). Planta 107: 161–170.

Gupta P.K., Timmis R., Pullman G., Yancey M., Kreitinger M., Carlson W. and Carpenter C. (1991) Development of an embryogenic system for automated propagation of forest trees. In: Vasil IK (Ed) Scale-up and Automation in Plant Propagation. Cell Culture and Somatic Cell Genetics of Plants Vol. 8. Academic Press, San Diego. pp 75–93.

Häggman H., Niemi K., Timonen H., Ylioja T. and Chiang V. (2006) Environmental aspects of lignin modified trees. In: Fladung M, Ewald D (eds) Tree Transgenesis. Springer, Heidelberg, pp 105-117.

Ide Y. and Yamamoto S. (1987) *In vitro* plantlet regeneration from axillary buds of juvenile seedlings of konara (*Quercus serrata*). J Jap For Soc 69: 109–112.

Johnson P.S., Shifley S.R. and Rogers R. (2002) The ecology and silviculture of oaks, CABI, New York, NY, pp. 503.

Khumbongmayum A.D., Khan M.L. and Tripathi R.S. (2005) Sacred groves of Manipur, northeast India: Biodiversity value, status and strategies for their conservation. Biodiv. and Cons. 14(7):1541-1582.

Kim Y. (2000) Somatic embryogenesis in *Quercus acutissima*. In Jain, S., Gupta, P. & Newton, R. (Eds), Somatic Embryogenesis in Woody Plants. Vol. 6. Dordrecht: Kluwer Academic Publishers. pp. 671-686.

Kleinschmit J. (1993) Interspecific variation of growth and adaptive traits in European oak species. Ann. Sci. For. 50 (1): 166-185.

Krahl-Urban J. (1959) Die Eichen. Paul Parey Verlag, Hamburg, 288 p.

Kramer A.T. and Pence V.C. (2012) The challenges of *ex situ* conservation for oaks. J Int. Oak Soc. 23:90-108.

Krussmann G. (1978) Hanbuch der Laubgehölze. Paul Parey Verlag, Berlin, 79-115.

Lloyd G. and McCown L.B. (1980) Commercially feasible micropropagation of mountain laurel, Kalmia latifolia, by use of shoot-tip culture. Comb Proc Intl Plant Prop. Soc. 30:421–427.

Lovato P.E., Schuepp H., Trouvelot A. and Gianinazzi S. (1999) In Mycorrhiza (eds Varma A and Hook B). Springer. Verlag. pp. 443-468.

Manzanera J.A. and Pardos J.A. (1990) Micro-propagation of juvenile and adult *Quercus suber* L. Plant Cell Tiss Org Cult 21:1-8.

Murashige T. and Skoog F. (1962) A revised medium for rapid growth and bioassays with tobacco tissue cultures. Physiol. Plant 15:473-497.

Neger F.T. and Münch E. (1950) Die Laubhölzer. Sammlung Göschen de Gruyter, Berlin Bd 718, pp.160.

Negi S.S. and Naithani H.B. (1995) Oaks of India, Nepal and Bhutan, Dehradun: International Book Distributors, India.

Pandey A. and Tamta S. (2012) Influence of kinetin on *in vitro* rooting and survival of banj oak (*Quercus leucotrichophora* L.) Afr. J. Biotech. 11(62):12538-12545. DOI: 10.5897/AJB12.161

Pandey A. and Tamta S. (2013a) Effects of different light conditions on *in vitro* seed germination in *Quercus serrata* Thunb. Assessment and conservation of forest genetic resources through biotechnological interventions. S. Singh and R Das ed(s) Next print, Kishoreganj, Road No.2 Ranchi-834001 (Jharkhand) India, pp.150-155.

Pandey A. and Tamta S. (2013b) Conservation of Banj and Kharsu Oak. Agroforestry and climate change management. A. Arunanchalam, J.C. Dagar and A.K. Singh ed (s) Pointer publishers, Jaipur 302003 (Raj.) India, pp.88-96.

Pandey A. (2013) Propagation of two species of Oaks using *in vitro* methods. Ph. D Thesis submitted to Kumaun University Nainital.

Pandey A., Brijwal L. and Tamta S. (2013) *In vitro* propagation and phytochemical assessment of *Berberis chitria*: An important medicinal shrub of Kumaun Himalaya, India J. Med. Plant Res. 7(15):930-937. DOI 10.5897/JMPR13.4435.

Pandey A. and Tamta S. (2014) *In vitro* propagation of the important tasar Oak (*Quercus serrata* Thunb.) by casein hydrolysate promoted high frequency shoot proliferation J. Sust. For. 33:590–603. DOI: 10.1080/10549811.2014.912587.

Purohit V.K., Palni L.M.S., Nandi S.K. and Rikhari H.C. (2002a) *In vitro* plant regeneration through cotyledonary nodes of *Quercus floribunda* Lindl. ex A. Camus (Tilonj oak), a high value tree species of central Himalaya. Curr. Sci. 833: 101–104.

Purohit V.K., Tamta S., Chandra S., Vyas P., Palni L.M.S. and Nandi S.K. (2002b) *In vitro* multiplication of *Quercus leucotrichophora* and *Q. glauca*: important Himalayan oaks. Plant. Cell Tiss. Org. Cult. 69:121–133.

Purohit V.K., Palni L.M.S., Nandi S.K., Vyas P. and Tamta S. (2002c) Somatic embryogenesis in *Quercus floribunda*, an important Central Himalayan oak. In: Role of plant tissue culture in biodiversity conservation and economic development. Nandi SK, Palni LMS, Kumar A (eds.) Gyanodaya Prakashan, Nainital, pp. 41-52.

Repo T., Mononen K., Alvila L., Pakkanen T.T. and Hänninen H. (2008) Cold acclimation of pedunculate oak (*Quercus robur* L.) at its northernmost distribution range Environ. Exp. Bot. 63:59-70.

San Jose M.C., Ballester A. and Vieitez A.M. (1988) Factors affecting *in vitro* propagation of *Quercus robur* L, Tree Physiol. 4:281-290.

Schenk R.U. and Hildebrandt A.C. (1972) Medium and techniques for induction and growth of monocotyledonous and dicotyledonous plant cell cultures. Can. J. Bot. 50:199 204.

Shrestha B.B. (2003) *Quercus semecarpifolia* Sm. in the Himalayan Region: ecology, exploitation and threats. Him. J. Sci. 1: 126-128.

Singh J.S. and Singh S.P. (1992) Forests of Himalaya: Structure, functioning and impact of man. Gyanodaya Prakashan, Nainital, pp.294.

Spurr S. and Barnes B.V. (1973) Forest Ecology. The Ronald Press Company, New York.

Tamta S., Palni L.M.S., Purohit V.K. and Nandi S.K. (2008) *In vitro* propagation of brown oak (*Quercus semecarpifolia* Sm.) from seedling explants. *In vitro* Cell. Dev. Biol. Plant 44(2):136-141.

Tamta S., Palni L.M.S., Vyas P. and Bisht M.S. (2009) Conservation through *in vitro* method: A case of plant regeneration through somatic embryogenesis in *Quercus semecarpifolia* Sm., J. Ame. Sci. 5(1):70-76.

Thomas F.M., Blank R. and Hartmann G. (2002) Abiotic and biotic factors and their interactions as causes of oak decline in central Europe. For Pathol 32:277–307.

Toppila R. (2012) *Ex situ* conservation of oak (*Quercus* L.) in botanic gardens: A North American perspective. MSc. Thesis, University of Delaware.

Vengadesan G. and Pijut P.M. (2009) *In vitro* propagation of northern red oak (*Quercus rubra* L.). *In vitro* Cell Dev Biol-Plant 45: 474–482.

Vieitez A.M., Corredoira E., Martinez M.T., San-Jose M.C., Sanchez C., Valladares S., Vidal N. and Ballester A. (2012) Application of biotechnological tools to *Quercus* improvement. Eur. J. Plant Res. 131(3):519-539.

Vieitez A.M., Corredoira E., Ballester A., Muñoz F., Durán J. and Ibarra M. (2009) *In vitro* regeneration of important North American oak species *Quercus alba*, *Quercus bicolor* and *Quercus rubra*. Plant Cell Tiss. Org. Cult. 98:135-145.

Wilhelm E. (2000) Somatic embryogenesis in oak (*Quercus* spp.). *In vitro* Cell Dev Bio Plant 36: 349-357

Yong-Wook K. (2000) Somatic Embryogenesis in *Quercus acutissima*. In Somatic Embryogenesis in Woody Plants pp. 671-685. Springer Netherlands.

Zobel B. and Talbert J. (1984) Applied forestry improvement, John Wiley & Sons. New York.

4

Micropropagation of Bamboo Towards Large Scale Production of Tissue Culture Raised Bamboo

Niladri Bag
Department of Horticulture
Sikkim University, 6th Mile, Gangtok – 737 102, India
E-mail: niladribag@rediffmail.com

Abstract

Bamboo is nature's unique gift to mankind. It has tremendous scope for use in eco-friendly agro-forestry programmes, to bring marginal lands into use. The requirement for bamboo is increasing for various traditional and industrial requirements. In India almost 99% of annual bamboo production comes from the natural forests and only about 1% is derived from plantations. Thus to cope up with the situation and demand new bamboo plantation is very important as increasing demand for bamboo has outstripped the supply, which has been steadily declining. Propagation of bamboo through seeds is less attractive due to infrequent flowering, monocarpic nature of the plant, poor seed set, low seed viability, poor survival of seedlings in nature and inherent phenotypic heterogeneity associated with seed raised plantations. The other traditional mode of bamboo propagation is by asexual methods using offsets and culm cuttings is being practiced for time immemorial but the process is slow, labour intensive and cannot meet the growing demand. In addition, there is also risk involved in the propagation of bamboo through vegetative means using material from plants of unknown age. Most bamboo species maintain a definite period of vegetative state, followed by synchronized flowering leading to death. In this regard *in vitro* approaches offer an attractive alternative for the rapid mass multiplication of selected elite clones of physiologically young age. The chapter describes in length micropropagation protocols developed for large scale multiplication of bamboo species. At the end it also cautioned the researchers, policy makers, and various thinkers, that during micropropagation selection of mother plant is very important as vegetatively propagated bamboos are known to 'remember' the age of its mother plant. Recent studies also suggest that long term observations should be carried out and assessed whether the 'age memory' of bamboo is retained when plantlets propagated via somatic embryogenesis using tissue culture technology.

Introduction

Bamboos are renewable and fast growing giant timber grass. Due to the wide adaptability and growth characteristics bamboos are now considered as a one of the important forest resources for biomass production. Like any angiosperm species and grasses, in general bamboo can be propagated through sexual or asexual means. In sexual method seeds are use as propagule. However, this method is unreliable and rarely known because seeds of bamboo are rarely available. Asexually bamboos are propagated by culm division, rhizome, offset, culm cutting, branch cutting, twig cutting and through plant tissue culture or micropropagation. In general natural propagations by sexual or asexual means are simple and best methods for bamboo multiplication. However, these methods are having inherent problems that limit them for large scale multiplication as this is highly labour demanding. Several reports (Bag 2015; Bag and Palni 2013) have described in detail about various conventional as well as microprogation methods of bamboo potential for large scale plantation.

Peculiar, unpredictable and uncertain flowering behavior of bamboos makes it difficult to get synchronous (in space and time) flowering in two or more species. This is one of the biggest natural hurdles for attempting hybridization among the different bamboo species, which, thereby, limits development of breeding methods in nature. In an attempt experiment were attempted for reciprocal crosses al National Chemical Laboratory (NCL), Pune by Nadgauda *et al.* (1997) to hybridize *B. arundinaria* with *D. strictus*, when both of these species flowered simultaneously under natural conditions in Maharastra, but it was an unsuccessful attempt and no hybrid seed was obtained out of the study. However, their observations are important for future research. Other standard methods, such as, bud pollination, washing and blotting of the stigmas, application of pollen germination medium on the stigma, and addition of pollen extract to the pollen germination medium etc. have also not given positive results. Presently, in various laboratories attempts are being made, with some success, to induce *in vitro* flowering in bamboos using tissue culture, the technique which has gained tremendous importance after 1990 when the first mile stone research was published in Nature (Nadguda *et al.* 1990). But so far flowering and hybridization of bamboo is still a challenge in research.

Plant cell and tissue culture has been an important tool for several decades for improving the understanding of cell biology, particularly in the areas of plant physiology and biochemistry. Micropropagation or plant tissue culture is one of the methods of clonal propagation. It is a specialized "high-tech" technique of producing a large number of genetically identical (true to type to mother plant or clonal) plants by culturing plant cells, tissues or organs in an artificial nutrient medium under germ-free or aseptic conditions. Maintenance of aseptic conditions in the laboratory, establishment of aseptic culture, standardization or development of the culture medium, culture conditions – i.e., temperature, light intensity in the culture room, photo period etc. are most important considerations for success of micropropagation. Now a day, tissue culture is one of the vital components of plant biotechnology and permits the regeneration of plants as clones and transgenics. It has gained popularity not only as a method for 'micropropagation', but also for 'conservation' of rare or endangered species that are on the verge of extinction. Other benefits derived from the use of this technology are: (1) rapid mass multiplication of disease-free-stock, new varieties and plants that are normally difficult to propagate, (2) possibility of plant propagation throughout the year without climatic constraints, (3) introduction of variability to generate new genotypes with disease/ pest resistance, tolerance to abiotic stresses, yield enhancement and quality improvement, etc., (4) production of pure breeding lines through haploid culture and chromosome duplication, (5) fusion of somatic cells resulting in unique hybrids, otherwise impossible to obtain through conventional sexual route, (6) short or long term storage of germplasm, and easy exchange of clean, uncontaminated plant material between different countries without normal quarantine restrictions, and (7) encapsulation of propagules to produce 'synthetic seeds'. Besides, tissue culture is used as a tool for somaclonal variation and for applying selection and mutagenic pressures. It is an important vehicle for regenerating plants following protoplast fusion and from transformed cells, and the success of recombinant DNA technology depends heavily on advances in tissue culture (Palni *et al.* 1993; Palni and Pandey 1994; Bag 2015).

Tissue culture studies of bamboo

The first report related to tissue culture of bamboo describing embryo culture and *in vitro* germination appeared in 1968 (Alexander and Rao 1968). Intensive research on bamboo tissue culture started with the work of Mehta *et al.* (1982), reporting regeneration and plantlet formation in *Bambusa arundinacea* through somatic embryogenesis. This was followed by several reports on different species and genera. Till date over 75 bamboo species (belonging to 22 genera) have been cultured under *in vitro* conditions with the aim of developing micropropropagation protocols; some of the important reports are summarised in. In most studies *in vitro* multiplication has been reported using either (i) juvenile explants (zygotic embryos, seeds, and seedlings), or (ii) nodal explants with healthy buds taken from mature culms.

A large number of papers on micropropagation of bamboos have been published so far. Many investigators have focussed on somatic embryogenesis of seedlings of tropical bamboos. Major research on micropropagation of bamboo focused on the clonal multiplication of elite genotypes, either juvenile or adult. The number of studies in this area however, is quite less, and success in this field is very difficult. It was observed that the propagation by explants taken from field- grown adult plants via axillary branching was much more problematic than with seedlings of bamboos.

Studies observed micropropagation of bamboo and its success depends upon the choice of starting material (seeds or adult plants). The advantages of using seedlings and seeds are that they will establish a new generation, and their establishment as explants in culture is easier. Whereas disadvantages are: (1) uncertainty of genetics of progeny, (2) availability of seeds for most species is difficult and seeds loose viability rapidly. In addition, wide variability of tissue response generate when juvenile explants are used in tissue culture for micropropagation (Saxena and Dhawan 1999). In contrast when explants are taken from adult bamboos problems are: (1) profound contaminations in explants, (2) hyperhydricity (3) initial establishment is difficult in culture (4) instability of multiplication rates, and (5) problems with rooting of micropropagated bamboos, etc. Literature says rooting percentages was very low and it was of 10% for *Bambusa vulgaris* to 73% for adult *Dendrocalamus longispathus*, when explants were taken from adult bamboos (Saxena and Dhawan 1994). Ramanayake and Yakandawala (1997) obtained 77% rooting in *Dendrocalamus giganteus*. So far low rooting frequencies and inconsistency in rooting are still major bottleneck for commercially viable protocols of bamboo micropropagation (Saxena *et al.* 1993). Recently few reports have demonstrated improved rooting percentages and successful lab to land transfer with great efficiency using explants taken from established mature mother plants and seeds (Bag *et al.* 2000, 2012).

Micropropagation using physiologically young and field tested elite clones offers a useful method of bamboo multiplication (Bag *et al.* 2000, 2012). Efficient methods of mass propagation now have been standardised using single node culm cuttings of mature bamboos (Sharma 1986; Sharma and Kaushal 1985; Woods *et al.* 1992). However, the traditional methods of propagation produce limited number of propagules, and are both labour intensive and time consuming (Kumar *et al.* 1992). Further, there is also risk involved in bamboo propagation through vegetative means using material from plants of unknown age due to the peculiar behaviour of mass flowering and death of flowered clumps. Thus the use of tissue culture (TC) techniques for rapid multiplication of bamboos has been recommended; it offers many advantages particularly when the explants are taken from physiologically young and field tested elite clones (Saxena and Bhojwani 1993). *In vitro* shoot multiplication of bamboos has been carried out using varied explants like zygotic embryo, seed, seedling explants (Nadgauda *et al.* 1990, Chambers *et al.* 1991; Bag *et al.* 2000), as well as from nodal segments taken from mature culms (Saxena and Bhojwani 1993; Chaturvedi *et al.* 1993; Woods *et al.* 1992). Plant regeneration through callus cultures obtained from nodal segments (Lin and Chang 1998; Sood *et al.* 2002), young leaves (Rout and Das 1994) has also been demonstrated. There are also reports of somatic embryogenesis and plant regeneration using inflorescence segments (Jullien and Tran Thanh Van 1994), anther (Yeh and Chang 1986b) and zygotic embryos (Tasy *et al.* 1990; Rao *et al.* 1985). Somatic embryogenesis using explants taken from field tested seedlings has also been reported (Sood *et al.* 1992, 2002).

However, TC raised plants have certain disadvantages. On transfer from laboratory to field conditions, even after undergoing a period of hardening, such plants take time to adapt to changed environmental conditions. Sometimes failure to acclimatize results in heavy mortality of TC raised plants. It is, therefore, essential that TC raised plants are gradually hardened and properly acclimatized prior to field transfer (Nadgauda 1990; Pandey *et al.* 2000; Bag *et al.* 2012).

Bamboo Micropropagation using Juvenile Explants

Most of the research on bamboo tissue culture has been aimed at developing shoot multiplication techniques using juvenile explants (seeds, excised zygotic embryos, or explants taken from aseptically grown seedlings (Bag 2015). Micropropagated cultures of bamboo can be readily initiated from zygotic embryos or *in vitro* grown seedlings. Similar to many other woody plants, explants from juvenile tissues are easier to establish. Tissue culture and micropropagation of bamboos by culturing juvenile explants comprise of multiple shoot formation and their subsequent rooting resulting in complete plantlets. Alexander and Rao (1968), respected as pioneer workers in the field of bamboo tissue culture, first reported *in vitro* germination of mature embryos from bamboo seeds. They used White's medium (White 1963) supplemented with 2% sucrose to obtain optimum and satisfactory growth (average of 8-10 shoots per explant) within 6-7 weeks.

Later, for the first time, Nadgir *et al.* (1984) gainfully exploited the potential of multiple shoot formation by using excised axillary branching from shoots derived from seedlings of *D. strictus.* They then estimated that about 10,000 plantlets could be obtained from a single embryo culture per year. Dekkers and Rao (1987) were also able to obtain multiple shoots in the same species but only few of those rooted. A simple micropropagation method of *D. strictus* by stationary culturing of nodes and coleoptile explants, derived from *in vitro* grown seedlings, in ½ MS liquid medium containing 2.2 mM BAP and 2 % sucrose, was first developed by Shirgurkar *et al.* (1996). Vongvijitra (1988) reported multiple shooting in *D. membranances* and *D. brandisii* from aseptically germinated seeds on MS medium supplemented with BAP. Yeh and Chang (1987) produced plantlets of *S. latiflorus* from *in vitro* germinated seeds by adventitious shoot formation and Saxena (1990) developed a complete micropropagation protocol for *B. tulda* through multiple shoot proliferation (at a rate of 4-5 fold per three week culture) by using 3 weeks old aseptically grown seedlings.

As a landmark discovery in 1990 by Nadgauda *et al.*, *D. brandisii* and *B. arundinaceae* were successfully micropropagated following aseptic seed germination on modified White's medium and multiplied in liquid MS medium supplemented with 0.5 ppm BAP and 5% coconut milk. Cultures were kept on a rotary shaker at 120 rpm and resulted in 15-20 shoots per vessel. Following the same method, after three sub-cultures, they were later able to induce *in vitro* flowering in both the species which later paved way for future breeding of bamboos. In 1991, Chambers *et al.* were successful in developing micropropagation protocol for *D. hamiltonii* by using intact epicotyls excised from aseptically germinated seeds and culturing and inducting of *in vitro* flowering on MS medium supplemented with BA either alone or in combination with coconut milk.

In 1999 an efficient and reproducible method for the large-scale propagation of *D. asper* was developed (Arya *et al.* 1999) by which a high frequency direct shoot proliferation was induced in aseptic seed cultures on MS medium supplemented with BA (1.0-10.0 mg/l). Multiple shoots (1-25 per explant) formed within 5 weeks with a multiplication rate of 15-16 fold following culture on MS medium supplemented with 3.0 mg /l BA. In another reports Bag *et al.* (2000, 2012) demonstrated efficient protocol for *in vitro* propagation of *Thamnocalamus spathiflorus*, an evergreen temperate bamboo, through multiple shoot formation from zygotic embryos excised from germinating seeds. In this study multiple shoot formation was standardized on MS medium supplemented with 5.0 μM BAP, 1.0 μM IBA and about 90% explants proliferated with 16-38 (average 28.6) shoots per explant in eight weeks. In an exceptional case, a maximum of 117 shoots was recorded from one clump (Bag *et al.* 2000).

Different media have been used for tissue culture of bamboos. In the beginning Alexander and Rao (1968) used White's Medium (White 1963). The other basal media used by different workers were MS (Murashigue and Skoog 1962), N_6 medium (Nitsh and Nitsh 1969), and B_5 medium (Gamborg *et al.* 1968). However, in the majority of studies, MS has been successfully used. Media are generally supplemented with different cytokinins (BAP, Kinetin, 2-ip, Zeatin), auxins (2,4-D, NAA, IBA), casein hydrolysate, coconut milk and sucrose. Nadgir *et al.* (1984) observed that liquid medium gave better results than semi-solid medium, and that a better culture aeration was maintained by placing the cultures on a rotary shaker at 120 rpm. Saxena (1990) also reported better shoot multiplication in liquid medium compared to semi-solid medium. He used 10 ml liquid medium in 150 ml flask for better aeration and to avoid submerization of explants. He omitted the use of a rotary shaker for aerating the liquid medium to reduce the cost of culturing (i.e., as shakers are costlier and it consumes electricity so when shakers are being used recurring cost will be higher). Micropropagation of *D. hamiltonii* was reported by Chambers *et al.* (1991) who were also able to induce *in vitro* flowering by culturing intact epicotyl tissue of 2-week-old seedlings germinated on MS medium containing 4.4 μM BA.

Bamboo Micropropagation using explants from field grown plant

It is now evident that since the last two decades most of the research on bamboos has focussed on developing micropropagation protocols through multiple shoot formation, and organogenesis or embryogenesis following culture of juvenile explants. Due to tremendous progress made, use of tissue culture for mass propagation of bamboos is now being strongly recommended (Varmah and Bahadur 1980; Saxena and Bhojwani 1993; Saxena and Dhawan 1999; Bag *et al.* 2000, 2012; Bag 2001; Agnihotri *et al.* 2007). However, successful reports on cloning of adult bamboos are limited. A few reports are available on bamboo propagation using nodal explants with axillary buds derived from field grown culms (Huang *et al.* 1989; Prutpongse and Gavinlertvatana 1992; Sood *et al.* 1992; Chaturvedi *et al.* 1993; Saxena and Bhojwani 1993; Jullien and Tran Thanh Van 1994; Das and Rout 1994; Hirimburegama and Gamage 1995; Bisht *et al.* 1998; Lin and Chang 1998; Arya and Sharma 1998; Mukunthakumar *et al.* 1999 Bag *et al.* 2000; Bag 2001; Mishra *et al.* 2001; Fig. 1A). In *D. strictus*, shoot proliferation (2-3 shoots from one node) on semisolid MS medium within 2-4 weeks was successfully demonstrated by Nadgir *et al.* (1984). Banik (1987) reported growth and proliferation of shoots in *D. strictus* on MS medium supplemented with BAP and NAA from dormant culm bud explants. In the same species, Dekkers and Rao (1987) and Chaturvedi *et al* (1993) obtained multiple shoots in nodal segments taken from precocious branches. In another report (Ravikumar *et al.* 1998) multiple shoot induction in *D. strictus* was obtained from axillary buds of mature plants on MS medium supplemented with BA and kinetin. In this case maximum shoot multiplication frequency of 35-45 shoots within 20-25 days of culture was obtained in liquid medium containing BA (0.5 mg/l). An interesting report by Mishra *et al.* (2001) appeared recently which demonstrated a comparatively higher rate of multiplication (about 4.5 fold per 5 weeks) by a single nodal culture of a 5-6 year old *D. strictus* following incorporation of 0.5 ml/l Vipul (traicontanol), a primary alkyl alcohol, to the medium in combination with BA. In case of *B. bamboos*, micropropagation techniques have been developed with the use of explants taken from 3-year-old plants raised under nursery conditions (Arya and Sharma 1998). Shoot multiplication was achieved in liquid and agar solidified MS medium supplemented with BA (3.0 mg/l) with a fivefold increment of shoots per subculture in four weeks. They also observed that the repeated subculture in liquid medium resulted in vitrification of shoots. During this experiment, flowering in two weeks accompanied by rooting of shoots when the shoots were shifted from 3.0 mg/l BA containing medium to rooting medium containing 3.0 mg/l NAA was interesting. In an another study (Banik, 1987) reported propagation of *B. glauceus* from dormant bud explants following a series of exploratory experiments with young branch nodes, dormant culm buds, non-dormant culm buds and emerging shoot tips on MS medium supplemented with BA and NAA. Sood *et al.* (1992) obtained direct shoot regeneration and plantlet production from sprouted nodal explants (occasionally 3-4) of 7-year-old plant. For further multiplication axillary shoots were excised and cultured on MS liquid medium containing same quantity (4.52 mM BAP and 2,4-D on a gyratory shaker (80 rpm). Prutpongse and Gavinlertvatana (1992) reported micropropagation of 54 species of bamboo belonging to 15 genera. In almost all the species, multiple shoots were produced by culturing nodal cuttings on MS medium containing 22.0 mM BAP. In some of the species addition of 5.4 mM NAA in the medium enhanced shoot multiplication and only very few plantlets could be regenerated adventitiously from callus. They were also able to store many species on half strength MS medium at room temperature for more than 15 months. Saxena and Bhojwani (1993) reported *in vitro* clonal multiplication of 4-year-old *D. longispathus*. They cultured single node segments from young lateral branches on a semi-solid MS medium containing 12.0 mM BAP and 3.0 mM kinetin and obtained multiple shoots. The shoots multiplied for 15 passages in liquid medium and then over 5 passages on semisolid MS medium containing 15.0 mM BAP, 1.0 mM IBA and 10% coconut water. They also found that the nature of explant and different seasons strongly influenced culture initiation. Jullien and Tran Thanh Van (1994) reported micropropagation of *B. glaucescens* (golden goddess) by plantlet regeneration from dormant nodal buds taken from a field grown plant on MS medium supplemented with NAA. Hirimburegama and Gamage (1995) reported plantlet production in *B. vulgaris* by culturing single nodal segments with axillary buds collected from field grown plants on MS medium supplemented with plant growth regulator. Like other species of bamboo, cytokinin was found to be essential for initiating bud-break in *B. vulgaris*. Bud break was further enhanced by the presence of GA_3 in the medium. Position of the bud on culms also affected bud-break; middle nodes responded better than the basal or terminal ones. Field grown 10-year-old *B. edulis* was micropropagated on MS medium supplemented with 0.1 mg/l of thidiazuron (TDZ; Lin and Chang 1998). Some green and albino shoots flowered on TDZ containing medium. Multiple shoots proliferated well from nodal explants on medium containing 0.1 mg/l TDZ alone, and the shoots appeared to be healthy and vigorous. The range of 0.01-6.0 mg/l TDZ promoted shoot multiplication, with an average of 3-4 shoots from one shoot in each subculture of 21 days. Another important bamboo species, viz., *D. brandishii* was also been micropropagated using culture of nodal explants taken from 3-5 years old plants (Mukunthakumar *et al.* 1999). They used MS media, modified as 825.0 mg/l NH_4NO_3, 950.0 mg/l KNO_3, 220.0 mg/l $CaCl_2$ $2H_20$ and 5.0 mg/l thiamine HCl and supplemented with BAP for shoot proliferation. Maximum (about 16 shoots per explant) were obtained within 45 days of culture on medium containing 5.0 mg/l BAP. For multiplication of a temperate bamboo (*T. spathiflorus)* a successful multiplication of a 2-year-old plant was demonstrated by Bag *et al.* (2000). In this report nodal segments, without leaf sheath, cultured on 1/2 MS medium sprouted within 4-5 days. Generally, 4-16 buds were present on each nodal explant and almost all the buds sprouted, but occasionally one or two remained suppressed. After 2-3 weeks, sprouted buds of 30-35 mm length were excised and cultured on MS medium supplemented with 5.0 μM BAP and 1.0 μM NAA for further multiplication. Maximum multiplication rate (6-11 shoots per explant) was recorded in eight weeks. The degree of proliferation also depended on the inoculum size. Propagules containing minimum 3-4 shoots proliferated at a maximum rate (6-11 shoots per explant) where as single shoots proliferated at a much slower rate (1-3 shoots per explant).

In nature it is quite common that 6-7 shoots initially sprout from one axil but only one of them grows into a branch. Under *in vitro* conditions, with the help of certain combinations of plant growth regulators, all these multiple shoots can be forced to grow and each shoot can be rooted and developed into a complete plantlet. Use of explants from mature clumps permits selection for field tested plus clumps but being grown in a vegetative propagation method the plantlets produced may retain the age of their parent clumps. Therefore, the limitation of this method is that the total yield depends on the time left for their flowering. In species having shorter intermediate periods of flowering (15-30 years) this can be a serious limitation. However, for species having longer intermediate periods of flowering (60-120 years), it may be acceptable, provided the explants are collected at an early stage in the vegetative growth phase.

After critical review of the available reports, the major problem encountered for developing a micropropagation protocol for any mature bamboo is the culture initiation. Bud dormancy and break varies with the position of buds in a plant, seasons and species (McClure 1966). In case of

D. giganteus bud- break was better just before the onset of the two major rainy seasons (the south-west and the north-east monsoons) in Kandy, India (Ramanayake and Yakandawala 1997). In *B. vulgaris*, bud- break was induced more frequently throughout the year, and was positively correlated with rainfall. In the Indian Himalaya, Saxena and Bhojwani (1993) found that *in vitro* bud-break in *D. longispathus* took place during the monsoon period. Another major limitation of this method was that a few regenerated shoots from mature nodes senesced and turned brown even after transfer to fresh medium (Nadgir *et al.* 1984; Chaturvedi *et al.* 1993). The problem may be overcome with frequent transfer following removal of the browned leaves and leaf sheaths and culturing on liquid medium (Ramanayake and Yakandawala 1997) or on medium solidified with 0.2% phytagel (Bag *et al.* 2000).

To reduce the risks of somaclonal variability during the multiplication of material, apical and axillary meristems as source of new plants are usually preferred. For bamboos, juvenile explants for micropropagation have been largely used as they are easy to propagate and can be multiplied easily under appropriate conditions. The main limitation of this method is that the seeds are heterozygous due to out breeding, whereas, in the other method nodal explants can be taken from single elite field tested adult plant thus assuring production of better quality plants.

Somatic Embryogenesis

Somatic embryogenesis can be defined as the process of embryo formation from somatic cells or non-gametic cells which differentiate as bipolar structures, and mimic zygotic embryos which can be germinated into plantlets (Street 1977). Although much progress has been made since the phenomenon was first observed in carrot in the late fifties (Steward *et al.* 1958), we are still far from understanding the process. Generally grasses, particularly the bamboos, are among those groups which are considered recalcitrant to regenerate in culture. In the past two decades tremendous progress has been made in the cell culture technology of many crops and horticultural plants, especially in all the major species of grasses, with clear evidence of regeneration of plants through somatic embryogenesis (Vasil 1987). Yet, till date, efficient protocols of somatic embryogenesis in bamboos are limited.

Fig. 1 : Somatic embryogenesis of bamboo: (A) A field grown mother plant. (B) Sprouted buds from nodal explants. (C) Somatic embryos showing germination under in vitro conditions. (D) Plantlets derived from somatic embryos. (E) *In vitro* propagated bamboo showing profuse roots. (F) Hardened TC raised plants in pots.

Micropropagation through somatic embryogenesis offers considerable advantages over other methods of clonal propagation. The embryos are bipolar and germinate to form both root and shoot, and, therefore, do not require rooting stage like in multiple shoots. Multiplication is quite rapid and more than a hundred embryos can be obtained per culture as compared to only a few shoots from nodes and shoot cultures. Maintaining and manipulating embryogenic cultures is easier and hence less labour-intensive and cost effective in contrast to shoot culture. This has potential for production of largest number of plantlets in a limited time, and the technology is amenable to mechanization to make the production process cheaper. Furthermore, somatic embryos are ideal for long term storage. In addition, scientists anticipate that the juvenile period and growth habit of regenerated plants through somatic embryogenesis is similar to plants derived from seeds. Thus, bamboo plants regenerated through somatic embryogenesis can have a complete vegetative life-span (Rao *et al.* 1990).

However, till date, the success of somatic embryogenesis in bamboo is very limited and in all the reported studies tissue of juvenile explants have been used. In fact, embryogenesis from tissue of mature plants is difficult. So far, there is only one report of somatic embryogenesis by explant taken from mature 45 years old bamboo is available (Bag *et al.* 2012; Fig. 1B, C). In another report, shoot bud regeneration of *D. hamiltonii* through callus phase derived from nodal segments of 7-year-old elite plant has been reported (Sood *et al.* 1992). Somaclonal variation and genetic instability created by an intermediate callus phase is another major disadvantage towards clonal propagation. Although somaclonal variants are not desirable when uniformity of propagules is required, there is always a possibility of getting mutant(s) through such methods.

Somatic embryogenesis and regeneration of plantlets of *B. arundinaceae* was reported for the first time in early eighties (Mehta *et al.* 1982) and was a landmark achievement which enhanced research in tissue culture of bamboos. According to published reports, somatic embryogenesis has been obtained in *B. arundinacea* by several workers (Mehta *et al.* 1982, Rao and Rao 1988), *B. balcoa* (Rao *et al.* 1990), *P. viridis* (El Hassan and Debergh 1987), *D. strictus* (Rao *et al.* 1985; Rao and Rao 1988; Mascarenhas *et al.* 1988; Mukunthakumar and Mathur 199; Rout and Das 1994; Saxena and Dhawan 1999); *S. latiflora* (Yeh and Chang 1987, Samora *et al.* 1988; Tsay *et al.* 1990); *B. beecheyana* and *B. oldhamii* (Yeh and Chang 1986a, b); *O. acuminata aztecorum* (Woods *et al.* 1992); *B. glaucescens* (Jullien and Tran Thanh Van 1994); *B. vulgaris* and *D. giganteus* (Rout and Das 1994).

Somatic embryogenesis and *in vitro* regeneration of plants from field grown elite 'maggar' bamboo (*Dendrocalamus hamiltonii*) belonging to two populations of different age (10 and 45-year-old) and locations have been established (Bag 2001). In this study, MS medium supplemented with BAP and 2,4-D (5.0 mM each) was found to produce cream-white callus with approximately 80.0% and 73.0 % efficiency in explants obtained from 10 and 45-year-old plants, respectively. Efficient embryogenesis was obtained from the callus when cultured on MS medium containing 5.0 μM BAP and 7.5 μM 2,4-D. In addition to the presence of hormone(s) in the medium, age of the mother plant was found to influence callus formation, subsequent embryogenesis and plantlet regeneration. The explants obtained from 10-year-old mother plant performed better in comparison to explants obtained from a 45-year-old mother plant (Fig. 1D).

In the majority of studies on somatic embryogenesis use of zygotic embryos as initial explants has been reported (Mehta *et at.* 1982; Rao *et al.* 1985, 1990; Yeh and Chang 1987; El Hassan and Debergh 1987; Dekkers and Rao 1989; Samora and Gruezo 1990; Woods *et al.* 1992; Mukunthakumar and Mathur 1992; Rout and Das 1994; Saxena and Dhawan 1999). In addition, regeneration from young leaves (Jullien and Tran Thanh Van 1994), haploid formation from anther culture (Tsay *et al.* 1990) and immature inflorescence explants (Yeh and Chang 1986a, b) have also been reported.

The basal nutrient medium is reported to influence induction of somatic embryogenesis in bamboos (Das and Rout 1994). Different workers have used various basal media for the induction of embryogenic callus and somatic embryogenesis. Among these Murashige and Skoog's basal medium (MS) was used by the majority of workers (Nadgauda *et al.* 1990; Saxena 1990; Woods *et al.* 1992; Saxena and Bhajwani 1993; Rout and Das 1994). In *D. giganteus*, *B. vulgaris* and *D. strictus* the MS basal medium supplemented with 3% sucrose and plant growth regulator was found to be effective for embryogenic callus initiation, embryogenesis and plantlet regeneration (Rout and Das 1994). However, for anther culture N_6 (Nitsh and Nitsh 1969) basal medium was found to be more effective than MS (Tsay *et al.* 1990). In some cases B_5 (Gamborg *et al.* 1968) basal medium has also been successfully used (Rao *et al.* 1985; Rao and Rao 1988) for various responses, but the reports lacked a comparative analysis with other medium. The effects of different auxins on somatic embryo development in bamboo species have been reported by several researchers. The most commonly used PGRs for induction of somatic embryogenesis are 2,4-D and BAP (Yeh and Chang 1987; Woods *et al.* 1992; Rout and Das 1994). Woods *et al.* (1992) and Rout and Das (1994) found 2,4-D to be essential for induction of somatic embryogenesis in different species of bamboo. In a somatic embroyogenesis study using explant of mature *D. hamiltonii* plant (Bag 2001; Bag *et al.* 2012), it was found that 2,4-D concentration was positively correlated to callus initiation and subsequent embryogenesis and in medium with little or nil 2,4-D embryogenesis did not occur. It seems that 2,4-D is essential for somatic embryogenesis in bamboo.

In vitro Flowering

Now we are living in the era of technology. With the progress of sciences in the 1940s, 1950s and 1960s exciting development of techniques and technologies took place all over the world. Eventually we have achieved crop improvement mainly through systemised breeding, resulting in high yielding hybrid production and the best example is green revolution, white revolution, etc. However, in several plants hybridization could not be achieved due to several inherent limitations within these plants. Therefore, in such cases further improvement cannot be achieved easily by these methods. Unpredictable flowering and poor seed formation in most bamboo species, and a long, species-specific vegetative phase, ranging from 3-120 or more years (Janzen 1976) make it very difficult to implement a systematic breeding programmes for bamboos. It is also difficult to get two or more species of bamboo those flower concurrently, in space and time which is a prerequisite for any breeding programme.

Monocarpic flowering, common among grasses, is a feature of bamboos also (John *et al.* 1995). However, the perennial (prolonged) vegetative phase lasting several years is quite interesting and a feature peculiar to bamboos. In India pioneering work on bamboo was initiated in the year 1968 on bamboo tissue culture (Alexander and Rao 1968), resulting in the development of a complete micropropagation protocol of bamboo (Mehta *et al.* 1982). Rao and Rao 1988 obtained *in vitro* flowering of bamboo (Rao and Rao 1988). There are a few mile stone researches done so far towards bamboo improvement. So far *in vitro* induction of flowering has been well established in several bamboo species, i.e., *B. arundinacea, B. vulgaris, D. brandisii, D. gigianteus, D. hamiltonii* and *D. strictus* (Rao and Rao 1988; Nadgauda *et al.* 1990, 1997; Chambers *et al.* 1991; Rout and Das 1994; Lin and Chang 1998; Ramanayake *et al.* 2001). Nadguda *et al.* 1990, 1997 and Chamber *et al.* 1991 had induced *in vitro* flowering using culture of juvenile explants obtained from seedlings. In other report Rout and Das (1994) had used nodal explants taken from plantlets regenerated through *in vitro* somatic embryogenesis. Further, Ramanayake *et al.* (2001) reported successful *in vitro* multiplication and flowering of *D. giganteus* (giant bamboo) following *in vitro* multiplication using single node culture of a plant of more than 70 years old. They cultured nodal explants and multiplied on MS medium supplemented with 26.6 mM BA and 2% sucrose.

Among these reports basic similarity were the use of seed or seedling tissue as the initial explants. Other similarities observed were use of plant growth regulators (PGR), i.e. benzyladenine (BA) - a cytokinin; and adenine sulphate (Ads) - a compound having cytokinin-like activity were supplemented in the medium. Nadgauda *et al.* (1997) reported enhanced accelerated growth and multiplication of flowering cultures of bamboo by using bioreactor. They (Nadgauda *et al.* 1997) also reported very interesting observation, demonstrating that the pH of the medium dropped dramatically from 5.6 to 4.2 after about 2 weeks of culture. They also found a positive correlation between this drops in the pH of the medium with maximum growth of the inflorescences. Joshi and Nadgauda (1997) noticed cytokinins, particularly BA, to be essential for induction of *in vitro* flowering.

Induction of *in vitro* flowering in different species of bamboo, which is unpredictable and take several years to flower in nature, are of immense importance towards improvement of bamboo. The phenomenon has potentiality of obtaining controlled, synchronous and concurrent flowering in more than a bamboo species, and also has opened up possibilities for interspecific/ intergeneric crosses and changes in genetic makeup, thus inherent improvement. However, it should be mentioned here, after about 15 years of first report on *in vitro* flowering of bamboos, it still seems to be an uphill task for the plant scientists and so far not even a single hybrid achieved. So, concerted efforts and scientific inputs are required to standardise and perfect this technique for its wider application, e.g., for effective cross-pollination and production of "hybrid" bamboos. The phenomenon of precocious flowering in bamboos has many interesting facets to it, and some of these, involving *in vitro* studies, have been fluently discussed by in various scientific forum Nadgauda *et al.* (1997). It must, however, be noted that there have been several instances of precocious flowering under natural conditions, and therefore reports of *in vitro* flowering must also be seen in the light of such observations (Raste and Bhojwani 1998).

In vitro Rooting

For the development of any successful *in vitro* micropropagation protocol, root induction of excised shoots and subsequent survival of plantlets in the soil are essential and crucial. Generally either a single or two step rooting procedure is adopted for *in vitro* rooting for plants propagated by methods other than somatic embryogenesis. Nadgir *et al.* (1984) recorded rooting in only 20% shoots of *D. strictus* following incubation of shoots for 96 h in the dark in MS medium containing 4.9 mM IBA and thereafter transferring to solid medium containing 0.25% activated charcoal. Sometimes coumarin (a phenol) was applied in conjunction with PGR for better rooting efficiency. *In vitro* propagated shoots of *B. tulda* were rooted in bunches of three on modified MS medium (NH_4NO_3 content reduced to half) containing 10.0 mM IAA and 68.0 mM coumarin (Saxena 1990). In the study it was also observed that rooting was equally good in liquid, solid and semi-solid medium. However, root growth was better with less vitrification of shoots on semi-solid medium (gelled with 0.2% gelrite). Saxena and Bhojwani (1993) showed high percentage of rooting (73%) in *D. longispathus* in a medium supplemented with IBA, IAA and coumarin. Other phenolic compounds, such as, caffeic acid and phloroglucinol were reported to regulate rooting and shoot proliferation (Jones 1976; James and Thurbon 1979; Nemeth 1986). These have been used for induction of rooting in bamboo species that are difficult to root (Chaturvedi and Sharma 1989; Chaturvedi *et al.* 1993). Chaturvedi *et al.* (1993) reported that the rooting response was better in *D. strictus* cultures initiated from explants collected in July-August than in those collected during September-October. They also observed that prolonged incubation in the rooting medium resulted in abnormal knotting of roots caused by the inhibitory effects of auxins on further development. Improved rooting (up to 92%) has been reported with combination of auxins and phenols (Saxena 1990; Sood *et al.* 1992). Bag *et al.* (2000, 2012) achieved 100% rooting following treatment with IBA for two weeks, and subsequent culture of shoots on auxin-free medium in case of a temperate bamboo (*T. spathiflorus*) and maggar bamboo (*Dendrocalamus hamiltonii* (Bag *et al.* 2012, Fig. 1E).

Acclimatization

Since last two decades, micropropagation has contributed in a big ways towards rapid mass multiplication of many crop, ornamental, and tree plants. However, extensive use of this technique has been restricted on account of (i) difficulty encountered in rooting of micropropagated shoots of some clones, and (ii) often high percentage of mortality during or following lab to land transfer. During *in vitro* culture, plantlets grow under special conditions of synthetic medium containing high nutrients, relatively high humidity, lower irradiance than ambient and exhibit abnormal morphology, anatomy and physiology (Ziv 1995; Pospiselova *et al.* 1999; Bag *et al.* 2000; Bag 2015). On immediate transfer to field conditions, these plantlets can easily be damaged due to changed environmental condition. It is, therefore, essential that the plantlets are hardened prior to transfer to fields. In case of micropropagated bamboos well rooted plantlets can be transferred to soil following simple hardening process and little care by controlling the relative humidity and temperature. Most of the *in vitro* micropropagation studies of bamboos claim successful transfer of micropropagated plants to the field but few of them reflect survival percent. Based on studies by different workers it is found that the survival rate varies from 20-100%. The survival of plantlets derived from somatic embryogenesis in *D. strictus* was reported to be 78% in the green house and 90-95% in a growth chamber (Rao and Rao 1988). *In vitro* raised plants with developed rhizomes after hardening record better survival rate after transplantation to the field. The survival and growth in the nursery are influenced by the quality of root system of the micropropagated plants type of potting mixture (Samora *et al.* 1988). Plants raised through tissue culture were established faster than seed derived plants. This is an advantage which can be utilised in reforestation and plantation programmes (Rao and Rao 1988). Saxena and Bhojwani (1993) demonstrated a simple process for hardening and better survival rate of tissue culture raised bamboo plantlets on their transfer to soil in which the plantlets were preconditioned under aseptic/ semi-aseptic environment conditions before transferring to polythene bags. The survival rate also varied when micropropagated plants were hardened inside different growth chambers and transferred during different seasons. Plantlets transferred during the monsoon season from glass house and polyhouse recorded a survival per cent 86%-92% as compared to those transferred during summer (80-85%) and winter (25-30%; Bhojwani and Sharma 1999). Rao and Rao (1998) designed a simple and low cost chamber for hardening and acclimatization of micropropagared bamboo plants.

The bamboo tissue culture progressed considerably towards mass propagation and large number of saplings has been produced by few laboratories (Mascrehans *et al.* 1988; Saxena and Dhawan 1999; Arya and Sharma 1998; Agnihotri *et al.* 2007). A major bottleneck in large scale application of this technology is high mortality during or following lab to land transfer (Ziv 1995, 1991a, b; Gaspar *et al.* 1987; Kevers *et al.* 1987). It needs to be emphasised that all the reports deal with the mechanical or seasonal aspects during acclimatization of bamboo and survival varied from 25-90%, but attention has not been paid to the sudden exposure of such aseptically raised plants to microbial communities, including minor and major pathogens that are present in the soil. Recently, besides many abiotic aspects, beneficial bacterial inoculation during transplantation as the first line of defence and for stimulating growth of micropropagated plants (Pandey *et al.* 2000, 2002) and mycorrhization (Varma and Schuepp 1996; Hernandez-Sebastia *et al.* 1999) have been used which have opened new avenues for better survival and stimulation of growth of micropropagated plants (Fig. 1F). This novel approach may help to improve and simplify transplantation of micropropagated bamboo with better establishment and growth in the field.

Conclusion

Now a days, bamboos are a major source of raw material for the pulp and paper industries and several other industrial products. Although it was once called poor man's timber, it is no longer cheap. Annually about 0.84 million tonnes of bamboo is being used by the Indian paper mills, clearly indicating an indiscriminate exploitation of bamboo forests well beyond sustainable levels of harvest (Anonymous 1992; Martin 1996). It has been projected that by the year 2015 there will be shortage of about 10 MMt of bamboo (Anonymous 1990). In order to meet this demand the pressure is going to be on the natural bamboo forests which will result in a reduction in the area under bamboo forests. Hence bamboo production needs to be highly encouraged and maximized by using techniques for mass multiplication of high yielding and fast growing bamboos for sustainable development; it has been proposed that bamboo should be promoted as a plantation crop in the country (Anonymous 2000). In this context, *m*icropropagation of bamboo would be useful as it will augment the supply of a large amount planting material. It would be ideal to micropropagate using tissue taken from adult plant as it will help in multiplying field tested superior clone on a large scale. Micropropagation through somatic embryogenesis offers considerable advantages over other methods of clonal propagation as this route has the potential for producing greatest number of plantlets of uniform genetic make up, and the technology is amenable to mechanization. Furthermore, due to their bipolar nature, rooted plants are easily obtained, and somatic embryos would seem ideal for encapsulation and long term storage. Beside infrastructural support and abiotic components, knowledge on plant-microbe interactions or the rhizospheric aspects should be given due emphasis that may be helpful in improving the overall success in large scale production of bamboos (Bag 2015; Bag *et al.* 2000, 2012; Pandey *et al.* 2000, 2002; Shroti *et al.* 2012).

Regeneration through somatic embryogenesis followed by callus phase offers the possibility of selecting superior somaclonal variants for enriching the existing genetic pool. Flowering in bamboos has always been an enigma to plant physiologists since ages (Janzen 1976). Therefore *in vitro* flowering, though achieved only in a few species so far, has the potential to unravel the intricacies of the flowering behavior of bamboo in future (Nadgauda *et al.* 1990, 1993, 1997; Mascarenhas *et al.* 1988; Rout and Das 1994). Once this method becomes standardized and reproducible in a large number of bamboo species, it would facilitate a planned breeding strategy so that useful traits can be combined through hybridization. Cryopreservation of bamboo pollen and female gametes can be another method towards hybridization of bamboo to overcome the barrier of asynchronous flowering of different species.

Another critical research goal should be directed towards protoplast technology. Though some preliminary work has been done (Tseng *et al.* 1975; Huang 1988; Rao and Rao 1988), a number of difficulties still remain to be solved for an effective technology to be developed. Another area of research in future which needs be addressed is the development of efficient protocols for suspension culture. Some efforts are being made in this area (Rao and Rao 1988) but there is still scope for further studies and applications of biotechnological methods for bamboo improvement (Saxena 1990; Saxena and Dhawan 1999; Saxena and Bhojwani 1993). It is believed that more and more exploration in the areas of hybridization of somatic cells, induction of positive somaclonal variations or genetic engineering will take place when the differentiation and regeneration skills of bamboos are fully solved. Bamboos are monocarpic and flowers only once after long vegetative phase, while plants propagated by asexual means, *e.g.*, cuttings and offsets are known to maintain the age of mother plant. It is predicted that plants regenerated by embryogenesis or through callus phase the juvenile period and growth habit of regenerated plants would be starting from day zero phase. But it remains to be seen when mature explant would be taken and plants regenerated by somatic embryogenesis or through callus phase would also remain the age of the mother plant.

The need for increasing bamboo plantation is very important as ever increasing demand for bamboo in the modern society has outstripped the supply, which has been steadily declining. One of the most important contributions of bamboo to the modern society is production of paper. It has several other well known uses, including used as industrial raw material, house construction, agricultural implements and many other traditional uses and provides employment opportunities to the rural population mainly from the bamboo growing regions. Besides its importance in the national 'economy', bamboo plantations are being considered particularly important in afforestation programmes for which production of healthy and physiologically young planting materials is a prerequisite. This can be done by establishing nurseries by juvenile explants from known sources and age since the bamboo propagules are known to retain the age of the mother plant and exhibit terminal flowering.

The demand for bamboos, because of its multiple utilities, will continue to increase as the time progresses. Therefore bamboo is no longer a 'poor mans timber', now it is fast emerging as a 'Green Gold'. With the advancement in bamboo research, and technology bamboo markets are growing very fast and offers new opportunities for promotion of bamboo as alternative to wood as well as other traditional uses. Therefore, it is a golden opportunity for the farmers, entrepreneurs, industries, banks, govt. agencies, etc. to tap the potential of economic, ecological and social gains. In the countenance of this situation, there is an urgent need for improved techniques of propagation to supply large number of plants. The current knowledge about bamboo tissue culture shows that rapid propagation of elite bamboo clones is now feasible. Concerted efforts are now required to take this technology to the field for commercial production of large number of plants for plantation programmes. It is hoped that in the future this technique will prove beneficial for the development of bamboos and bamboo based industries. When new varieties/ hybrids will be obtained by conventional or advanced biotechnological techniques, or if the mother culms become scarce in number, micropropagation will prove to be an extremely effective and applicable method for proliferation of valuable germplasm in a limited time.

References

Agnihotri R.K., Bag N., Nandi S.K. and Plani L.M.S. (2007) Clonal propagation and field evaluation of "Maggar" bamboo (*Dendrocalamus hamiltonii* Nees et Arn. Ex Munro). In: Arya and Arya (Eds) Utilization of Biotechnology in Plant Science, Forest Research Institute, Dehradun (India) pp. 145-154.

Alexander M.P. and Rao, T.C. (1968) *In vitro* culture of bamboo embryos. Curr. Sci. 37: 415.

Anonymous (1992) Action Plan for Himalaya. G.B. Pant Institute of Himalayan Environment and Development, Kosi-Katarmal, Almora, India, 68p.

Anonymous (2000) Role of science and technology in the integrated development of Uttarakhand. G.B. Pant Institute of Himalayan Environment and Development, Kosi-Katarmal, Almora, India.

Arya S. and Sharma S. (1998) Micropropagation technology of *Bambusa bambos* through shoot proliferation. Indian For. 124 (9): 725-731.

Bag N. (2001) Mass propagation of tea, maggar bamboo and dev-ringal. D. Phil. Thesis submitted to the H.N.B. Garahwal University, Srinagar, India.

Bag N. (2015) Biotechnology of Bamboo – Present and Future. Kunal Books, Ansari Road, Daryaganj, New Delhi, 136p.

Bag N. and Palni L.M.S. (2013) Seed germination studies of Dev-ringal *Thamnocalamus spathiflorus* (Trin.) Munro - a temperate bamboo. Indian Forester 139 (7): 610-614.

Bag N., Palni L.M.S., Chandra S. and Nandi S.K. (2012) Somatic embryogenesis in 'Maggar' bamboo (*Dendrocalamus hamiltonii*) and field performance of regenerated plants. Curr. Sci. 102(9): 1279-1287.

Banik R.L. (1987) Techniques of bamboo propagation with special reference to pre-rooted and pre-rhizomed branch cuttings and tissue culture. In: Rao, A.N., Dhanarajan G. and Sastry C.B. (eds.) Recent Research on Bamboos. The Chinese Academy of Forestry and IDRC, Canada, pp. 160-169.

Begum F., Paul S., Bag N., Sikdar S.R. and Sen S.K. (1995) Somatic hybrids between *Brassica juncea* (L). Czern. and *Diplotaxis harra* (Forsk.) Boiss and the generation of backcross progenies. Theor. Appl. Genet. 91: 1167-1172.

Bisht M.S., Vyas P., Bag N. and Palni L.M.S. (1998) Micropropagation of some plants of Indian Himalayan Region. in: Srivastava P.S. (ed.) Plant tissue culture and molecular biology – applications and prospective. Narosa Publishing House, New Delhi, pp. 126-169.

Chambers S.M., Heuch J.H.R. and Pirrie A. (1991) Micropropagation and *in vitro* flowering of the bamboo *Dendrocalamus hamiltonii* Munr., Plant Cell Tiss. Org. Cult. 27: 45-48.

Chang W.C. and Lan T.H. (1995) Somatic embryogenesis and plant regeneration from roots of bamboo (*Bambusa beecheyana* Munro var. *beecheyana*). J. Plant Physiol. 145: 535-538.

Chaturvedi H.C. and Sharma M. (1989) *In vitro* production of cloned plans of jojoba [*Simmondsia chinenis* (Link) Schneider] through shoot proliferation in long term culture. Plant Sci. 63: 199-207.

Chaturvedi H.C., Sharma M. and Sharma A.K. (1993) *In vitro* regeneration of *Dendrocalamus strictus* Nees through nodal segments taken from field-grown clumps. Plant Sci. 91: 97-101.

Dekkers A.J. and Rao A.N. (1987) In vitro culture of three bamboo genera. In: Seminar on Tissue Culture of Forest Trees, Kuala Lumpur, June 15-18.

El Hassan A.A. and Debergh P. (1987) Embryogenesis and plantlet development in the bamboo *Phyllostachys viridis* (Young) McClure. Plant Cell Tiss. Org. Cult. 10: 73-77.

Gamborg O.L., Miller R.A. and Ojima K. (1968) Nutrient requirements of suspension cultures of soybean root cells. Expt. Cell Res. 50: 151-158.

Gaspear T., Kevers C., Debergh P., Maene L., Paques M. and Boxus P. (1987) Vitrification, Morphological, Physiological and Ecological aspects: In: Bonga J.M and Durzan (eds.) Cell and tissue culture in forestry, Martinus Nijhoff Publishers, Dordrecht. Vol. 1, pp. 152-166.

Glimelius K., Fahlesson J., Landgren M., Sjodin C. and Sundberg E. (1991) Gene transfer via somatic hybridization in plants. Trends Biotech. 9: 24-30.

Hernandez-Sebastia C., Piche Y. and Desjardins Y. (1999) Water relations of whole strawberry plantlets *in vitro* inoculated with *Glomus intraradices* in a tripartite culture system. Plant Sci. 143: 81-91.

Huang L.C. (1988) Bamboo callus, cells in liquid suspension, protoplasts and plant regeneration in vitro, In: Int. Conf. Applied Biology and Biotechnology, Hongkong, pp. 4.

James D.J. and Thurbon I.J. (1979) Rapid *in vitro* rooting of the apple root stock M-9. J. Hort. Sci. 54: 309-311.

Janzen D.H. (1976) Why bamboo wait so long to flower. Ann. Rev. Ecol. Syst. 7: 347-391.

John C.K., Joshi M.S., Nadgauda R.S. and Mascarenhas A.F. (1994) Selection – a valuable method for bamboo improvement. Curr. Sci. 66 (11): 822-824.

Joshi M. and Nadgauda R.S. (1997) Cytokinins and *in vitro* induction of flowering in bamboo: Bambusa arundinacea (Retz) Wild. Curr. Sci. 73: 523-526.

Jullien F. and Tran Thanh Van K. (1994) Micropropagation and embryoid formation from young leaves of *Bambusa glaucescene* 'Golden goddess'. Plant Sci. 98: 199-207.

Kevers C., Prat R., and Gaspar Th. (1987) Vitrification of carnation *in vitro*: changes in cell wall mechanical properties, cellulose and lignin content. Plant Growth Regul. 5: 59-66.

Lin C.S. and Chang W.C. (1998) Micropropagation of *Bambusa edulis* through nodal explants of field-grown culms and flowering of regenerated plants. Plant Cell Rep. 17: 617-620.

Martin M. (1996) Where have the groves gone? *Down To Earth*. 5 (2): 27-31.

Mascarenhas A.F., Nadgir A.L., Thengane S.R., Phadke C.H., Khuspe S.S, Shirgurkar M.V., Parasharami V.A. and Nadgauda R.S. (1988) Potential application of tissue culture for propagation of Dendrocalamus strictus, in Rao I.V.R., Gnanaharan R. and Sastry C.B. (eds.) Bamboo: Current Research. For. Res. Inst., Kerala, India, and IDRC, Canada, pp. 159-166.

Mehta U., Rao I.V.R. and Mohan Ram H.Y. (1982) Somatic embryogenesis in bamboo, in Fujiwara, A. (ed.) Proc. Int. Cong. Plant Tissue Culture, Tokyo, Japan. pp. 109-110.

Mishra Y, Rana P.K., Shirin F. and Ansari S.A. (2001) Augmenting *in vitro* shoot multiplication by vipul (triacontanol) and adventitious rhizogenesis by rice bran extract in *Dendrocalamus strictus*. Indian J. Exp. Biol. 39: 165-169.

Mukunthakumar S., Mathur J., Nair P.K.K. and Mathur S.N. (1999) Micropropagation of *Dendrocalamus brandisii* Kurz. using *in vitro* nodal explants. Indian For. 125 (12): 1239-1243.

Murashige T., and Skoog F. (1962) A revised medium for rapid growth and bioassays with tobacco tissue cultures. Physiol. Plant. 15: 473-497.

Nadgauda R.S., John C.K., Parasharami V.A., Joshi M.S. and Mascarenhas A.F. (1997) A comparison of *in vitro* with *in vivo* flowering in bamboo: *Bambusa arundinacea.* Plant Cell Tiss. Org. Cult. 48: 181-188.

Nadgauda R.S., Parasharami V.A. and Mascarenhas A.F. (1990) Precocious flowering and seeding behavior in tissue-cultured bamboos. Nature 344: 335-336.

Nadgir A.L., Phadke C.H., Gupta P.K., Parsharami V.A., Nair S. and Mascarenhas A.F. (1984) Rapid multiplication of bamboo by tissue culture. Silvae Genet. 33: 219-223.

Nemeth G. (1986) Introduction of rooting, in Bajaj Y.P.S. (ed.) Biotechnology in Agriculture and forestry, Trees I. Springer, Berlin, Heidelberg, New York, Tokyo, vol. 1, pp. 49-64.

Nitisch J.P. and Nitsch C. (1969) Haploid plants from pollen grains. Science 163: 85-87.

Palni L.M.S. and Pandey A. (1994) Plant Biotechnology: Opportunities and Challenges. In: Somvanshi R. and Lokeshwar R.R. (eds.), Current Advances in Science and Animal Production in India. Int. Book Dist. Co. Lucknow, pp. 174-188.

Palni L.M.S., Bag N., Nadeem M., Tamta S., Vyas P., Bisht M.S., Purohit V.K., Kumar A., Nandi S.K., Pandey A. and Purohit A.N. (1998) Micropropagation: Conservation through tissue culture of selected Himalayan plants. In: Anonymous (ed.) Research for mountain development: some initiatives and accomplishments. Gyanodaya Prakashan, Nainital, pp. 431-452.

Pandey A., Bag N., Chandra B. and Palni L.M.S. (2002) Biological hardening: a promising technology for tissue culture industry. In: Nandi S.K., Palni L.M.S. and Kumar A. (eds.) Role of plant tissue culture in biodiversity conservation and economic development, Gyanoda Prakashan, Nainital, India, pp. 565-577.

Pandey A., Palni L.M.S. and Bag N. (2000) Biological hardening of tissue culture raised tea plants through rhizosphere bacteria. Biotech. Let. 22: 1087-1091.

Pospisilova J., Ticha I., Kadlecek P., Haisel D. and Plzakova S. (1999) Acclimatization of micropropagated plants to *ex vitro* conditions. Biol. Plant. 42 (4): 481-497.

Prutpongse P. and Gavinlertvatana P. (1992) *In vitro* micropropagation of 54 species from 15 genera of bamboo. Hort. Sci. 27: 453-454.

Ramanayake S.M.S.D. and Yakandawala K. (1997) Micropropagation of a giant bamboo *Dendrocalamus giganteus*) from nodal explants of field grown culms. Plant Sci. 129: 213-223.

Rao I.V.R. and Rao I.U. (1988) Tissue culture approaches to the mass propagation and genetic improvement of bamboos. In: Rao I.V.R., Gnanaharan R. and Sastry C.B. (eds.) Bamboo: Current Research. For. Res. Inst., Kerala, India and IDRC, Canada, pp. 151-158.

Rao I.V.R., Yusoff A.M., Rao A.N. and Sastry C.B. (1990) Propagation of bamboo and rattan through tissue culture. Bamboo and Rattan Research Network, IDRC, Canada, 60 p.

Rao U., Rao I.V.R. and Narang V. (1985) Somatic embryogenesis and regeneration of plants in the bamboo *Dendrocalamus strictus.* Plant Cell Rep. 4: 191-194.

Raste A.P. and Bhajwani S.S. (1998) *In vitro* clonal propagation propagation of bamboo. In: Srivastava, P.S. (ed.) Plant tissue culture and molecular biology – applications and prospective. Narosa Publishing House, New Delhi, pp. 538-553.

Rout G.R. and Das P. (1994) Somatic embryogenesis and *in vitro* flowering of 3 species of bamboo. Plant Cell Rep. 13: 683-686.

Samora A.B. and Gruezo S.S. (1990) Embryo culture of bamboo (*Dendracalamus strictus* Nees), *Philippine Ag.* 73: 199-206.

Samora A.B., Gruezo S.S. and Damasco O.P. (1988) Callus induction and plant regeneration from internode tissues of Bamboo (*Dendrocalamus latiflorus* ev. *Machiku*), *Philippine Ag.* 71: 76-84.

Saxena S. (1990) *In vitro* propagation of the bamboo (*Bambusa tulda* Roxb.) through shoot proliferation. Plant Cell Rep. 9: 431-434.

Saxena S. and Bhojwani S.S. (1993) *In vitro* clonal multiplication of 4-year-old plants of bamboo, *Dendrocalamus longispathus* Kurz. In Vitro Cell. Dev. Biol. 29: 135-142.

Saxena S. and Dhawan V. (1999) Regeneration and large-scale propagation of bamboo (*Dendrocalamus strictus* Nees) through somatic embryogenesis. Plant Cell Rep. 18: 438-443.

Sharma O.P. (1986) Mass multiplication of *Dendrocalamus hamiltonii* Munro - a critical evaluation. Indian For. 112: 517-523.

Shirgurkar M.V., Thengane S.R., Poonawala I.S., Jana M.M., Nadgauda R.S. and Mascarenhas A.F. (1996) A simple *in vitro* method of propagation and rhizome formation in *D. strictus* Nees. Curr. Sci. 70: 940-941.

Sikdar R.S, Chatterjee G., Das S. and Sen S.K. (1990) "Erussica" the intergeneric fertile somatic hybrid developed through protoplast fusion between Eruca sativa and *Brassica juncea* (L.) Czern. Theor. Appl. Genet. 79: 561-567.

Sood A, Ahuja PS, Sharma M, Sharma OP & Godbole S (2002) *In vitro* protocols and field performance of elites of an important bamboo *Dendrocalamus hamiltonii* Nees et Arn. Ex Munro. Plant Cell Tiss. Org. Cult. 71: 55-63.

Sood A., Palni L.M.S., Sharma M. and Sharma O.P. (1994) Improved methods of propagation of maggar bamboo, *Dendrocalamus hamiltonii* Nees et Arn. ex Munro, Biotechnology in India, pp. 199-212.

Sood A., Palni L.M.S., Sharma M., Chand G. and Sharma O.P. (2002) Micropropagation of *Dendrocalamus hamiltonii* Munro (Maggar bamboo) using explants taken from seed raised and field tested plus plants. J. Plant Biol. 29(2): 125-132.

Sood A., Palni L.M.S., Sharma M., Rao, D.V., Chand G. and Jain N.K. (1993) Micropropagation of tea using cotyledon culture and encapsulated somatic embryos. J. Plant. Crops 21: 295-300.

Sood A., Sharma O.P. and Palni L.M.S. (1992) Improved methods of propagation of maggar bamboo (*Dendrocalamus hamiltonii* Nees et Arn. ex Munro) using single node cuttings taking from juvenile culms of elite seedlings. J. Am. Bamboo Soc. 9: 17-24.

Steward F.C., Mapes M.O. and Mears K. (1958) Growth and organized development of culured cells. II Organization in cultures grown freely suspended cells. Am. J. Bot. 45: 705-708.

Street H.E. (1977) *Plant Tissue and Cell Culture*, Blackwell Scientific Publications, Oxford, UK.

Tsay H.S., Hsu J.Y. and Yeh C.C. (1996) Haploidy in bamboo (*Sinocalamus latiflora*) by anther culture. In: Jain S.M., Sopory S.K. and Velleux R.E. (eds.) In vitro haploid production in higher plants. Kluer Academic Publishers, The Netherlands, Vol. 3, pp. 337-347.

Tsay H.S., Yeh C.C. and Hsu J.Y. (1990) Embryogenesis and plant regeneration from anther culture of bamboo [*Sinocalamus latiflora* (Munro) McClure]. Plant Cell Rep. 9: 349-351.

Tseng T.C., Lin D.F. and Shaio S.Y. (1975) Isolation of protoplasts from crop plants. Bot. Bull. Acad. Sincia 16: 55-60.

Varma A. and Schuepp H. (1996) Influence of mycorrhization on the growth of micropropragated plants. In: Mukerji K.G. (ed.) *Concepts in Mycorrhizal Research*, Kluwer Academic Publishers, Netherlands, pp. 113-132.

Varmah J.C. and Bahadur K.N. (1980) Country report: India. In: Lessard, G. and Chouinard, A. (eds.) Bamboo Research in Asia, I.D.R.C., pp. 19-46.

Vasil I.K. (1987) Developing cell and tissue culture systems for the improvement of cereal and grass crops. J. Plant Physiol. 128: 193-218.

Vasil V. and Vasil I.K. (1981) Somatic embryogenesis and plant regeneration from suspension cultures of pearl millet (*Pennisetum americanum*). Ann. Bot. 47: 669-678.

Vongvijitra R. (1988) Traditional vegetative propagation and tissue culture of some Thai bamboos. In: Rao I.V.R., Gnanaharan R. and Sastry C.B. (eds.), *Bamboo: Current Research*, For. Res. Inst., Kerala, India and IDRC, Canada, pp. 148-150.

White P.R. (1963) *The cultivation of animal and plant cells*, The Ronald Press, New York, 228 p.

Woods S.H., Phillips G.C., Woods J.E. and Collins G.B. (1992) Somatic embryogenesis and plant regeneration from zygotic embryo explants in Mexican weeping bamboo, *Otatea acuminata aztecorum*. Plant Cell Rep. 11: 257-261.

Yeh M.L. and Chang W.C. (1986a) Plant regeneration through somatic embryogenesis in callus culture of green bamboo (*Bambusa oldhamii* Munro), Theor. Appl. Genet. 73: 161-163.

Yeh M.L. and Chang W.C. (1986b) Somatic embryogenesis and subsequent plant regeneration from inflorescence callus of *Bambusa beecheyana* Munro var. *beecheyana*. Plant Cell Rep. 5: 409-411.

Yeh M.L. and Chang W.C. (1987) Plant regeneration via somatic embryogenesis in mature embryo-derived callus of *Sinocalamus latiflora* (Monro) McClure. Plant Sci. 51: 93-96.

Ziv M. (1991a) Vitrification: Morphological and physiological disorders of in vitro plants. In: Debergh P.C. and Zimmerman R.H. (eds.) Micropropagation technology and application, Kluwer Academic Publishers, Dordrecht, pp. 45-69.

Ziv M. (1991b) Quality of micropagated plants – Vitrification. In Vitro Cell Devel. Biol. – Plant 27: 64-69.

Ziv M. (1995) *In vitro* acclimatization, in: Aitken-Christie J.; Kozai, T.; Smith M.L. (eds.) *Automation and Environmental Control in Plant Tissue Culture*, Kluwer Academic Publishers, Netherlands, pp. 493-516.

5

Bacilli as Bioinoculants for Sustainable Agriculture and Its Challenges

Pankaj Kumar[1]* and R. C. Dubey[2]
[1]Department of Microbiology, Dolphin (PG) College of Science & Agriculture Chunnikalan, Fatehgarh Sahib-140307, Near Chandigarh, Punjab, India
[2]Department of Botany and Microbiology
Gurukul Kangri Vishwavidyalaya, Haridwar-249404, Uttarakhand, India
**E-mail: guptapankaj23@gmail.com*

Abstract

Bacilli, a group of endospore forming bacteria present in soil, contribute directly or indirectly to crop productivity. Bacilli have many physiological unique traits like multilayered cell wall, stress resistant endospore formation, secretion of peptide antibiotics, peptide signal molecules, extracellular enzymes, which make them ubiquitous and assist to their survival under adverse environmental conditions for long time. The principal mechanisms for plant growth promotion by bacilli include production of several growth stimulating phytohormones such as Indole acetic acid, gibberalic acid, cytokinin, solubilization of phosphate, potassium, zinc, production of iron chelating compound siderophore, nitrogen fixation, inhibition of plant ethylene synthesis by producing 1-aminocyclopropane-1-carboxylate (ACC) deaminase, inhibition of phytopathogens by antibiosis (production of antibiotics) as well as by secreting cell wall degrading enzymes like chitinase, β-1,3-glucanase, β-1,4-glucanase, β-1,6-glucanase, cellulases, chitinase, chintosanses, amylase and toxins (bacteriocins). The advantage of Bacilli over the non-spore forming bacteria is that the spores are more vigorous and resistant to the extreme conditions like temperature, pH and high concentrations of chemicals. This enhances the shelf-life of biofertilizers which have bacterial spores for up to 3 years. This chapter discusses about bacilli as potential bioinoculants for sustainable agriculture.

Keywords : Bacilli, Endospore, Plant Growth Promotion, Bioinoculants.

Introduction

The genus *Bacillus* was originally named as *Vibrio* by Christian Gottfried Ehrenberg for the spore-forming, Gram-positive, rod-shaped bacteria. Later on in 1872, it was renamed as *Bacillus* by Ferdinand Cohn, a student of Robert Koch. The genus *Bacillus* comprises 266 species. They are ubiquitous, with the largest 16S diversity and environmental diversity. It is a member of the division Firmicutes and *Bacillaceae* family. *Bacillus* species are catalase positive, obligate aerobes or facultative anaerobes (Turnbull 1996). Under stressful environmental conditions, the cells produce oval endospores that can stay dormant for long time, which is distinguishing feature of the family. These characteristics originally defined the genus, but not all such species are closely related, and many have been moved to other genera (Madigan *et al.* 2005).

Bioinoculants or microbial inoculants are artificially multiplied cultures of certain soil microorganisms that can improve soil fertility and crop productivity by increasing the supply or availability of primary nutrients to the host plant. Although the beneficial effects of legumes in improving soil fertility was known since ancient times and their role in biological nitrogen fixation was discovered more than a century ago, commercial exploitation of such biological processes is of recent interest and practice. Bioinoculants add nutrients through the natural processes of nitrogen fixation, solubilizing phosphorus, potassium, zinc etc and stimulating plant growth through the synthesis of growth-promoting substances such as Indole acetic acid, gibberellins, cytokinin etc. It reduces the use of chemical fertilizers and pesticides. The microorganisms in bioinoculant restore the soil's natural nutrient cycle and build soil organic matter. Through the use of bioinoculant, healthy plants can be grown, while enhancing the sustainability and the health of the agricultural soil.

The first successful application and commercial production of Plant Grwoth Promoting Rhizobacteria (PGPR) is a *B. subtilis* strain A13, which was isolated more than 25 years ago in Australia based on *in vitro* inhibitory activity to all of nine pathogens tested and was subsequently shown to promote plant growth. Microbial inoculation of PGPR into agriculture system, used either to increase crop health or to manage plant diseases, depends on the development of formulations with suitable carriers that support the survival of bacteria for a long time. Development of formulations with increased shelf-life and broad spectrum of action with consistent performance under field conditions could pave the way for commercialization of the technology at a faster rate. However, successful inoculation of PGPR often is limited by several factors including environmental conditions, the number of infective cells applied, presence of competing indigenous rhizobacteria and the inoculation method (Smith 1992). Production of biofertilizer in India increased from 25065.0352 tons in 2008-09 to 46836.82 tons in 2012-13. Mostly *Bacillus*, *Rhizobium*, *Azotobacter*, *Azospirillum*, *Acetobacter*, PSB (phosphate solubilizing bacteria) are being used as biofertilizers. Although *Rhizobium* inoculants have been in the market for a century, yet several free-living rhizobacteria and other microorganisms have also been commercialized as bioinoculant and bio-pesticides. These bioinoculants play an important role in enhancing crop productivity through nitrogen fixation, phosphate solubilization, plant hormone production, ammonia excretion and controlling various plant diseases (Smith 1992).

Bacilli as Plant Growth Promoting Rhizobacteria (PGPR) and biocontrol agents

Bacillus, an endospore forming bacteria occurs in agricultural fields and contribute to crop productivity directly or indirectly. Physiological traits, such as multilayered cell wall, stress resistant endospore formation, and secretion of peptide antibiotics, peptide signal molecules, and extracellular enzymes, are ubiquitous to these bacilli and contribute to their survival under adverse environmental conditions for extended periods of time. Several species of *Bacillus* are reported to enhance growth of crop plants (Tilak and Reddy 2006; Minaxi *et al.* 2011). The principal mechanisms of growth promotion include production of growth stimulating phytohormones such as IAA, gibberalic acid, cytokinin, solubilization of phosphate, potassium, zinc, production of siderophore, antibiosis, *i.e.*, production of antibiotics, inhibition of plant ethylene synthesis by producing ACC deaminase, and induction of plant systemic resistance to pathogens (Idris *et al.* 2007; Chung *et al.* 2008; Richardson *et al.* 2009; Sunar *et al.* 2015). Recently, some workers reported nitrogen fixation in different species of *Bacillus* (Kumar *et al.* 2012). It is very likely that plant growth promotion by rhizosphere bacilli may be a result of combined action of two or more of these mechanisms. Pathogenic microorganisms affecting plant health are a major threat to food production, and traditional methods, viz., crop rotation, breeding for resistant plant cultivars, and application of chemical pesticides, seem to be insufficient to control root diseases of important crop plants (Johri *et al.* 2003). Biological control of pathogens has been increased in the past few years partly due to change in the public concern over the use of chemicals and the need to find alternative of chemicals for disease control. *Bacillus* species express antagonistic activities by suppressing the pathogens (Chen *et al.* 2009a, b; Arrebola *et al.* 2010 and Kumar *et al.* 2012; Oslizlo *et al.* 2015). Enhancement of plant growth by root-colonizing species of *Bacillus* is well documented. Spore-forming bacilli have an advantage over the non-spore formers such as *Pseudomonas*. Spores are more robust and resistant to the elevated temperature and high concentrations of chemicals. Moreover, the shelf-life of biological products based on bacterial spores can be up to 1-3 years. A disadvantage of using the spores is the requirement of time to return to the metabolic active stage of a vegetative cell after their application. In the past few years, research has been directed more toward the induced systemic resistance (ISR). Induced resistance may be defined as a physiological state of enhanced defensive capacity elicited in response to specific environmental stimuli and consequently the plants innate defenses are potentiated against subsequent biotic challenges (van Loon 2000). Choudhary and Johri (2008) have reviewed ISR by *Bacillus* spp. in relation to crop plants and emphasized on the mechanisms and possible applications of ISR in the biological control of pathogenic microbes. Various strains of species *B. amyloliquefaciens*, *B. subtilis*, *B. pasteurii*, *B. cereus*, *B. pumilus*, *B. mycoides*, and *B. sphaericus* are known as potential elicitors of ISR and exhibit significant reduction in the incidence or severity of various diseases on different crops (Kloepper *et al.* 2004; Choudhary and Johri 2008; Yang *et al.* 2009, 2011; Kamal *et al.* 2014). It is believed that plants have the ability to acquire enhanced level of resistance to pathogens after getting exposed to biotic stimuli provided by many PGPRs and this is known as rhizobacteria mediated ISR (Choudhary *et al.* 2007).

Very recently Sunar *et al.* (2015) reported *Bacillus altitudinis* isolated from Darjeeling hills, India having good plant growth promoting activities such as phosphate solubilization, Hydrogen Cyanide (HCN), siderophore, (IAA), chitinase, protease production and significantly inhibited phytopathogens. They found enhancement of all the growth parameter in *Vigna radiata*, *Cicer arietinum*, and *Glycine max* by application of *B.*

altitudinis. Beneduzi *et al.* (2008) investigated several strains of bacilli, mainly species of the genera *Bacillus* and *Paenibacillus*, displaying important plant growth promoting (PGP) characteristics isolated from seven distinct rice production zones of the Rio Grande do Sul State, south Brazil. Of those 296 isolates, 94 and 148 produced between 0.1 and 30 mg of IAA ml^{-1} *in vitro* after 72 and 144 h of incubation, respectively. Twenty two isolates were able to solubilize phosphate and 32 isolates produced siderophores. *Paenibacillus* and *Bacillus* genera were the most prominent groups in the rhizosphere and soil populations analyzed. The isolate SVPR30, identified by 16S rRNA gene sequence analysis as a strain of *Bacillus* sp., was chosen for *in vivo* greenhouse experiments and proved to be very efficient in promoting a significant increase in the roots and shoot parts of rice plants. *Bacillus* spp. isolated from the rhizosphere of *Phaseolus vulgaris* produced significant amounts of IAA (Kumar *et al.* 2012). Several strains of *Bacillus subtilis* and *B. cereus* have been isolated and selected in China, due to their ability to promote plant growth and control fungal diseases that affect wheat roots (Ryder *et al.* 1999). Among these strains is *B. cereus* A47, which has been used in several cultures in that country, stimulating wheat production by around 11%. In turn, *B. subtilis* B908 was used to control rice sheath blight, caused by *Rhizoctonia solani*. Both strains were also tested in two different soils in Australia, in experiments conducted in a green house. These bacteria considerably reduced the severity of diseases caused by *R. solani* and *Gaeumannomyces graminis* var. *tritici*, stimulating also the growth of wheat seedlings (Ryder *et al.* 1999). *Bacillus megaterium* strain B388, isolated from rhizosphere soil of pine belonging to a temperate Himalayan location has been characterized by Trivedi and Pandey (2008). In the same year Singh *et al.* (2008) reported enhancement of growth parameter of chirpine by application of *B. subtilis* BN1. The carrier based formulations of the bacterium resulted in increased plant growth in bioassays. The rhizosphere colonization and the viability of the cells entrapped in alginate beads were greater in comparison to coal or broth based formulations. In our laboratory, we have also reported *Bacillus* spp. having good plant growth promoting activities and served as good biocontrol agent against various phytopathogens such as *Macrophomina phaseolina*, *Fusarium oxysporum*, *F. solani*, *Sclerotinia sclerotiorum*, *Rhizoctonia solani* and *Choletotricum* sp. (Kumar *et al.* 2012; Chen *et al.* 2013; Abdalla *et al.* 2014; Oslizlo *et al.* 2015; Xiong *et al.* 2015). To develop a successful microbial bioinoculant, PGPR must possess: (i) high rhizosphere competence,

(ii) high competitive saprophytic ability, (iii) ability to enhance plant growth, (iv) ease for mass multiplication, (v) broad spectrum of action, (vi) excellent and reliable control (vii) safe to environment, (viii) compatible with other rhizobacteria, and (ix) ability to tolerate desiccation, heat, oxidizing agents and UV radiations.

Carriers

Several types of materials are used as carrier for seed or soil inoculation. For preparation of seed inoculant, the carrier material is milled to fine powder with particle size of 10-40 μm. For soil inoculation, carrier material in granular form (0.5-1.5 mm) is generally used. Granular forms of peat, perlite, talcum powder, charcoal or soil aggregates are suitable for soil inculation. According to the "Handbook for Rhizobia" (Somasegaran and Hoben 1994), the properties of a good carrier material for seed inoculation are: (*i*) non-toxic to inoculant bacterial strain, (*ii*) good moisture absorption capacity, (*iii*) easy to process and free of lump-forming materials, (*iv*) easy to sterilize by autoclaving or gamma-irradiation, (*v*) available in adequate amounts, (*vi*) inexpensive, (*vii*) good adhesion to seeds, (*viii*) good pH buffering capacity, (*ix*) non-toxic to plant, (*x*) organic matter should be 40%, and (*xi*) should have high water holding capacity and surface area. There are several carrier materials commercially available for different crops having diverse physical properties (Table 1).

Table 1 : Physical properties of carriers

Carriers	pH	Water holding capacity (%)	Inherent moisture content (%)
Soil	7.1	70	1.3
Sand	7.2	230	0.3
Lignite	7.0	250	0.3
Charcoal	7.0	620	5.1
Sawdust	6.8	350	13.8
Soil + Sawdust (1:1)	7.1	320	6.9

Besides above parameters some other criteria must be considered during carrier selection. (i) Survival of the inoculant bacteria on seed. Seeds are not sown immediately after seed coating with the inoculant bacteria. The bacteria have to survive on seed surface against drying condition until placed into soil. (ii) Survival of the inoculant bacteria during the storage period. (iii) Survival of the inoculant bacteria in soil. After being introduced into the soil, the inoculant bacteria have to compete with native soil microorganisms for the nutrient and habitable niche, and have to survive against grazing protozoa.

Carrier materials that offer the available nutrient and/or habitable micro-pore to the inoculant bacteria will be desirable. In this sense, materials with micro-porous structure, such as soil aggregate and charcoal, will be good carrier for soil inoculant. Many of the microbial inoculants all over the world are based on solid peat formulations for seed inoculation and a number of information is available on the properties and effect of the inoculant. The typical peat is not available in India in sufficient amount but peat like material is available in Nilgiris (Ootacamund or Udagamandalam) has been found to be very good carrier. Whereas lignite is available in Neyveli lignite mines of south India. Several materials have been tested as an alternative carrier source for bacterial inoculants, such as soil, coal, vermiculite, perlite, attapulgite, sepiolite, amorphous silica, cork compost, grape bagasse, talc based powder and wood flour and other plant composts (Ferreira and Castro 2005; Khavazi *et al.* 2007; Albareda *et al.* 2008; Sallam *et al.* 2013). Charcoal, filter mud, Nile silt, bagasse and groundnut shells were used as carries for two phosphorus solubilizing bacteria, *Streptomyces albus* and *Bacillus megatherium* var *phosphaticum*. The results showed that the four materials can be used as carriers. However, Charcoal was found to be the most efficient carrier (Osman and Mohamed 2006). Arora *et al.* (2008) reported sawdust as a superior carrier for production of multipurpose bioinoculant using plant growth promoting bacteria which enhanced the productivity of *Trifolium repense.*

Bioinoculants

Developing countries like India are experiencing difficulties regarding their economic, agricultural and industrial development. The aim of today's scientific agricultural practices is to produce enhanced grain yield to supply food to rapidly increasing population. Thus the present scenario demands the use of chemical fertilizers and other agrochemicals. However the production cost of these chemical products is too high as it increases pressure on the fossil fuel reserves of the country. These fossil fuel reserves are finite or unsustainable in the long-term and give hazardous oxidized products posing threat to environmental and human health. Moreover the inorganic fertilizer application tremendously debilitates the soil's physical and chemical status as well as drastically alters microbial diversity including microbial flora and fauna vital to improve soil fertility. In 1906, when the first production unit of chemical fertilizer started in India was producing single super phosphate in Ranipet (Near Chennai) with annual capacity of 6000 MT reached a level of 120.41 lakh MT of nitrogen and 56.19 lakh MT of phosphatic nutrient on 31.03.2012, making India the 3rd largest fertilizer producer in the world (Annual report, Department of fertilizers, India 2012-2013). India is the second largest consumer of fertilizers in the world after China (FAI 2010). More than half of the total fertilizer is consumed by only two crops i.e. rice and wheat. Sustainable agriculture is configured on use of a variety of prophecies, phenomena and products stressing on land reclamation and awareness towards hazards of over-industrialization and pollution on health of ecosystem. This concept urges the utilization of an array of techniques as organic farming, biofertilizers, biocontrol agents/ biopesticides, mixed/inter-cropping.

Biofertilizers is still an unclear term. It can be easily found that biofertilizers are identified as plant extract, composted urban wastes and various microbial mixtures with unidentified constituents and chemical fertilizer formulations supplemented with organic compounds. However biofertilizer is most commonly referred to the use of soil microorganisms to increase the availability and uptake of mineral nutrients from plants. So it is necessary to define the term biofertilizer. There is a proposal that biofertilizer be defined as a substance which contains living microorganisms which colonizes the rhizosphere or the interior of the plant and promotes growth by increasing the supply or availability of primary nutrient and/or growth stimulus to the target crop, when applied to seed plant surfaces or soil. Biofertilizer is a natural product carrying living microorganisms derived from the root or cultivated soil. They don't have any ill effect on soil health and environment. Besides their role in atmospheric nitrogen fixation and phosphorous and other mineral solubilisation, these also help in stimulating the plant growth hormones providing better nutrient uptake and increased tolerance towards drought and moisture stress (Vessey 2003; Nakkeeran *et al.* 2005). A small dose of biofertilizer is sufficient to produce desirable results because each gram of carrier of biofertilizers contains at least 10 million viable cells of a specific strain. Marginal farmers in India cannot afford the chemical fertilizers but the use of biofertilizers is affordable and economical for wide variety of crops such as cereals, millets, oilseeds, spices, vegetables, fruits and plantation crops. The cost: benefit ratio of biofertilizers is always higher, so their use is acceptable in all places of cultivation.

Ideal bioinoculants should: (i) have increased shelf life, (ii) not to be phytotoxic (iii) dissolve well in water and should release the bacteria (iv) tolerate adverse environmental conditions, (v) be cost effective and give reliable control of plant diseases, (vi) be compatible with other agrochemicals, and (vii) be cheap and readily available for formulation development. In the carrier-based solid bio-fertilizers, the microorganisms have a shelf-life of only six months. They are not tolerant to UV (ultra violet) rays and temperature more than 30°C.

Solid based Bioinoulants

Solid carrier-based microbial inoculants used for natural supply of phosphorus, potassium, zinc etc to the plants, biological control of soil borne phytopathogens, biological degradation of organic contaminants, soil life and fertility improvement.

In solid based bioinoculant, the microorganisms have a shelf life of only six months. They are not tolerant to UV rays and temperatures more than 30°C. The population density of these microbes is only 10^8(10 crores) cfu/ml at the time of production. This count gradually reduces day by day. In the fourth month it reduces to 10^6(10 lakhs) cfu/ml and at the end of 6 months the count is almost nil. That's why the solid based bioinoculants are not much effective and not popular among the farmers.

Liquid based Bioinoculants

The defects of solid based bioinoculants can be rectified and fulfilled by Liquid bioinoculants. The shelf life of the microbes in these liquid bioinoculants is approximately two years. They tolerate high temperatures (55°C) and UV radiations. The count is as high (10^9 cfu/ml), which is maintained constant up to two years. So, the application of 1 ml of liquid bioinoculants is equivalent to the application of 1 kg of 5 months old solid carrier based bioinoculants. Since these are liquid formulations the application in the field is also very simple and easy. They are applied using hand sprayers, power sprayers, fertigation tanks and as basal manure mixed along with farmyard *manure* etc (Ngampimol and Kunathigan 2008).

Other advantages of Liquid Bio-fertilizers

(i) Contains special cell protectants or substances that encourage formation of resting spores or cysts.

(ii) Contains special nutrients that ensure longer shelf life, Better survival on seeds and soil and tolerance to adverse conditions.

(iii) Liquid formation ensures that the product is easy to handle and apply.

(iv) Since the organisms are stabilized during production, distribution and storage, the activity is enhanced after the contact and interaction with the target crops.

(v) No loss of properties due to storage up to 55 degrees.

(vi) Greater potentials to fight with native population.

(vii) Dosages are 10 times lesser than the carrier-based bio-fertilizers.

(viii) Very high enzymatic activity since contamination is nil.

Bureau of Indian Standards (BIS) norms for Bioinoculants

The Bureau of Indian Standards (BIS) is the National Standards Body of India working under the aegis of Ministry of Consumer Affairs, Food & Public Distribution, Government of India. It is established by the Bureau of Indian Standards Act, 1986 which came into effect on 23 December 1986. The Minister in charge of the Ministry or Department having administrative control of the BIS is ex-officio President (Emaad Amin) of the BIS. The organization was formerly the Indian Standards Institution (ISI), set up under the Resolution of the then Department of Industries and Supplies No. 1 Std.(4)/45, dated 3 September 1946. The ISI was registered under the Societies Registration Act, 1860. It has 25 members drawn from Central or State Governments, industry, scientific and research institutions, and consumer organizations. Its headquarters is in New Delhi, with regional offices in Kolkata, Chennai, Mumbai, Chandigarh and Delhi, and 20 branch offices. It also works as WTO-TBT enquiry point for India. BIS is a founder member of International Organisation for Standardization (ISO). Bureau of Indian Standard (BIS) have developed quality standards for Bioinoculants/ Biofertilizers especially for *Rhizobium*, *Azotobactor Azospirilum* and Phosphate solubilizing Bacteria (PSB) (Table 2). No standards are available for Blue Green Algae (BGA), Azolen, VAM.

Table 2 : BIS norms for bioinoculants

Base		Carrier based or liquid based
Viable Cell Count (CFU)**		Minimum 10^7 cells/g of carrier material or 10^7 cells/ml of liquid material
Contamination level		No contamination of 10^5 dilution
Particle size in case of carrier based material		All Material shall pass through 0.15 – 0.212mm IS Sieve
pH	6.5 – 7.5	
Moisture percent by weight, maximum in case of carrier based material	30-40%	
Efficiency Character	PSB	The strain should have minimum 30% phosphate solubilizing capacity, when tested spectrophotometrically. In terms of zone formation, minimum 5 mm solubilization zone in prescribed media having at least 3 mm thikness.
	Rhizobium	Should Show effective nodulation on all the legume species (Crops) listed on the packet.
	Azotobacter	The strain should be capable of fixing at least 10 mg of N per g of sucrose consumed
	Azospirillum	Formation of white pellicle in semi-solid Nitrogen–free bromothymol blue media

Source - Bureau of Indian Standard (BIS)

Some other general norms of BIS for biofertilizers

- Bio-fertilizers are not included in FCO (*Fertilizer Control Order*) and as such no penal action can be taken for supplying sub-standard biofertilizers.
- Government does not control the prices of Organic Manures and Bio-fertilizers.
- Import of Organic manures & biofertilizers must be banned. However, export is considered on case to case basis by DGFT

(Directorate General of Foreign Trade) on the basis of recommendation of the Ministry of Agriculture.

- Import of biofertilizers should be restricted to avoid Competition with indigenous unit and also to avoid entry of pathogens through bio-fertilizer. However DGFT can allow imports.

Prospects of Use of Biofertilizers

Positive aspects

- Cheap source to augment plant nutrient supply.
- Easy to treat seeds.
- Effect of Rhizobial inoculation is established.
- Work well under low fertilizer use.
- Work well in combination with organics.

Negative aspects

- Bio-fertilizers other than Rhizobia have limited use so far.
- Manufacture is difficult.
- Not available in Blocks/ Panchayats/ Villages.
- Only limited quantity is available to farmers.
- Outside agencies often supply spurious materials which affect farmers considerably.
- Not much residual value.

General awareness for popularization of biofertilizers

- Training of farmers on quality composting.
- Farmer participating demonstrations on use of quality composts.
- Awareness creation among farmers quality Vs quantity.
- Crop demonstrations on vegetables, pulses, oilseeds, cereals on use of quality manures + bio fertilizers depending on crop removal.
- Emphasis on residue management.

- Soil health cards – Soil Testing, identifying critical issues – soil acidity, phosphorus, boron etc.
- Developing indicators to determine maturity of composts.

Constraints in promotion of biofertilizer

- No policy on soil test based nutrient use.
- Soil testing network in the State- very poor.
- State Govt. should promote vermi composting, leaf composting, phospho composting, crop residue management through farmers groups, WIG, SHGs in KVKs/ ATMAs.
- SAUs, ZRS, KVKs need to be encouraged for biofertilizer production. State Govt. should decide on targets well in advance and ensure that it reaches the farmers in time.
- Mass Media support is meagre in this field.
- Absence of a sound & scientific agricultural policy in the State.
- No quality control lab for organics/ Bio-fertilizers.

To produce powder and granular formulations, spore-forming strains such as *Bacillus* have an advantage over non spore-forming strains such as *Pseudomonas* which have to be formulated as vegetative cells. Spores are more robust and resistant to elevated temperature and high concentrations of chemicals that are part of the spray-drying process. Moreover, the shelf-life of biological products based on bacterial spores can be up to 1-3 years. A disadvantage of the use of spores is that, after application, they need time to return to the metabolic active stage of a vegetative cell. The poor performance of biofertilizers can be attributed to (i) stains which do not survive well in adverse climatic and soil conditions; (ii) inefficient production technologies resulting in the production of contaminated product with poor nitrogen fixing ability; (iii) the use of inappropriate carriers, and storage and transport facilities which reduce the shelf life, and (iv) lack of knowledge and understanding among the farmers of the correct procedures for the storage and application of biofertilizers. Gaps exist between India and some of the leading countries in all areas of biofertilizer production and application technology. These gaps are particularly important in the field of strains, techniques used for sterilization, fermentation, and carriers. The effects of these are aggravated further by poor quality control maintained by most producers. In the circumstance, a significant increase in the level of penetration and demand of biofertilizers is possible only if these gaps are reduced and strict quality control is maintained. There is a need to identify and develop superior strains. At present only single strains of *Rhizobium*, *Azotobacter* and *Azospirillum, Bacillus etc.* are used. It is necessary that composite strains of these biofertilizers are produced and used. At present 122 biofertiliser production units are in operation, other than NBDC/RBDCs (capacity 18500 tonne out of which 83 are established with the central financial assistance. Total production of biofertiliser is 10000 tonne. (www.nbdc.com). National Biofertiliser Development Centre (NBDC), Ghaziabad, a unit of Government of India produce many biofertilizers based on *Azotobacter, Azospirillum, Rhizobium,* P-solubiliser and BGA, but there is no any biofertilizers based on *Bacillus* spp. which has many advantage (as mentioned above) comparing to other PGPR. While promotional efforts are important, the success of these efforts will depend on the availability of biofertilizers of high and consistent quality. A system by which the quality is monitored by the central and state level authorities may be devised and enforced. Application of *Bacillus* spp. in combination with other potent plant growth promoting strains performed better result comparison to alone (Sivaramaiah *et al*. 2007; Ramirez and Kloepper 2009). The well-known Kodiak *B. subtilis* strain GB03 is thought to survive on seeds until planted and then uses seed exudates during seed germination, directionally multiplying to reach young roots, and maintaining a robust population in the presence of field crops via plant–microbial interactions (Kloepper *et al*. 2004). Some of the other commercially available bioinoculants based on *Bacillus* spp. and their PGP attributes are listed in Table 3.

Table 3 : Commercially available Bioinoculants based on *Bacillus* spp.

Trade Name	*Bacillus* spp.	PGP Properties	Formulation	References
BioP-Plus	*Bacillus coagulans*	P-solubilization	Powder	www.soo.co.in/biopplus.htm
Ballad Plus (Approved only for USA)	*Bacillus pumilus* QST 2808	Biocontrol	Liquid	www.agraquest.com
BioYield	*B. subtilis* GB122 and *B. amyloliquefaciens* GB99	Biocontrol	Gustafson	www.gustafson.com
Companion	*B. subtilis* GB03, *B. lichenformis*, *B. megaterium*	Biocontrol	Liquid	www.growthproducts.com
Companion-Dry super concentrate	*B. subtilis* GB03	Biocontrol	Powder	www.growthproducts.com
Eco Guard	*B. licheniformis* SB3086	Biocontrol	Liquid spore concentrate	www.novozymes.com
Forgreen-BS (ISP)	*B. subtilis*	Biocontrol	Powder	http://www.biocentury.com
Kodiac/GUS 2000	*B. subtilis* GB03	Biocontrol	Powder	www.gustafson.com
Phosphomax	*Bacillus* sp.	P-Solubilization	Powder	www.alibaba.com
Rhizo-Plus	*B. subtilis* FZB24	Biocontrol	Granule	www.home.fhtw-berlin.de
Serenade	*B. subtilis* QST713	Biocontrol	Powder	www.agraquest.com
Shakti	*Bacillus* sp.	Biocontrol	Powder	www.amitbiotech.com
SUKAHAN	*B. subtilis*	N-fixation	Powder	www.alibaba.com

Sonata (Approved for five country including USA)	*Bacillus pumilus* QST2808	Biocontrol	Liquid	www.agraquest.com
Subtilix	*B. subtilis* MBI600	Biocontrol	Powder	www.beckerunderwood.com
Teeka	*Bacillus* sp.	P-solubilization, Biocontrol	Liquid, granular	www.sribio.com
YieldShield	*B. subtilis* GB34	Biocontrol	Powder	www.gustafson.com
	Bacillus subtilis ME488	Biocontrol		Chung et al. 2008
***Bacillus* and Other PGPRs**				
Bio Speed	*Bacillus coagulans* and (*A. chroococcum, Torulospora globosa,* sea weed extract, humic acid)	P- solubilization and N-fixation	Liquid	
Bevenovo	*Bacillus coagulans B. megaterium* and *Azotobacter spp.*	Provide nutrients	Powder	www.alibaba.com
HiStick N/T	*B. subtilis* and *Rhizobia*	Biocontrol	Slurry	www.beckerunderwood.com
Phosfert	*B. megaterium var. phosphaticum* and (*Azotobacter chroococcum, A. vinelandii*)	P- solubilizationand N-fixation	Powder	www.krishimitra.net
Phospho-bacterin	*B. megaterium* var. *phosphaticum* and *Pseudomonas* spp.	P-solubilization	Powder	www.iari.res.in
Seawinner	Bacilli and Seawead	Biocontrol , K, P- solubilization	Powder	ww.seawinner.ell.alibaba.com
Vitagro	*B. firmus*, *B. polymyxa*, *B. megaterium, B. licheniformis* and (*P. striata, Trichoderma viride*)	P- solubilization, Biocontrol	Powder	www.microbax.com

Challenges in Bioinoculants

Challenges in selection

During selection of plant growth promoting rhizobacteria following point must be considered:

- Host plant specificity
- Adaptation to the particular rhizosphere
- Climatic condition
- PGPR must be a good root colonizer
- Pathogens of such host plant/soil during isolation and screening of PGPR

In case of chemical fertilizer industry thousands of prospective compounds are screened annually to protect the plant and climate for further development, but this type of approaches are not in the case of PGPR. During selection of organisms with the potential to control soil borne phytopathogens must isolate from soil that are suppressive to that pathogens (Kumar and Dubey 2012).

Challenges in application

1. Achieving consistent performance in the field where there is heterogeneity of abiotic and biotic factors and competition with indigenous organisms is more difficult.
2. Knowledge of these factors can aid in determination of optimal concentration timing and placement of inoculants and of soil and crop management strategies to enhance survival and proliferation of the inoculants.
3. The concept of engineering or managing the rhizosphere to enhance PGPR function by manipulation of the host plant substrate for PGPR or through agronomic practice is gaining increasing attentions.
4. Development of better formulation to ensure survival and activity in the field and compatibility with chemical and biological seed treatment is another area of focus.

Challenges in commercialization

Before registration and commercialization of PGPR product a number of hurdles must be overcome

1. Scale up and production of the organisms under commercial fermentation condition while maintaining quality, stability and efficacy of the product.
2. Formulation development must consider factors such as shelf life, compatibility with current application practices, cost and ease of application.
3. Health and safety testing may be required to address such issue as non target effects on other organisms including toxigenicity, altergenicity, pathogenicity, persistence in the environment and potential for horizontal gene transfer.
4. The product claim whether as a fertilizer supplement or for biological control will determine to which federal agency application for registration should be addressed in Canada and the USA.
5. Capitalization cost and potential market must be considered in the decision to commercialize $1 million US required to register a biopesitcide product in North America.

Conclusion

Bioformulation development increases the shelf life of bacteria for a longer period and sustained direct and indirect plant growth promoting activities which makes it potential for further use. To promote adoption of biofertilizer production and use following suggestions should be taken into consideration: Support of applied research in the field of biofertilizer through financial support, training of personnel and providing logistics. Continuous improvement of skills and capacity building of the agricultural extensionists in the areas of biofertilizer use and strengthen collaboration between the different research institutions, the central and state ministries of agriculture. Establishment of technical committees for development of production and use of biofertilizers technology. Adoption of international quality control standards for production of microbial inoculants. Establishment of extension programmes among farmers and agricultural companies for dissemination of biofertilizers and encouragement of bioferilizers for commercial production. Close supervision from the government authorities to the imported biofertilizers is highly needed.

References

Abdalla S.A., Soad AAA, Ibrahim EA, Naim AME (2014) *In Vitro* screening of *Bacillus* isolates for biological control of early blight disease of Tomato in Shambat Soil. World J. Agri. Res. 2(2): 47-50.

Albareda M, Rodriguez-Navarro DN, Camacho M, Temprano FJ (2008) Alternatives to peat as a carrier for rhizobia inoculants: Solid and liquid formulations. Soil Biol. Biochem. 40: 2771-2779.

Arora NK, Khare E, Naraian R, Maheshwari DK (2008) Sawdust as a superior carrier for production of multipurpose bioinoculant using plant growth promoting Rhizobial and pseudomonad strains and their impact on productivity of *Trifolium repense*. Curr. Sci. 95(1): 90-94.

Arrebola E, Jacobs R, Korsten L (2010) Iturin A is the principal inhibitor in the biocontrol activity of *Bacillus amyloliquefaciens* PPCB004 against postharvest fungal pathogens. J. Appl. Microbiol. 108: 386–395.

Beneduzi AD, Peres LK, Vargas MH, Bodanese-Zanettini, Passaglia LMP (2008) Evaluation of genetic diversity and plant growth promoting activities of nitrogen-fixing Bacilli isolated from rice fields in South Brazil. Appl. Soil Ecol. 39: 311-320.

Chen XH, Koumoutsi A, Scholz R, Schneider K, Vater J, Sussmuth R, Piel J, Borriss R (2009a) Genome analysis of *Bacillus amyloliquefaciens* FZB42 reveals its potential for biocontrol of plant pathogens. J. Biotechnol. 140: 27–37.

Chen XH, Scholz R, Borriss M, Junge H, Mogel G, Kunz S, Borriss R (2009b) Difficidin and bacilysin produced by plant-associated *Bacillus amyloliquefaciens* are efficient in controlling fire blight disease. J. Biotechnol. 140: 38–44.

Chen Y, Yan F, Chai Y, Liu H, Kolter R, Losick R, Guo JH (2013) Biocontrol of tomato wilt disease by *Bacillus subtilis* isolates from natural environments depends on conserved genes mediating biofilm formation. Environ. Microbiol. 15(3): 848–864.

Choudhary DK, Johri BN (2008) Interactions of *Bacillus* spp. and plants with special reference to induced systemic resistance (ISR). Microbiol. Res. 164: 493-513.

Choudhary DK, Prakash A, Johri BN (2007) Induced systemic resistance (ISR) in plants: mechanism of action. Ind. J. Microbiol. 47: 289-297.

Chung S, Kong H, Buyer JS, Lakshman DK, Lydon J, Kim SD, Roberts D (2008) Isolation and partial characterization of *Bacillus subtilis* ME488 for suppression of soilborne pathogens of cucumber and pepper. Appl. Microbiol. Biotechnol. 80(1): 115-23.

Ferreira EM, Castro IV (2005) Residues of the cork industry as carriers for the production of legumes inoculants. Silva Lusitana 13: 159-167.

Fertiliser Association of India (2010) "Fertiliser Statistics 2009-10 and earlier issues", The Fertiliser Association of India, New Delhi.

Idris EES, Iglesias DJ, Talon M, Borriss R (2007) Tryptophan dependent production of indole-3-acetic acid (IAA) affects level of plant growth promotion by *Bacillus amyloliquefaciens* FZB42. Mol. Plant Microb. Interact. 20: 619–626.

Indian Fertilizer Scenario (2013) Department of Fertilizers, Ministry of Chemicals and Fertilizers, Government of India.

Johri BN, Sharma A, Virdi JS (2003) Rhizobacterial diversity in India and its influence on soil and plant health. Adv. Bioch. Engin. Biotechnol. 84: 49-89.

Kamal R, Gusain YS, Kumar V (2014) Interaction and symbiosis of AM fungi, Actinomycetes and Plant Growth Promoting Rhizobacteria with plants: Strategies for the improvement of plants health and defense system. Int. J. Curr. Microbiol. App. Sci. 3(7) 564-585.

Khavazi K, Rejali F, Seguin P, Miransari M (2007) Effects of carrier, sterilization method, and incubation on survival of *B. japonicumin* soybean (*Glycine max* L.) inoculants. Enzy. Microbial Technol. 41: 780-784.

Kloepper JW, Ryu CM, Zhang S (2004) Induced systemic resistance and promotion of plant growth by *Bacillus* spp. Phytopathology 94: 1259–1266.

Kumar P, Dubey RC (2012) Plant growth promoting rhizobacteria for biocontrol of phytopathogens and yield enhancement of *Phaseolus vulgaris* L. J. Curr Persp. Appl Microbiol. 1: 6-38.

Kumar P, Dubey RC, Maheshwari DK (2012) *Bacillus* strains isolated from rhizosphere showed plant growth promoting and antagonistic activity against phytopathogens. Microbiol. Res. 167(8): 493– 499.

Madigan MT, Martinko JM, Dunlap PV, Clark DP (2005) Brock Biology of Microorganisms (11th ed.), Prentice Hall.

Minaxi, Nain L, Yadav RC, Saxena J (2011) Characterization of multifaceted Bacillus sp. RM-2 for its use as plant growth promoting bioinoculant for crops grown in semi arid deserts. Appl. Soil Ecol. doi:10.1016/j.apsoil.2011.08.001.

Nakkeeran S, Fernando WG, Dilantha, Siddiqui ZA (2005) Plant growth promoting rhizobacteria formulations and its scope in commercialization for the management of pests and diseases. PGPR: Biocontrol and Biofertilization, 257-296.

Ngampimol H, Kunathigan V (2008) The Study of Shelf Life for liquid biofertilizer from vegetable waste. AU J.T. 11 (4): 204-208.

Oslizlo A, Stefanic P, Vatovec S, Beigot GS, Rupnik M, Mandic-Mulec I (2015) Exploring ComQXPA quorum-sensing diversity and biocontrol potential of *Bacillus* spp. isolates from tomato rhizoplane. Microbial Biotechnol. DOI: 10.1111/1751-7915.12258.

Osman AG, Mohamed SS (2006) Testing the suitability of some materials as carriers for phosphorus solubilizing bacteria. J. Sci. Technol. 7: 52-61.

Ramirez CA, Kloepper JW (2009) Plant growth promotion by *Bacillus amyloliquefaciens* FZB45 depends on inoculum concentration and P-related soil properties. 8th International PGPR workshop. Proceedings book. Portland, Oregon.

Richardson A, Peter A, Hocking P, Simpson R, George T (2009) Plant mechanisms to optimise access to soil phosphorus. Crop Pasture Sci. 60: 124–143.

Ryder MH, Yan Z, Terrace TE, Rovira AD, Tang W, Correll RL (1999) Use of strain of *Bacillus* isolated in China to suppress takeall and rhizoctonia root rot, and promote seedling growth of glasshouse grown wheat in Australian soils. Soil Biol. Biochem. 31: 19–29.

Sallam NA, Riad SN, Mohamed MS, El-eslam AS (2013) Formulations of *Bacillus* spp. and *Pseudomonas fluorescens* for biocontrol of Cantaloupe root rot caused by *Fusarium solani* J. Plant Prot. Res. 53(3): 395-300.

Singh N, Pandey P, Dubey RC, Maheshwari DK (2008) Biological control of root rot fungus *Macrophomina phaseolina* and growth enhancement of *Pinus roxburghii* (Sarg.) by rhizosphere competent *Bacillus subtilis* BN1. World J. Microbiol Biotechnol. 24(9): 1669-1679.

Sivaramaiah N, Malik DK, Sindhu SS (2007) Improvement in symbiotic efficiency of chickpea (*Cicer arietinum*) by coinoculation of *Bacillus* strains with *Mesorhizobium* sp. Ind. J. Microbiol. 47: 51–56.

Smith RS (1992) Legume inoculant formulation and application. Can. J. Microbiol. 38: 485-492.

Somasegaran P, Hoben HJ (1994) Handbook for Rhizobia, Methods in Legume-Rhizobium technology. Springer Verlag, New York. pp. 332-341.

Sunar K, Dey P, Chakraborty U, Chakraborty B (2015) Biocontrol efficacy and plant growth promoting activity of *Bacillus altitudinis* isolated from Darjeeling hills, India. J. Basic Microbiol. 55(1): 91-104.

Tilak KVBR, Reddy BS (2006) *B. cereus* and *B. circulans* novel inoculants for crops. Curr. Sci. 90(5): 642–644.

Trivedi P, Pandey A (2008) Plant growth promotion abilities and formulation of *Bacillus megaterium* strain B 388 (MTCC 6521) isolated from temperate Himalyan location. Ind. J. Microbiol. 48: 342-347.

Turnbull PCB (1996) *Bacillus.* In: Barron's Medical Microbiology (eds. Baron S *et al.*) (4th ed.), University of Texas, Medical Branch.

van Loon LC (2000) Systemic induced resistance. In: Slusarenko AJ, Fraser RSS, van Loon LC (eds) Mechanism of resistance to plant diseases. Kluwer, Dordrecht, pp 521- 574.

Vessey JK (2003) Plant growth promoting rhizobacteria as bio-fertilizers. Plant Soil 255: 571-586.

Xiong J, Zhou Q, Luo H, Xia L, Li L, Sun M, Yu Z (2015) Systemic nematicidal activity and biocontrol efficacy of *Bacillus firmus* against the root-knot nematode *Meloidogyne incognita.* World J. Microbiol. Biotechnol. 31(4): 661-667.

Yang JW, Yu SH, Ryu CM (2009) Promising of defense-related genes confers root-colonizing Bacillielicited induced systemic resistance in pepper. Plant Pathol. J. 25: 389–399.

Yang SY, Park MR, Kim IS, Kim YC, Yang JW, Ryu CM (2011) 2-Aminobenzoic acid of *Bacillus* sp. BS107 as an ISR determinant against *Pectobacterium carotovorum* subsp. carotovotrum SCC1 in tobacco Eur. J. Plant Pathol. 129: 371–378.

6

Application of Biosurfactants in Agriculture and Bioremediation: An Indian Perspective

Parthasarathi R[2], Abitha B[3], Shalini D[1], Gomathi M[1], Allen John Henry[1], Kathireshan A.K[1], Tongmin Sa[4], and Melvin Joe M[1,4*]
[1]Department of Microbiology, School of Life Sciences, VELS University, Velan Nagar, Pallavaram Chennai-600117, Tamilnadu, India
[2]Department of Microbiology, Faculty of Agriculture, Annamalai University, Annamalainagar-608002, Tamilnadu, India
[3]School of Biosciences and Technology, VIT University, Vellore-632014, Tamilnadu, India
[4]Department of Environmental and Biological Chemistry, Chungbuk National University, Cheongju, Chungbuk, 361-763, Republic of Korea
**Email: micromelvin@gmail.com*

Abstract

Biosurfactants (BS) are amphiphilic compounds produced by microorganisms mainly bacteria and are known to contain hydrophobic and hydrophilic moieties that play a major role in reducing surface tension (ST) and interfacial tensions between individual molecules. Chemically, BS is a complex molecules consisting of lipopeptides, glycolipids, polysaccharide protein complex, fatty acids and phospholipids. They are known for their biodegradability, low toxicity, and renewable nature. They are produced from cheaper raw materials, including agro-industrial wastes. Rhamnolipids (RLs), that belong to glycolipid type biosurfactant is one of the most intensively studied biosurfactant produced extensively by *Pseudomonas aeruginosa*. Biosynthesis of rhamnolipid involves three major reactions: biosynthesis of the lipid moiety, followed by the biosynthesis of the sugar moiety, and finally, the enzymatic dimerization and the rhamnosyl transfer, which yield the final products. *rhlA* gene, is the major gene responsible for rhamnolipid synthesis, however, its exact function is still unknown. Several attempts have been made to improve biosurfactant production through mutation, genetic engineering, and Response Surface Optimization (RSM) studies. Biosurfactant usage finds application in environmental and agricultural sectors such as i) bioremediation of hydrocarbons, and heavy metals ii) enhanced oil recovery and iii) pesticide biodegradation and biopesticides production in agriculture.

In this chapter we reviewed the work done on biosurfactant production and applications in agriculture and industry in India under the following headings.

1) Biosurfactant nature, structure and chemical properties.
2) Metabolic pathways and genes involved in the biosurfactant synthesis
3) Genetic and metabolic engineering for enhanced biosurfactant production
4) Response surface optimization studies for Industrial production
5) Potential application of biosurfactants in Environment and Agriculture

Keywords: Biosurfactants, Rhamnolipids, rhlA gene, Bioremediation, Environment, Agriculture

Introduction

Surfactants have broad applications ranging from washing the dishes to advanced oil recovery operations. Primarily the chemical surfactants obtained from petrochemical industries were widely used world over with the value of approximately $ 9.4 billion tons/annum (Desai and Bannat, 1997). However as the chemical surfactants impart negative impacts on our environment, the use of alternative sustainable surfactants from biological sources *i.e.*, biosurfactants is the need of the hour. Biosurfactants are surface-active biomolecules produced by a variety of microorganisms. Similar to chemical surfactants, biosurfactants have unique properties of lowering the surface and interfacial tensions and Critical Micelle Concentration (CMC). The biosurfactants are now acquiring a preference-potential status, because of their biodegradability, biocompatibility, low toxicity, specificity

(Sanket and Yagnik 2013), their ability to withstand high temperature and tolerate high salt concentration (Banat *et al.* 2010). Biosurfactants have potential industrial applications including cosmetics (Brown 1991); textile (Kesting *et al.* 1996); crude oil recovery and bioremediation of contaminated sites (Finnerty 1994; Pekdemir *et al.* 1999) petrochemical industries (Banat *et al.* 2000); mining processes (Zouboulis *et al.* 2003) pharmaceutical industries (Singh and Cameotra 2004) food and beverage industries (Nitschke and Costa 2007); soap/detergent making process (Mukherjee 2007; Helmy *et al.* 2011)

According to the report published by Transparency Market Research the global market for biosurfactant will reach US$ 2,210.5 million by the year 2018, with annual growth rate of 3.5% from 2011 to 2018. As per the Grand View Research Inc., 2014, the global biosurfactants market was 344,068.40 tons in 2013 and is expected to reach 461,991.67 tons by 2020, growing at a CAGR of 4.3% from 2014 to 2020.

Biosurfactant nature, structure and chemical properties

Biosurfactants are amphiphilic and polyphilic polymers that tend to interact with the phase boundary between two phases in heterogenous system, defined as the interface. For all interfacial systems it is known that organic molecules from the aqueous phase they tend to immobilize at the solid interface. They eventually form a film known as a conditioning film, which will change the properties like wettability and surface energy of the original surface (Neu 1996).

Biosurfactants are produced on living surfaces, mostly on microbial cell surfaces or excreted extracellularly and contain hydrophobic and hydrophilic moieties that confer the ability to accumulate between fluid phases, thus reducing the surface tension and interface respectively (Karnath *et al.* 1999). These characters confer excellent detergency, emulsifying, foaming and dispersion (Finnerty 1994; Desai and Banat 1997; Lang 2002).

Types of biosurfactant

Rosenberg and Ron (1999) suggested that biosurfactants can be divided into low and high molecular mass molecules, which efficiently can lower surface and interfacial tension and molecular mass polymers, which are more effective as emulsion stabilizing agents. The major classes of low mass surfactants include glycoloipids, lipopeptides and phospholipids whereas high mass surfactants include polymeric surfactants and particulate surfactants. Types of biosurfactants produced by various microorganisms are tabulated in Table 1.

Table 1 : Types of biosurfactants produced by various Microorganisms

S.No	Biosurfactant class	Microorganisms	Reference
1.	**Glycolipids**		
	Rhamnolipids	*Pseudomonas* sp. *Pseudomonas aeruginosa*	Parra *et al.* (1989) Lang and Wullbrandt (1999)
	Trehalose lipids	*Rhodococcus erythropolis*, *Arthrobacter* sp. *Mycobacterium* sp.	Lang and Philip (1998) Li *et al.* (1984)Cooper *et al.* (1989)
	Sophorolipids	*Torulopsis bombicola*, *Torulopsis apicola*	Gobbert *et al.* (1984) Hommel *et al.* (1987)
2.	**Lipopeptides**		
	Surfactin/Iturin/ Fengycin	*Bacillus subtilis*	Arima *et al.* (1968) Bernheimer *et al.* (1970)
	Lichenysin	*Bacillus licheniformis*	Yakimov *et al.* (1995)
3.	**Fatty acids, neutral lipids and phospholipids**		
	Fatty acids	*Corynebaterium lepus*	Cooper *et al.* (1978)
	Neutral lipids	*Nocardia erythropolis*	MacDonald *et al.* (1981)
	Phospholipids	*Thiobacillus thiooxidans*	Beeba and Umbreit (1971)
4.	**Polymeric biosurfactants**		
	Emulsan/ Biodispersan	*Acinetobacter calcoaceticus*	Rosenberg *et al.* (1988)
	Alasan	*Acinetobacter radioresistens*	Navon-Venezia *et al.* (1995)
	Liposan	*Candida lipolytica*	Cirigliano and Carman (1984)
	Carbohydrate protein-lipid	*Pseudomonas fluorescens*	Desai *et al.* (1988)
	Protein PA	*Pseudomonas aeruginosa*	Hisatsuka *et al.*(1972)
	Mannan-lipid-protein	*Candida tropicalis*	Kappeli *et al.*(1979)
5.	**Particulate biosurfactant**		
	Vesicles and fimbriaes	*Acinetobacter calcoaceticus*	Kappeli *et al.*(1979)
	Whole microbial cells	Variety of bacteria	Rosenberg (1986)

Adapted and modified from Karanth *et al.* (1999)

Low molecular-weight biosurfactant

Low molecular-weight biosurfactants are generally glycolipids and lipopeptides. The best studied glycolipids are rhamnolipids, trehalolipids and sophorolipids, which are disaccharides that are articulated with long-chain fatty acids. These type of biosurfactants are efficient in lowering surface and interfacial tension at the air/water interface (Ron and Rosenberg 2001; Salihu *et al.* 2009).

Glycolipids

Rhamnolipids from *P. aeruginosa* is currently commercialized by Jeneil Biosurfactant, USA, mainly as a fungicide for agricultural purposes or an additive to enhance bioremediation activities. Trehalolipids are produced by a number of different microorganisms, such as *Mycobacterium*, *Nocardia* and *Corynebacterium*. However, the most extensively studied compounds in this class are trehalose dimycolates produced by *Rhodococcus erythropolis* (Rapp *et al.* 1979).

Glycolipids are carbohydrates whose constituents mainly mono, di, tri and tetrasaccharides include glucose, mannose, galactose, rhamnose, galactose sulfate and glucoronic acid. These carbohydrates combine with long-chain aliphatic acids or hydroxy aliphatic acids. This linkage is by means of either ether or an ester group. In this glycolipid type the best known are rhamnolipids, trehalolipids and sophorolipids (Desai and Banat 1997; Muthusamy *et al.* 2008). The structure of glycolipid is provided in Fig.1.

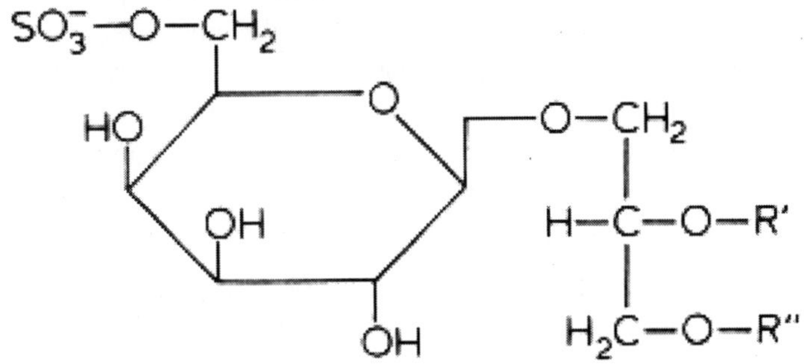

Fig. 1 : Structure of glycolipids

Rhamnolipid

Jarvis and Jhonson (1949) first described the production of rhamnose containing glycolipids in *Pseudomonas aeruginosa.* Edward and Hayashi (1965) reported the formation of glycolipid type R-1 containing two rhamnose and two hydroxydecanoic units in *Pseudomonas aeruginosa.* Rhamnolipids from *Pseudomonas* spp. have been demonstrated to lower the interfacial tension against n-hexadecane to 1 mN/m and the surface tension to 25 to 30 mN/m. They also emulsify alkanes and stimulate the growth of *Pseudomonas aeruginosa* on hexadecane (Hisatsuka *et al.* 1971; Parra *et al.* 1989). Typical structure of rhamnolipid is given in Fig. 2.

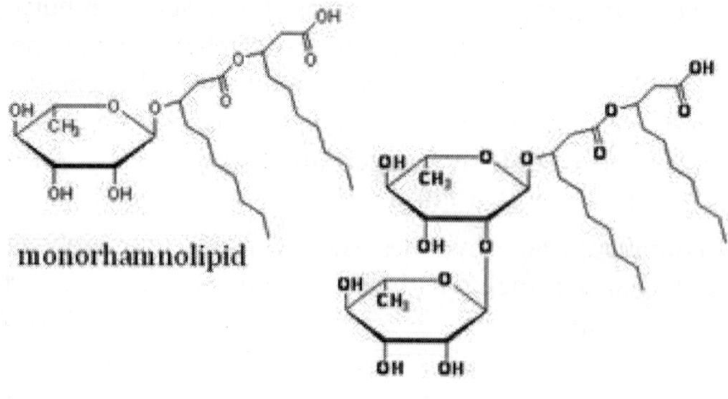

Fig. 2 : Structure of mono and di-rhamnolipid

Rahman *et al.* (2006) suggested that, rhamnolipid produced by *Pseudomonas aeruginosa* strains are among the most effective surfactants when applied for the removal of hydrophobic compounds from contaminated soils. Their ability to form emulsions and solubilise water-immiscible compounds makes them suitable for various industrial and environmental applications (Costa *et al.* 2010).

Pacheco *et al.* (2012) reported the production of rhamnolipid under oxidative stress condition by *Pseudomonas aeruginosa* PA1 and identified the proteins which are potentially involved in the rhamnolipid production and responsible for the oxidative stress.

Trehalolipids

Trehalolipids from different organisms differ in the size, structure of mycolic acid, the number of carbon atoms and the degree of unsaturation

(Cooper *et al.* 1989). Disaccharide trehalose linked at C-6 and C-6' to mycolic acid is associated with most species of *Mycobacterium*, *Nocardia* and *Corynebacterium.* Mycolic acids are long-chain, α-branched-β-hydroxy fatty acids (Desai and Banat 1997). Structure of trehalolipid is provided in Fig. 3.

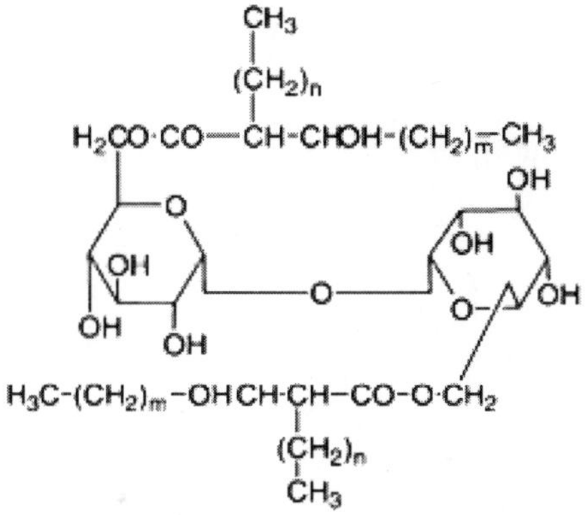

Fig. 3 : Structure of Trehalolipids

Trehalose lipids are mainly produced by *Rhodococcus* sp. and present interesting physicochemical and biological properties (Lang and Philip, 1998). Li *et al.* 1984) reported that the Trehalose lipids from *Rhodococcus erythropolis* and *Arthrobacter* sp. lowered the surface and interfacial tension from 25 to 40 and 1 to 5mN/m respectively. The production of trehalose lipids from alkanotrophic *Rhodococcus ruber* on gaseous alkanes propane and butane was reported by Philip *et al.* (2002).

Marques *et al.* (2009) characterized the physicochemical properties, chemical structure and cellular toxicity of trehalose lipids produced by the strain *Rhodococcus erythropolis* 51T7 in order to determine the optimum conditions for its applications.

Sophorolipids

Sophorolipids consist of a dimeric carbohydrates sophorose attached with a long chain hydroxyl fatty acid (Fig. 4). They are mainly produced by yeasts such as *Torulopsis bomicola*, *Torulopsis apicola* and *Wickerhamiella domericqiae* (Chen *et al.* 2006). The Critical Micelle Concentration (CMC) and the solubilization ratio of the sophorolipids biosurfactant were found to be in a good range compared with synthetic surfactants (Schippers *et al.* 2000). It has been shown that *T. petrophilum* produces sophorolipids on water insoluble substrates such as alkanes and vegetable oils (Rau *et al.* 2001).

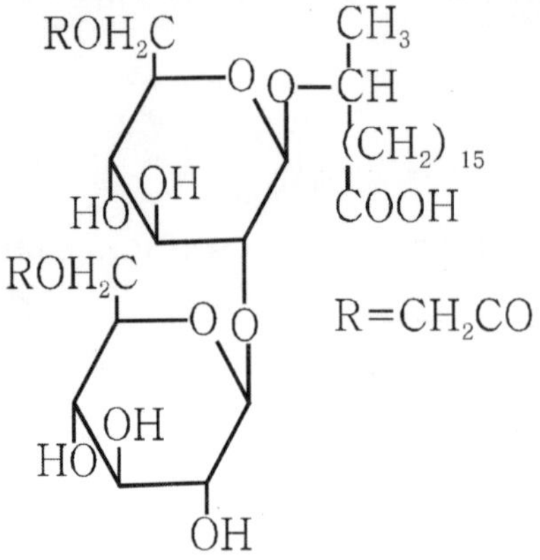

Fig. 4 : Structure of Sophorolipids

Generally, sophorolipids occur as a mixture of macrolactones and free acid form (Fig. 3). Sophorolipids have the capacity to lower the surface tension of water from 72.8 mN/m to 40 to 30 mN/m, with a Critical Micelle Concentration of 40 to 100 mg/l (Van Bogaert *et al.* 2007).

Lipopeptides and Lipoproteins

Lipopeptide biosurfactants are cyclic compounds and they are mostly isolated from *Bacillus* and *Pseudomonas* type bacteria Lipopeptides mainly consist of hydrophilic peptides, generally they 7 and 10 amino acids long, linked to a hydrophobic fatty acid structure. *Bacillus* produce cyclic lipopeptides consist of three major groups known as the surfactin, iturin and fengycin families. Surfactin is the most commonly studied and it contains 7 amino acid cyclic sequences connected to a C13–C16 fatty acid (Kakinuma *et al.* 1969). Structure of aminolipopeptide and cyclic aminolipopeptide is provided in Fig. 5.

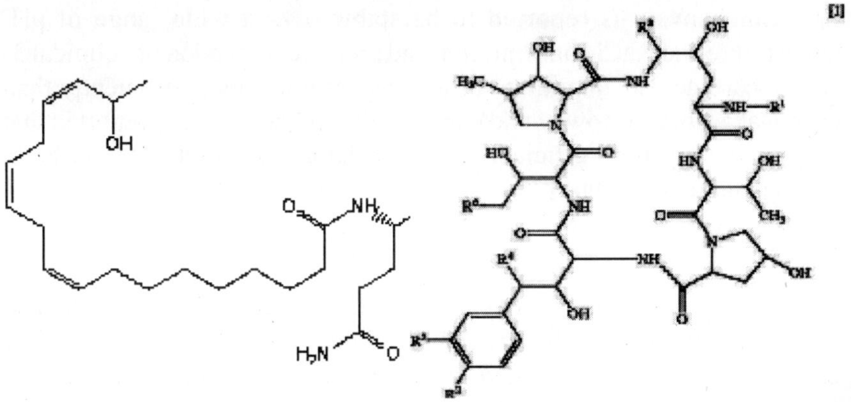

Fig. 5 : Structure of aminolipopetide and cyclic aminolipopetid

Surfactin

Surfactin, cyclic lipopeptide is one of the most powerful biosurfactant which is composed of seven amino acid ring structure coupled to a fatty acid chain via lactone linkage. It lowers the surface tension of water from 72 to 27mN/m at concentrations as low as10 mg/l (Osman *et al.* 1994; Peypoux *et al.* 1999; Mulligan 2005). Structure of surfactin is provided in Fig 6.

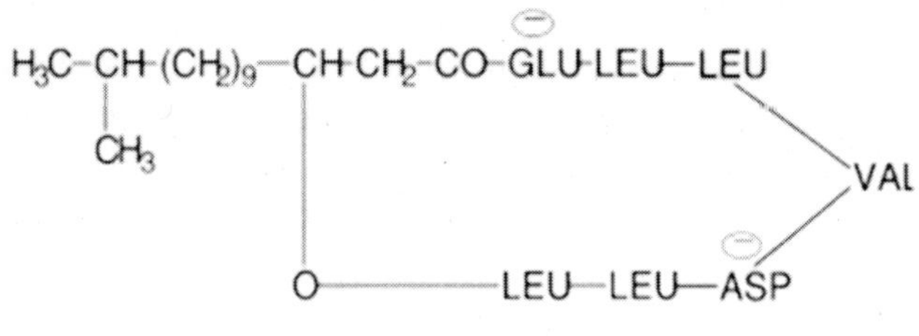

Fig. 6 : Structure of surfactin

Lichenysin

Lichenysin produced by *Bacillus licheniformis* exhibits similar structure and physiochemical properties to that of surfactin (McInerney *et al.* 1990). Lichenysin A produced by *Bacillus licheniformis* strain BAS50 characterized to contain a long chain b-hydroxy fatty acid molecule (Yakimov *et al.* 1996).

Linchenysin is reported to be stable over a wide range of pH, temperature and NaCl concentration and promote dispersion of colloidal 3-silicon carbide and aluminum nitride slurries much more efficiently than chemical agents (Horowitz and Currie 1990). It has also been reported that lichenysin is a more efficient cation chelator compared with surfactin (Grangemard *et al.* 2001).

Iturin

Iturin group of compounds are cyclic lipoheptapeptides, which contain a b-amino fatty acid in its side chain and the first compound of this type Iturin A was isolated from *Bacillus subtilis* (Peypoux *et al.* 1978). Lipopeptides belonging to the itruin family are potent antifungal agents which can also be used as biopesticides for plant protection (Mizumoto *et al.* 2007).

Fengycin

Fengycin is a lipodecapeptide containing b-hydroxyl fatty acid in its side chain and comprises of C_{15} to C_{17} variants which have a characteristic Alanine-Valine dimorphy at position 6 in the peptide ring (Schneider *et al.* 1999). Wang *et al.* (2004) demonstrated that, the identification of fengycin homologues produced by *Bacillus subtilis* by using Electrospray Ionization Spectrometry (ESI-MS) techniques.

Fatty acids, Phospholipids and Neutral lipids

Several bacteria and yeast produces large quantities of fatty acids and phospholipids surfactant during growth on n-alkanes. Thehydrophilic-lipophilic balance (HLB) is directly related to the length of the hydrocarbon chain in their structure. Fatty acids produced from alkanes as a result of microbial oxidations have been considered as surfactant. Apart from these straight chain acids, microorganisms produce complex fatty acid containing OH groups and alkyl branches; example of such complex acids includes corynomycolic acids (Rahman and Gakpe, 2008).

Myroides sp. SM1 was found to produce bile acids, cholic acid, deoxycholic acid and their glycine conjugate, when cultivated in Marine Broth (Maneerat *et al.*, 2004). *Acinetobacter* sp. Strain HO1-N on Hexadecane produces the phospholipids which is mainly phosphatidylethanolamine (Muthusamy *et al.* 2008).

Phospholipids have been qualitatively produced from *Thiobacillus thiooxidans* that are responsible for wetting elemental sulphur necessary for growth (Beeba and Umbriet 1971). Phosphatidylethanolamine produced by *Rhodococcus erythropolis* grown on n-alkane resulted in the lowering of interfacial tension between water and hexadecane to less than 1 mN/m and CMC of 30 mg l^{-1} (Kretschmer *et al.* 1982).

High molecular-weight biosurfactant

Ron and Rosenberg (2001) reported that a large number of bacterial species from different genera produce exo-cellular polymeric surfactant composed of polysaccharides, proteins, lipopolysaccharides, lipoproteins or complex mixtures of these polymers. They are more effective in stabilizing oil in water emulsions. They are highly efficient emulsifiers that work at low concentrations and exhibit considerable substrate specificity and it was referred to as bioemulsans (Dastgheib *et al.* 2008; Salihu *et al.* 2009).

Polymeric microbial surfactants

This type of biosurfactant comprises a polymeric heterosaccharide containing proteins. The best studied polymeric biosurfactants are emulsan, liposan, mannoprotein and polysaccharide protein complexes (Desai and Banat 1997). *Acinetobacter calcoceticus* RAG-1 produces a potent polyanionic amphipathic heteropolysaccharide bioemulsifier that belongs to this type of surfactant (Kappeli and Finnerty 1979).

Emulsan

Emulsan is a very effective emulsifying agent for hydrocarbons in water even at a concentration as low as 0.001 to 0.01per cent. It is one of the most powerful emulsion stabilizers known today and resists inversion even at a water-to-oil ratio of 1:4. On long standing, this emulsion separates into two layers. The upper cream layer, which is known as emulsanosol, contains 70 to 75 per cent oil (Zosim 1982).

Liposan

Liposan is an extracellular water-soluble emulsifier synthesized by *Candida lipolytica* and is composed of 83 per cent carbohydrate and 17 per cent protein (Cirigliano and Carman 1984).

Manno protein

Cameron *et al.* (1988) recently reported the production of large amounts of manno protein by *Saccharomyces cerevisiae*, this protein showed excellent emulsifying activity towards several oils, alkanes and organic solvents.

Alsan

Alasan is an anionic alanine-containing heteropolysaccharide protein biosurfactant produced by *Acinetobacter radioresistens* KA-53. The protein component of alasan appears to play an important role in both the structure and activity of the complex and it lowers interfacial tension from 69 to41 mN/m at 20 °C and has a CMC value of 200 μg/ml (Navon-Venezia 1995; Barkay *et al.* 1999).

Particulate biosurfactant

This type of biosurfactant is produced by *Acinetobacter* sp. when grown on hexadecane and it is accumulated in the extracellular vesicles of 20-50 mm diameter with a buoyant density of 1.158 g/cm^3. They are composed of protein, phospholipid and lipopolysaccharide. Extracellular membrane vesicles partition hydrocarbons forming micro emulsion which plays an essential role in alkane uptake by microbial cells. The purified vesicles are composed of proteins, phospholipids and lipopolysaccharide (Kappeli and Finnerty 1979).

Biosurfactant producing microorganisms

Microorganisms utilize a variety of organic compounds as a source of carbon and energy for their growth. When the carbon source is an immiscible substrate like hydrocarbon, microorganisms facilitate their diffusion and uptake into the cell by producing biosurfactant. Hydrocarbon-degrading bacterial populations are generally dominated by a few main bacterial genera such as *Pseudomonas*, *Bacillus*, *Sphingomonas* and *Actinobacteria* in soils and sediments and *Pseudoalteromonas*, *Halomonas*, *Alcanivorax* and *Acinetobacterin* from marine ecosystems. It is not surprising that a lot of biosurfactant or bioemulsifier producers belong to these same genera (Bodour *et al.* 2003).

Pseudomonas and Vibrionaceae strains with the capacity to produce biosurfactants were observed by Persson and Molin (1987). Biosurfactants are difficult to produce in an economic manner for several reasons. An improvement of production yield is hampered by the strong foam formation. Consequently diluted media have to be applied and only immobilized systems provided an increased productivity of about 3 $gl^{-1}h^{-1}$ as described by Fiecher (1992).

El-Sheshtawy and Doheim *et al.* (2014) demonstrated that biosurfactants are generally microbial metabolites with the typical amphiphilic structure of a surfactant. This study investigated potential biosurfactants production of *Pseudomonas aeruginosa ATCC*-10145 and *Bacillus subtilis* NCTC-1040 using glucose and n-hexadecane as substrates separately and compared it with the production in conventional medium. *Pseudomonas aeruginosa* growing in Bushnell has mineral salt (BHMS) medium with glucose as substrate decreased the surface tension from 72 mN/m of distilled water to 32 mN/m, this strain had higher reduction than *Bacillus subtilis* among all the substrates tested. The crude biosurfactant showed good action as antimicrobial activity against different bacterial and fungal species.

Gunther *et al.* (2004) described rhamnolipids as naturally occurring biosurfactants constructed of rhamnose sugar molecules and β-hydroxyl alkanoic acids, having a wide range of potential commercial applications. Rhamnolipid production by *P. chlororaphis* was achieved by growth at room temperature in static cultures of mineral salts medium containing 2% glucose.

Pseudomonas aeruginosa SP4, isolated from petroleum-contaminated soil in Thailand was used to produce a biosurfactant from mineral medium (MM) with palm oil. The biosurfactant production was done in two identical lab-scale aerobic sequencing batch reactor (SBR) units. The results indicated that the SBR can potentially be adapted for the biosurfactant production and most of the biosurfactant was found to be excreted during the aeration step of the SBR operation (Pornsunthorntawee *et al.* 2008). It has been reported that 2-3 per cent of screened populations in uncontaminated soils are biosurfactant producing microorganisms which increases to 25 per in hydrocarbon polluted soils (Bodour *et al.* 2003).

Kumar *et al.* (2006) reported that hydrocarbon degrading microorganisms having increased capacity to produce biosurfactant due to their co-existance and it can be effectively used for speedy metabolic activity. Thermophilic *Geobacillus pallidus* XS2 and XS3 were isolated from oil contaminated soil sample in Yumen oil field, China and it could be utilized for bioemulsifier production using different hydrocarbons (Zheng *et al.* 2011).

Genes involved in biosurfactant biosynthesis

Lang and Wullbrant (1999) reported that *rhl*R and *rhl*I genes are arranged sequentially and regulate *rhl*AB gene that encode rhamnosyltransferase, that play a major role in rhamnolipid biosynthesis. N-acylhomeserine lactones influence the Rh1r regulator proteins. These authors reported that if the rhII gene is defective rhamnolipid production is only possible if synthetic N-acylhomoserine lactone is added.

Olvera *et al.* (1999) reported that *algC* gene participates in rhamnolipid production by catalyzing the first step in the deoxy-thymidine-diphospho-L-rhamnose (dTDP-L-rhamnose) pathway, i.e., the conversion of glucose-6-phosphate to glucose-1-phosphate. Furthermore, these authors also reported that the *Alg*C protein encoded by *alg*C gene plays a central role in the production of the three *P. aeruginosa* virulence associated saccharides: alginate, LPS and rhamnolipid. According to Fulcher (2010) genes *rhlA*, *rhlB* and *rhlC* encode enzymes *Rhl*A that catalyze the terminal steps of the pathway. *Rhl*A is involved in synthesis of a 3-(3-hydroxyalkanoyloxy) alkanoate from a (3*R*)-3-hydroxyacyl-[acyl-carrier protein]. RhlB is a rhamnosyltransferase predicted to be membrane-associated, that produces a mono-rhamnolipid. This mono-rhamnolipid acts as a substrate for the RhlC rhamnosyltransferase, a inner-membrane associated enzyme, which is predicted to produce a di-rhamnolipid. On the other hand, Zhu *et al.* (2008) reported that *Rhl*C has no role in rhamnolipid biosynthesis. These authors showed that *Rhl*A alone is necessary and sufficient to form the acyl moiety of rhamnolipid, and that flux through the fatty acid synthesis pathway would accelerate to support both rhamnolipid and fatty acid biosynthesis.

Schmidberger *et al.* (2014) reported strong increase in relative expression of the genes for rhamnolipid synthesis, *rhlA* and *rhlC*, as well as the genes of the *pqs* quorum sensing regulon under iron-limiting conditions. Iron repletion on the other hand caused a down-regulation of those genes. Expression of *rhlA*, *rhlC* and *pvdS*, as well as the gene for the 3-oxo-C12-HSL synthetase, *lasI*, increased until late stationary growth phase. After this time point, their expression steadily decreased.

Ismail *et al.* (2014) investigated the expression of biosurfactants production genes, *rhl*ABC, in cultures of *Pseudomonas* sp. AK6U grown with inorganic or organic sulfur sources. AK6U grew with either inorganic sulfate (MgSO4), dibenzothiophene (DBT), or DBT-sulfone as a sole sulfur source in the presence of glucose as a carbon source. The AK6Ucultures produced variable amounts of biosurfactants depending on the sulfur source. RT-qPCR revealed variations in the expression levels of the *rhl*ABC genes depending on the sulfur source. The DBT cultures had higher expression levels for the three genes as compared to the DBT-sulfone and inorganic sulfate cultures. The increased expression of *rhl*C in the DBT cultures is indicative for production of higher amounts of dirhamnolipids compared to the DBT-sulfone and inorganic sulfate cultures.

Genes involved in Sophorolipid biosynthesis

Inge *et al.* (2013) reported that sophorolipid gene cluster consisting of five genes directly involved in sophorolipid synthesis in a yeast *S. bombicola*. These five genes include cytochrome P450 monooxygenase, two glucosyl-transferases, an acetyltransferase and a transporter these enzymes are directly involved in the core biosynthetic pathway.

Genes involved in Surfactin biosynthesis

Sfp gene of *Bacillus* species that encodes phosphopantetheinyl transferase, required for the non-ribosomal biosynthesis of surfactin. This gene is a member of the *srf*A operon that codes for a nonribosomal peptide synthetase complex also known as surfactin synthetase. Cosima *et al.* (1998) reported that the *sfp* gene of *Bacillus* species encodes phosphopantetheinyl transferase, required for the nonribosomal biosynthesis of surfactin. Nakane *et al.* (1992) reported that the *srf*A operon houses 4 genes (*srf*A, *srfAB*, *srfAC*, and srfAD), which together code for the surfactin synthetase subunits.

Porob *et al.* (2012) amplified and cloned genomic DNA of the isolate *Bacillus tequilensis* NIOS11. The cloned gene has an open reading frame of 675 bp coding for 224 amino acids with a theoretical molecular weight of 26,166.33 Da and a predicted isolectric point (pI) of 5.51. BLASTP analysis revealed 99% similarity to the earlier reported *sfp* gene of *B. subtilis*.

Genes involved in Putisolvin biosynthesis

Surfactant activity of putisolvins appeared in the culture medium at the end of the exponential growth phase (Kuiper *et al.* 2003). These authors reported Putisolvins I and II have important functions such as (i) to reduce the surface tension of the medium, (ii) to increase the formation of an emulsion with toluene, (iii) to stimulate swarming motility, and (iv) to inhibit biofilm formation and to degrade existing biofilms. Furthermore, these authors reported that *Pseudomonas putida* PCL1445 produces two cyclic lipopeptides, putisolvins I and II, which possess surfactant activity and play an important role in inhibition of biofilm formation and degradation.

Metabolic pathways involved in Biosurfactant synthesis

Rhamnolipids are synthesized by diverting precursors for the rhamnosyl and lipid moieties from central metabolic pathway of rhamnolipid biosynthesis. The donor of the rhamnosyl moiety is dTDP-L-rhamnose, produced from D-glucose by a series of reactions catalyzed by the *Alg*C and *Rml* enzymes (Maier and Soberon-Chavez 2000). It has been suggested that the lipid moiety precursor is diverted from *de novo* fatty acid synthesis since the stereochemistry of the β-hydroxy fatty acid in the lipid moiety matches that of the intermediates in *de novo* fatty acid synthesis (Zhu and Rock 2008).

Choi *et al.* (2011) reported the metabolic pathways for PHA and rhamnolipid synthesis using octanoic-1-^{13}C acid using ^{13}C NMR analysis. NMR analysis revealed that the monomer-unit (R)-3-hydroxyoctanoate-1-^{13}C being converted from the octanoic acid substrate and effectively incorporated into PHA. In the first step in rhamnolipid synthesis, the (R)-3-hydroxyoctanoate-1-^{13}C is suggested to be firstly converted to (R)-3-hydroxydecanoate-1,3-^{13}C via fatty acid de novo biosynthesis pathway and then further processed into (R)-3-((R)-3-hydroxyalkanoyloxy) alkanoic acids (HAAs) via *Rhl*A.

Strategies for enhancement of biosurfactin production

The science of manipulating and improving microbial strains, in order to enhance their metabolic products for biotechnological applications is termed as strain improvement. The microbial production strain is considered as the heart of fermentation industry, so improvement production of metabolites favours the cost reduction of capital outlay (Barrios-Gonzalez *et al.* 2003). Strain improvement can be carried out either by induced mutation or by gene cloning techniques. Different strategies involved in microbial strain improvement for enhanced biosurfacatant production in given below (Table 2).

Table 2: Enhanced production of biosurfuctant / production by mutant and recombinant strain

Microorganism	Methods of strain improvement	Status of improve strain	References
Bacillus subtilis ATTC 21332	UV mutation	3 times more than parent strain	Mulligan *et al.* (1989)
Bacillus licheniformis	Random mutagenesis with nitrosoguanidine	391 mgl^{-1}	Sung-chyrlin *et al.* (1998)
Pseudomonas aeuroginosa PTCC 1637	Random mutagenesis with nitrosoguanidine	10 times more than parent strain	Abbas Tahzibi *et al.* (2003)
Pseudomonas aeuroginosa	Gamma ray induced mutation	8.50 gl^{-1}	Raza *et al.* (2007)
Recombinant strain – *Pseudomonas putida*	Cloning rhlAB-rhamnosyltransferase gene from *P. aeuroginosa* into *P. putida* and *P. fluroscenes*	60 mg/l	Ochsner *et al.* (1995)
Recombinant *Bacillus subtilis* M113	Cloning of plasmid PC112 containing lpa-14 gene.	8 times more enhanced production	Ohno *et al.* 1995
Recombinant *Bacillus subtilis* ATCC 21332	Carries genetic engineered peptide synthetase	Surfactin produced with less toxicity	Symmank *et al.* (2002)
Recombinant *Pseudomonas aeuroginosa* strain	Cloning of *E.coli* lac Z Y genes into *P. aeuroginosa* strains.	Ability to utilize glucose & lactose	Koch *et al.* (1988)
Recombinant E.*coli* strain	Cloning of *rhl*AB and rm 1BDAC operons from P. *aeuroginosa*	Attains ability to produce membrane on lipids.	Carbrera- valladares *et al.* (2006)
Recombinant E.*coli* strain	Expression of only *rhl*AB gene in *E. coli.*	Ability to produce RL	Wang *et al.* (2007)
Recombinant P.*putida* strain	Cloning of both *rhl*AB genes and *rhl*RI quorum sensing system	Higher production of RL	Cha *et al.* (2008)
Recombinant E.*coli* strain	Cloning of genes *sfp*O, *srf*A of *B. subtilis* SK320	Two fold increased production	Sekhon *et al.* (2011)

Mutation

Efficient production of any biotechnological products depends on the use of hyper producing microbial strains. Even though, the yields of biosurfactins are enhanced by using cheap raw materials, optimized production conditions and efficient downstream processing, employing naturally hyper producing microbial strains determines the viability of the commercial production. Genetics of the biosurfactin producing organisms are considered as vital factor as they influence the yield of biosurfactins because the ability to produce this metabolite is conferred by genes only.

In this strategy the microbial production strains of biosurfactins are subjected to mutation, followed by random screening. Generally the mutation occurs spontaneously at the rate of 10^{-10} and 10^{-15} per generation and per gene. As this spontaneous mutation occurs at low frequency, the spontaneous mutation is not used much in industrial strain improvement. However induced mutation is a good strategy for strain improvement. The induced mutation can be carried out with either physical mutagens like U-V light or chemical mutagens like N-methyl-N-nitro-N-nitrosoguanidine or ethyl methane sulphonate.

Many researchers attempt to improve the biosurfactin production by mutation. A UV mutated strain of *Bacillus subtilis* ATCC 21332 for producing increased amounts of surfactin was developed by exposing approximately 3000 cells to short wave length UV. This stable mutant produced three times more of the biosurfactant, surfactin, than the parent strain (Mulligan *et al.* 1989). Lin *et al.* (1998) used a *Bacillus licheniformis* mutant strain designated KGL11 to produce higher levels of lipopeptide biosurfactants (391 mg $l^{"1}$). The mutant was derived by random mutagenesis with n-methyl-n2 -nitro-n-nitrosoguanidine treatment and compared to parent strain it produced 12 times more biosurfactins. Tahzibi *et al.* (2003) reported that a mutant strain of *Pesudomonas aeruginosa* PTCC1637 produced 10 times more production of rhamnolipid biosurfactants compared to parent strain. The mutant strain was developed by random mutagenesis with n-methyl-n2 -nitro-n-nitrosoguanidine and was selected on Siegmund-Wagner (SW) plates. A mutant of *Bacillus subtilis*-E8 was developed by low energy nitrogen ion beam implantation (Qingmei *et al.* 2006). The optimum condition of ions implantation used was20 keV of energy and $2.6 \times 10^{15} N^+/cm^2$ dosage to develop the mutant strain of *B.subtilis*-E8. Raza *et al.* (2007) developed a gamma ray induced *Pesudomonas aeruginosa* mutant EBN-8 for the higher production of rhamnolipids. They used wastes from canola, soybean and corn oil refineries as substrates for rhamnolipids production by this mutant strain. The mutant strain produced higher amount of rhamnolipids (8.50 g l^{-1}) in canola oil refinery waste (COD=20 g l^{-1}) supplemented with sodium nitrate (at COD/N=20) other than the substrates.

Recombinant DNA technology

Besides using mutant hyper producing biosurfactant strains, many scientists tried to improve the biosurfactant production by using genetic engineering technology. In most case of the mutational studies, *Pesudomonas aeruginosa* has been used because of its substrate versatility and its ability to degrade the environmental contaminants *in situ*. The production of biosurfactants by recombinant strains offers few advantages as compared to the production with mutant *P. aeruginosa. P. aeruginosa* is an opportunistic pathogen, with implied increased safety when handling large quantities of cultures. However in the recombination technology, the heterologous hosts like *E. coli –a nonpathogenic* may be used to produce biosurfactants of *P. aeruginosa.* Another advantage is the possibility of constitutive production of biosurfactants in heterologous host, in contrast to the tightly regulated production in *P. aeruginosa* (Abdel-Mawgoud *et al* 2011). However strain improvement by genetic engineering technique needs sophisticated technique, skill and apparatus, thus, increasing cost in research and development. In spite of these disadvantages this technique is very efficient to boost the industrial biosurfactant production, making it more viable for commercialization.

Several attempts were made to produce biosurfactants of *Pseudomonas* in heterologous hosts. But, none produces biosurfactants in comparable levels as the best *P. aeruginosa* strains. The cloning of *rhlAB* rhamnosyltransferase gene from *P. aeruginosa* into various hosts like *Pseudomonas fluorescens*, *Pseudomonas oleovorans*, *Pseudomonas putida*, and *E. coli* was performed to check the biosurfactin production in non-pathogenic strains (Ochsner *et al.* 1995). Among these different hosts, high production was achieved with *Pseudomonas putida* which produced 60mg/L of rhamnolipid, but no production was observed in *E. coli.* In another attempt a recombinant strain, *Bacillus subtilis* MI 113, was developed with the cloning of plasmid *pC112*harbouring *lpa-14*, a gene related to surfactin production. This strain showed 8 times more enhanced production of surfactin in solid-state fermentation using soybean curd residue.

Recombinant technology favors the formation of new strains to survive in cheaper substrates. For example, rhamnolipid producing *P. aeruginosa* strains are naturally unable to utilize the glucose; hence a glucose utilizing strain of *P. aeruginosa* was developed by cloning the *E. coli lacZY* genes into the chromosomes of *P. aeruginosa* (Koch *et al.* 1988). This recombinant strain could now grow in a medium containing glucose and lactose. This paves the way for developing a cheaper fermentation medium using easily available raw materials like whey. In such a medium only a recombinant strain can survive; but not natural strain or revertants. The production of mono-rhamnolipids in *E. coli* strain was successfully done by Carbrera-Valladares *et al* (2006) by cloning *rhl*AB and *rhl*BDAC operons from *Pseudomonas aeruginosa.* They found that the availability of dTDP-L-rhamnose in *E. coli* restricts the production of mono-rhamnolipids in this species. Thus by using rDNA techniques and by cloning the above mentioned operons in *E.coli* they generated rhamnolipid producing *E.coli* strains which was not reported till then. As cited in A. Mohammed Abdel-Mawgoud *et al* 2011, Wang *et al.* (2007) claimed RL production in *E. coli* expressing only *rhl*AB. Cha *et al.* (2008) were able to successfully increase rhamnolipid production by cloning both the *rhlAB* rhamnosyltransferase genes and the *rhlRI* quorum sensing system into *P. putida* to enhance rhamnolipid production. A maximum yield of 7.2 g/L of RLs was achieved. *sfpO, srfA* of *Bacillus subtilis* SK320 into *E. coli.* Also it was observed that biosurfactants produced by *Bacillus subtilis* SK320 and recombinant strains *BioS a*, *BioS b*, *BioS c* were found to be effective emulsifiers, reducing the surface tension of water from 72 dynes/cm to as low as 30.7 dynes/cm.

Metabolic engineering

Metabolic engineering is generally considered as the targeted and purposeful alteration of metabolic pathways found in an organism in order to better understand and to achieve a particular goal. This goal may be to increase process productivity or to extend metabolic capability by the addition of extrinsic activities for better production. In essence, metabolic engineering is the practice of altering genes and metabolic pathways within a cell or microorganism to increase its production of a specific substance. Metabolic engineering strategy has been successfully applied in strain improvement for biosurfactant production. However, in this strategy only a selected group of genes encoding enzymes and regulatory proteins have been manipulated based on the available information and research experience. Now, the recent advances in bioinformatics provide data of genomics, proteomics and metabolomics of industrially useful strains. Based on this data, together with experimental observations *in vivo* and predictions of modeling and simulation *in silico* can make us to better understand the features and functions of biological systems (Lee *et al.* 2005). The biosynthesis of rhamnolipids proceeds via the transfer of dTDP-L-rhamnose. In the *de novo* fatty acids synthesis 3-(3-hydroxyalkanoyloxy) alcanoic acid – HAA is synthesized by RhlA enzyme. Then HAA is converted to mono-rhamnolipd by RhlB enzyme. Afterwards this mono-rhamnolipd is converted to di-rhamnolipids by RhlC enzyme. CoA-β-hydroxyacids are the precursors of rhamnolipids. *Pseudomonas sp.* metabolize glucose via the Entner–Doudoroff pathway (ED) and pyruvate, via the ED specific intermediate 2-keto-3-deoxy-6-phosphogluconate (KDPG). However, *Pseudomonas sp.* cannot degrade glucose via glycolysis, because all enzymes required are not present in the organisms. However, in *P. aeruginosa* and *P. fluorescens*, only a single enzyme is missing for a fully functional glycolysis, i.e., phosphofructokinase which catalyses the phosphorylation of fructose-6-phosphate (fru-6-P) to fructose-1,6-bisphosphate (fru-1,6-2P). *P. putida* also lacks a gene for fructokinase/hexokinase, which phosphorylates fructose (fru) to fru-6-P. Oxidation of glucose to pyruvate via the ED pathway compared to glycolysis yields only one ATP less per glucose molecule (Henkel *et al.* 2012).

The other ways for improvement of rhamnolipid yields may be achieved by replacing *de novo* synthesis of rhamnolipids with either shorter metabolic route (e.g. the synthesis of precursors in fewer steps) or processes where precursors are used as a feed, which can also directly be incorporated into the product (Henkel *et al.* 2012). The effective production of sophorolipids by *Candida bombicola* is done by this method. When the precursors are fed simultaneously into production, higher yield of biosurfactants could be achieved effectively, because energy-consuming *de novo* synthesis of precursors is no longer required. Generally medium sized fatty acids are required for synthesis of rhamnolipids. If these fatty acids are directly fed into the production medium, then the synthesis of this kind of fatty acids by production strains through *de novo* synthesis is not at all required. The production strain directly starts to utilize this as precursor for the production of rhamnolipids. However, to enter into the rhamnolipid biosynthetic pathway, the substrate fatty acids should be linked to an acyl-carrier protein (ACP). This linking is performed by ACP-ligases. Hence the recombinant strain of *C. bombicola* cloned with ACP-ligase enzyme may be well used in production medium with plant oils as substrate. The substrate fatty acids derived from plant oils by lipases can be easily linked by ACP-ligase and utilized directly in the biosynthesis of rhamnolipids.

Response Surface Methodology

Response surface methodology was introduced by G.E.P. Box and K.B. Wilson in the year 1956 to explore the relationship between several explanatory variables against one of the response variable. This method uses a series of designed experiments to obtain an optimal response using a second-degree polynomial model. Though this model is easy to estimate and apply the results obtained are based only on an approximation. Usually a first degree of polynomial model uses a factorial experiment or a fractional factorial design to determine the explanatory variable that have an impact on the response variables/variable of interest. However, if it is suspected that only one significant explanatory variable is left, then a more complicated design such as central composite design can be implemented to estimate a second- degree polynomial model that could be used for optimization.

Optimization of medium components for biosurfactant production

RSM finds a number of applications in chemical and biochemical experiments. Classical method for medium optimization involves changing one variable at a time, while keeping the other variables fixed. The major problem with the classical method is that they are laborious, time consuming and may not guarantee the determination of optimal conditions since they are single dimensional. Conversely, carrying out experiments with every possible factorial combination of the test variable is not possible due to the requirement for large number of experiments (Sen, 1997).

In RSM, for the first screening it is recommended to evaluate the result and to estimate the main effects according to a linear model. Based on the evaluation, the variable that have the highest influence on the results were selected for the further study. Using this method a large number of experimental variables can be investigated without the need to increase the number of experiments.

Optimization of experiments involving synthetic /chemically defined media

A chemically defined media is entirely free of animal/plant/microbe derived components and represents the purest form. The term chemically defined media according to Jayme and Smith (2000) a basal formulation, which may also be protein-free and is comprised of biochemically defined low molecular weight constituents. Joshi *et al.* (2007) optimized the medium components for biosurfuctant production by *Bacillus licheniformis* K51 and these authors identified the important media components as H_3PO_4, $CaCl_2$, H_3BO_3, and Na-EDTA based on Placket –Burman method. The optimum concentration for higher biosurfactant production was (g/l): Glucose 1.1, $NaNO_3$ 4.4, $MgSO_4.7H_2O$ 0.8, KCl 0.4, $CaCl_2$ 0.27, H_3PO_4 1.0ml/l and trace elements H_3BO_3 0.25, $CuSO_4$ 0.6, $MnSO_4$ 2.2, Na_2MOO_4 0.5, $ZnSO_4$ 6.0, $FeSO_4$ 8.0, $CuCl_2$ 1.0 and Na-EDTA 30.0. Using these statistical methods these authors improved the relative biosurfuctant yield as critical micelle dilution (CMO) from 10X to 15X, when compared to the non-optimized rich medium.

Sivapatnasekaran *et al.* (2010) enhanced reported biosurfuctant production in Modified Mineral Medium (MMM) optimized through RSM. Using the results of a single factor at a time optimization experiment Glucose, Urea, $MnSO_4$, and $MgSO_4$ were identified as the critical compounds that affected biosurfuctant production. Biosurfuctant production in MMM was increased by RSM for biosurfuctant production.

Optimization of experiments involving undefined media

An undefined medium consists of complex ingredients, such as yeast extract or casein hydrolysate, which consist of a mixture of many chemicals of unknown proportions. Manivasagan *et al.* (2013) optimized medium composition and culture conditions for the glycolipid biosurfuctant production by *Streptomyces* sp. MAB36. These authors applied Placket –Burman design to find out the key ingredients and conditions for higher glycolipid production by Central Composite Design (CCD) to optimize for significant variables such as starch, casein, crude oil and incubation time.

Based on the reports these authors reported that fructose and yeast extract as the best carbon and nitrogen sources for glycolipid production Praveen *et al.* (2014) based on the statistical analysis of coefficient in Placket-Burman design experiments reported that pH, saw dust and glycerol were found to have significant effect on biosurfurctant production. Keeping all the other factor as fixed and optimization of these three selected variable through Box-Behnken design shows maximum predicted biosurfuctant production was high at a pH of 7.37.

Optimization for biosurfactant production using Neem cake as a raw material RSM-CCD was attempted with *Pseudomonas* sp. PEFL4 (Joe *et al.* unpublished results). The specific coded values of neem oil cake, inoculum size, pH and temperature are provided in Table 2. Central composite design (CCD) matrix of independent variables and their corresponding experimental and predicted yields of emulsification activity (EI24%) is provided in Table 3. Analysis of variance (ANOVA) of main effects of factors for production of biosurfactant by *Pseudomonas sp.* PEFL4 was provided in Table 4 & 5. The three dimensional contour plots for the maximum emulsification index is provided in Fig 7. Based on the design the optimized pH, inoculation load (log Cfu/ml), neem cake quantity (g) and temperature ÚC were 6, 7, 4.75 and 30, respectively.

Table 3 : Specific coded values of Neem oil cake, Inoculum size, pH and Temperature values.

S.No	Independent variable	Coded values				
		-2	-1	0	1	2
1.	Neem Cake	1.25	3	4.75	6.5	8.25
2.	Inoculation load	3	5	7	9	11
3.	pH	2	4	6	8	10
4.	Temperature	10	20	30	40	50

Table 4 : Central composite design (CCD) matrix of independent variables and their corresponding experimental and predicted yields of emulsification activity (EI24%)

Std	Run	A: Neem Cake	B: Inoculation load	C: pH	D: Temperature	Experimental	Predicted
5	1	3	5	8	20	42	44
11	2	3	9	4	40	42	44
30	3	4.75	7	6	30	54	56
2	4	6.5	5	4	20	32	34
1	5	3	5	4	20	34	36
21	6	4.75	7	2	30	12	14
16	7	6.5	9	8	40	72	74
3	8	3	9	4	20	34	37
14	9	6.5	5	8	40	54	58

Contd...

18	10	8.25	7	6	30	68	72
4	11	6.5	9	4	20	54	56
25	12	4.75	7	6	30	72	73
20	13	4.75	11	6	30	76	74
12	14	6.5	9	4	40	48	49
17	15	1.25	7	6	30	34	38
24	16	4.75	7	6	50	44	47
15	17	3	9	8	40	54	56
23	18	4.75	7	6	10	32	34
27	19	4.75	7	6	30	56	58
26	20	4.75	7	6	30	54	56
6	21	6.5	5	8	20	52	54
19	22	4.75	3	6	30	22	26
10	23	6.5	5	4	40	42	44
29	24	4.75	7	6	30	72	74
22	25	4.75	7	10	30	34	36
8	26	6.5	9	8	20	29	34
13	27	3	5	8	40	48	50
28	28	4.75	7	6	30	54	756
9	29	3	5	4	40	34	36
7	30	3	9	8	20	37	39

Table 5 : Analysis of variance (ANOVA) of main effects of factors for production of biosurfactant by *Acinetobacter* sp. ACMS25

Source	Sum of Squares	df	Mean Sum of Squares	F value	P Value
Model	5348.78	14	382.06	3.09	0.0188*
A-Neem Cake	661.50	1	661.50	5.36	0.0352
B-Inoculation load	816.67	1	816.67	6.61	0.0213
C-pH	522.67	1	522.67	4.23	0.0575
D-Temperature	450.67	1	450.67	3.65	0.0754
AB	12.25	1	12.25	0.099	0.7571
AC	2.25	1	2.25	0.018	0.8944
AD	20.25	1	20.25	0.16	0.6912
BC	100.00	1	100.00	0.81	0.3824
BD	121.00	1	121.00	0.98	0.3379

Contd...

CD	196.00	1	196.00	1.59	0.2270
A^2	76.19	1	76.19	0.62	0.4444
B^2	128.76	1	128.76	1.04	0.3234
C^2	2060.19	1	2060.19	16.68	0.0010
D^2	663.05	1	663.05	5.37	0.0350
Residual	1852.42	15	123.49		
Lack of Fit	1441.08	10	144.11	1.75	0.2785**
Pure Error	411.33	5	82.27		
Cor Total	7201.20	29			

R-Squared -0.742; Adj R-Squared-0.502; Pred R-Squared -0.234 *Significant

**Non-significant

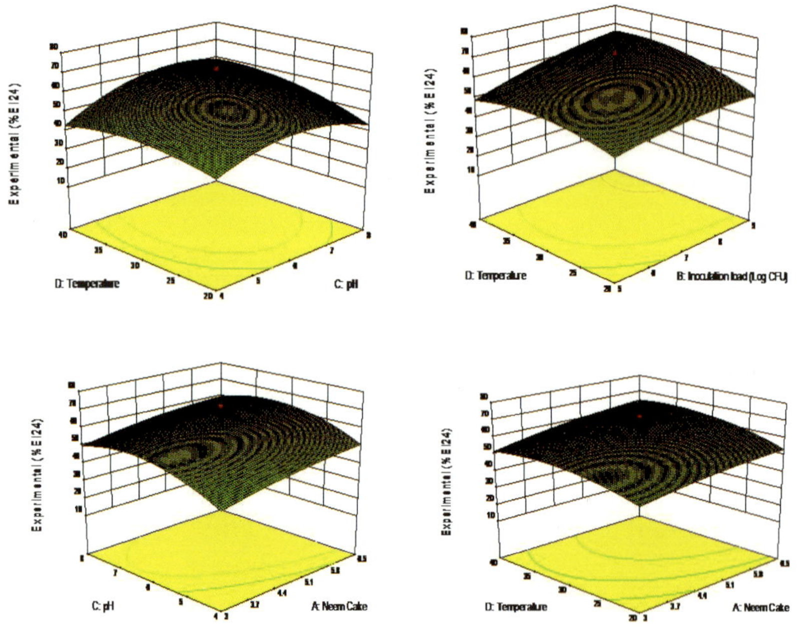

Fig. 7 : Three dimensional contour plots for the maximum emulsification index (% EI24) (maximum biosurfactant production) of Pseudomonas sp. PEFL4. Emulsification index (% EI24) as unction of temperature and pH (A), temperature and inoculation load (B), pH and Neem cake (C) temperature and neem cake (D) (Joe et al. unpublished results)

Application of biosurfactant in Agriculture

In Agriculture enormous use of harsh surfactants for soil application paved way for the use of biosurfactants since they are considered to be less toxic and eco-friendly (Dhara *et al.* 2013). When compared to their chemical counterparts, biosurfactants are biodegradable, and demonstrate excellent surface activity, and possess high specificity, show effectiveness under extreme conditions, and can be reused through regeneration (Koglin *et al.* 2010; Xu *et al.* 2011).

Biosurfactants facilitate biocontrol through various mechanisms such as parasitism, antibiosis, competition, induced systemic resistance, and hypovirulence (Singh *et al.* 2007). They play a major role in agriculture in enhancing the antagonistic activities of microbes and microbial products (Kim *et al.* 2004). Bais *et al.* (2004) reported that upon root colonization, *B. subtilis* 6051 forms a stable, extensive biofilm and secretes surfactin, which protects the plant from the pathogenic bacterial infection caused by *Pseudomonas syringae.* Perneel *et al.* (2008), reported that when phenazines and rhamnolipids were combined the metabolites significantly reduced the mycelial growth of *P. myriotylum*. In addition, microscopic analysis revealed substantial vacuolization and disintegration of *Pythium* hyphae after incubation in liquid medium amended with both metabolites. Arutchelvi and Doble (2010) isolated rhamnolipid-producing *Pseudomonas aeruginosa* CPCL strain from a petroleum-contaminated soil. Rhamnolipid of this strain was able to reduce surface tension of water from 72 to 44 mN/m at a critical micelle concentration of 11.27 +/- 1.85 mg^{-1}. Furthermore, these authors reported that this strain can be used for biocontrol of *Verticilllium microsclerotia*. Geetha and Manonmani (2010) reported that surfactin could be used as a mosquitocidal toxin and this toxin is stable to UV and varied temperature, active at acidic and basic pH and temperatures between 25 and 42 ÚC. Afsharmanesh *et al.* (2013) attempted different doses of irradiation from 100 Gray to 3000 Gray on *Bacillus subtilis* UTB1 and reported that six mutants are more effective against *A. flavus* in pistachio nuts than the parental strain. Tomar *et al.* (2014) tested 95 bacterial isolates for their biosurfactant as well as biocontrol activity against *Phytophthora infestans*. Results of this experiment revealed that only 15.8% isolates showed biosurfactant activity. Application of different types of biosurfactants as biocontrol agents is given in Table 6.

Table 6 : Application of different types of biosurfactants as biocontrol agents

Type of Biosurfuctant	M.O's Involved	Usefulness as a Biocontrol agent	References
Rhamnolipid	*Pseudomonas fluorescens* strain 6L10, 6ba6 and 3ss9	Invitro growth inhibition of *Phytophthora capsici*	Ozyilmaz and Benlioglu (2013)
Rhamnolipid	*Pseudomonas aeruginosa* 1/3	Phytopthora infestans 10-15% reduction in disease incidence	Tomar *et al.* (2014)
Surfactin	*Bacillus subtilis* ssp. *Subtilis* (VCRCB471)	Pupicidal activity against *Anopheles stephensi*	Geetha and Manonmani (2010)
Partially purified biosurfactant containing a fatty acid derivative	*Halomonas* sp. BS4	In vitro antibacterial activity (*Staphylococcus aureus*, *Klebsiella pneumonia*, *Streptococcus pyogenes*), and antiviral activity (White spot syndrome virus)	Donio *et al.* (2013)
Iturin A, surfactin A and fengycin	*Bacillus subtilis* CMB32	Control of anthracnose disease caused by *Colletotrichum gleospororoides*	Pyonggll *et al.* (2009)
Cyclic lipopeptide	*Pseudomonas* CMR12a	Rhizoctonia root rot of bean caused by *Rhizoctonia solani*	Daes *et al.* (2011)
Iturin and fengycin families	*Bacillus subtillis*	Antagonism of *Podosphaera fusa* on melon	Romero *et al.* (2007)
Biosurfactant	*Bacillus* spp. B1, B2, BW and OS17	Effective against alternaria tenius and *Fussarium oxysporium* var *lycopersici*	Violeta *et al.* (2011)
Iturin	*Bacillus amyloliquefaciens* PPCB004	Effective against post harvest pathogens of citrus, mango, orange and avocado	Arebola *et al.* (2009)

Application of biosurfactant in Environmental remediation

Biosurfactants play a role is promising role in environmental remediation due to their biodegradability, low toxicity and effectiveness in enhancing biodegradation and solubilization of low solubility compounds. Different applications in bioremediation is given in Table 7.

Table 7 : Application of different types of biosurfactants in environmental remediation

Type of Biosurfuctant	M.O's Involved	Usefulness as a Biocontrol agent	References
Rhamnolipid	Commercial Biosurfactant	Soil bound cadmium, zinc and lead	Herman *et al.* 1995; Han *et al* 1994
Rhamnolipid	Commercial Biosurfactant	Biosurfactant was utilized in micellar-enhanced filteration of heavy metals such as Cu, Zn, Ni, Pb, Cd, using two membranes	El-zeftawy and Mulligan (2011)
Rhamnolipid	Commercial Biosurfactant	Addition of 1 per cent to 0.5% rhamnolipid improved the removal of copper	Dahrazama and Mulligan (2007)
Rhamnolipid	*Pseudomonas aeruginosa* ATCC9027	Maximum complexion capacity of 0.2% cd^{2+}/ rhamnolipid on molar basis was obtained	Tan *et al.* (1994)
Crude biosurfactant	*Candida lipolytica* UCPO98	Crude biosurfactant removed 96% Zn and Cu and reduced concentration of Pb, Cd, and Fe from test specimen	Rufino *et al.* (2011)
Biosurfactant	*Fusarium* sp. BS8	For microbial enhanced oil recovery (EOR)	Quazi *et al.* (2013)
Biosurfactant	*Bacillus* spp. SH_2O/SH26	Cleaning of oil contaminated vessels/ enhanced biodegradation of oil sludge	Diab and Din (2013)

Contd...

Biosurfactant	*Pseudomonas aeruginosa* J4/ *Bacillus subtillis* ATCC21332	For enhanced biodegradation of diesel contaminated water and soil	Wang *et al.* (2008)
Surfactin	*Bacillus subtillis*	Sediments contaminated with Zn, Cu, Cd, oil and greese	Mulligan *et al.* (1999)
Rhamnolipid	Consortium of bacteria	Positive role in hydrocarbon degradation	Chen *et al.* (2013)

Mulligan *et al.* (1999; 2001) used biosurfactants to remove heavy metals from an oil-contaminated soil and was evaluated by batch washes with surfactin, a rhamnolipid and a sophorolipid. These authors reported that single washing with 0.5% rhamnolipid removed 65% of the copper and 18% of the zinc, whereas 4% sophorolipid removed 25% of the copper and 60% of the zinc from the contaminated soil. Cameotra and Singh (2008) reported that a mixture of rhamnolipid consortium degraded 91% of the hydrocarbon content of soil contaminated with 1% crude oil sludge in 5 weeks. Sinha and Paul (2010) reported that among three *Pseudomonas* strains (PGS1, PGS2 and PGS3), strain PGS1 was found to be effective in removing more than 50% of chromium and cadmium from the medium. Mulligan *et al.* (2010) evaluated the feasibility of using surfactin from *Bacillus subtilis*, for the removal of heavy metals from a contaminated soil. These authors reported that after one and five batch washings of the soil, 25 and 70% of the copper, 6 and 25% of the zinc, and 5 and 15% of the cadmium could be removed by 0.1% surfactin with 1% NaOH, respectively. Dahrazma and Mulligan (2007) evaluated the performance of rhamnolipid, in a continuous flow configuration (CFC) for removal of heavy metals (copper, zinc, and nickel) from the sediments taken from Lachine Canal, Canada. In this experiment, rhamnolipid solution with a constant rate was passed through the sediment sample within a column. These authors reported that 1% NaOH to 0.5% rhamnolipid improved the removal of copper by up to 4 times compared with 0.5% rhamnolipid alone. Singh and Cameotra (2013) described the potential application of lipopeptide biosurfactants in removal of petroleum hydrocarbons and heavy metals from the soil samples collected from industrial dumping site. These authors reported that soil washing with biosurfactant solution removed significant amount of petroleum hydrocarbon (64.5 %) and metals namely cadmium (44.2 %), cobalt (35.4 %), lead (40.3 %), nickel (32.2 %), copper (26.2 %) and zinc (32.07 %). Fu *et al.* (2014) reported that biosurfactant application improved the micro-environment and composting efficiency, by their the adsorption of heavy metals and humic acid (HA). These authors reported that Pb^{2+} adsorption was increased with increasing pH values and was weakened with rising temperature.

Conclusion

In spite of this tremendous advancement biosurfactants research, industrial production of biosurfactants is still low and the cost of production is high. The major problems faced in biosurfactant production is high cost of synthetic media, lack of process optimization studies with agro-based materials, and lack of commercially available genetically improved strains. This chapter is written in such a way to give in depth knowledge on the different types of biosurfactants with special reference to commercially important biosurfactants such as rhamnolipid and surfactin. Genes such as *rhl*ABC and *Sfp* gene involved in the biosynthesis of rhamnolipid and surfactin and different metabolic have been discussed in detail to explore possibilities for strain improvement through mutation and genetic engineering for higher production of commercially viable biosurfactants. Use of cheaper substrates like agro industry wastes, and process optimization statistical studies through RSM has also been discussed for promoting higher biosurfactant production with reduced costs.

References

Abdel-Banat B.M., Hoshida H., Ano A., Nonklang S and Akada R. (2010) High-temperature fermentation: how can processes for ethanol production at high temperatures become superior to the traditional process using mesophilic yeast. Appl. Microbiol. Biotech. 85:861-867.

Abdel-Mawgoud A.M., Hausmann R., Lépine F., Müller M.M and Déziel E. (2011) Rhamnolipids: detection, analysis, biosynthesis, genetic regulation, and bioengineering of production. Biosurfactants, Springer. pp. 13-55.

Arrebola E., Jacobs R., and Korsten L. (2010) Iturin A is the principal inhibitor in the biocontrol activity of *Bacillus amyloliquefaciens* PPCB004 against postharvest fungal pathogens. J. Appl. Microbiol. 108:386-395.

Arutchelvi J. and Doble M. (2010) Characterization of glycolipid biosurfactant from *Pseudomonas aeruginosa* CPCL isolated from petroleum-contaminated soil. Lett. Appl. Microbiol 51:75-82. DOI: 10.1111/j.1472-765X.2010.02858.x.

Bais H.P., Fall R., and Vivanco J.M. (2004) Biocontrol of *Bacillus subtilis* against infection of *Arabidopsis* roots by *Pseudomonas syringae* is facilitated by biofilm formation and surfactin production. Plant Physiol. 134:307-319.

Banat I.M., Franzetti A., Gandolfi I., Bestetti G., Martinotti M.G., Fracchia L., Smyth T.J., and Marchant R. (2010) Microbial biosurfactants production, applications and future potential. Appl. Microbiol. Biotech. 87:427-444.

Banat I.M., Makkar R.S., and Cameotra S. (2000) Potential commercial applications of microbial surfactants. Appl. Microbiol. Biotech. 53:495-508.

Beeba J., and Umbreit W. (1971) Microbial production of bio surfactants and their importance. J. Bacteriol. 108:612.

Bodour A.A., Wang J.M., Brusseau M.L., and Maier R.M. (2003) Temporal change in culturable phenanthrene degraders in response to long term exposure to phenanthrene in a soil column system. Environ. Microbiol. 5:888-895.

Brown M. (1991) Biosurfactants for cosmetic applications. Int. J. Cosmetic Sci. 13:61-64.

Cabrera-Valladares N., Richardson A.-P., Olvera C., Treviño L.G., Déziel E., Lépine F., and Soberón-Chávez G. (2006) Monorhamnolipids and 3-(3-hydroxyalkanoyloxy) alkanoic acids (HAAs) production using *Escherichia coli* as a heterologous host. Appl. Microbiol. Biotech. 73:187-194.

Cameotra S.S., and Singh P. (2008) Bioremediation of oil sludge using crude biosurfactants. Int. Biodeter Biodegr. 62:274-280.

Caspi R., Foerster H., Fulcher C.A., Kaipa P., Krummenacker M., Latendresse M., Paley S., Rhee S.Y., Shearer A.G., and Tissier C. (2008) The MetaCyc Database of metabolic pathways and enzymes and the BioCyc collection of Pathway/Genome Databases. Nucleic Acids Res. 36: D623-D631.

Cha M., Lee N., Kim M., Kim M., and Lee S. (2008) Heterologous production of *Pseudomonas aeruginosa* EMS1 biosurfactant in *Pseudomonas putida*. Bioresource Technol. 99:2192-2199.

Chen Q., Bao M., Fan X., Liang S., and Sun P. (2013) Rhamnolipids enhance marine oil spill bioremediation in laboratory system. Mar. Pollut. Bull. 71:269-275.

Choi M.H., Xu J., Gutierrez M., Yoo T., Cho Y.H., and Yoon S.C. (2011) Metabolic relationship between polyhydroxyalkanoic acid and rhamnolipid synthesis in *Pseudomonas aeruginosa*: comparative (1)(3)C NMR analysis of the products in wild-type and mutants. J. Biotechnol. 151:30-42. DOI: 10.1016/j.jbiotec.2010.10.072.

Cirigliano M.C., and Carman G.M. (1984) Isolation of a bioemulsifier from *Candida lipolytica*. Appl. Environ. Microbiol. 48:747-750.

Cosmina P., Rodriguez F., Ferra F., Grandi G., Perego M., Venema G., and Sinderen D. (1993) Sequence and analysis of the genetic locus responsible for surfactin synthesis in *Bacillus subtilis*. Mol. Microbiol. 8:821-831.

D'aes J., Hua G.K.H., De Maeyer K., Pannecoucque J., Forrez I., Ongena M., Dietrich L.E., Thomashow L.S., Mavrodi D.V., and Höfte M. (2011) Biological control of Rhizoctonia root rot on bean by phenazine-and cyclic lipopeptide-producing *Pseudomonas* CMR12a. Phytopathology 101:996-1004.

Dahrazma B., and Mulligan C.N. (2007) Investigation of the removal of heavy metals from sediments using rhamnolipid in a continuous flow configuration. Chemosphere 69:705-711. DOI: http://dx.doi.org/10.1016/j.chemosphere.2007.05.037.

Dastgheib S., Amoozegar M., Elahi E., Asad S., and Banat I. (2008) Bioemulsifier production by a halothermophilic *Bacillus* strain with potential applications in microbially enhanced oil recovery. Biotechnol. Lett. 30:263-270.

Desai J.D., and Banat I.M. (1997) Microbial production of surfactants and their commercial potential. Microbiol. Molecul. Biol. R. 61:47-64.

Diab A., and Din G. (2013) Application of the biosurfactants produced by *Bacillus* spp.(SH 20 and SH 26) and *P. aeruginosa* SH 29 isolated from the rhizosphere soil of an Egyptian salt marsh plant for the cleaning of oil-contaminataed vessels and enhancing the biodegradat. Afr. J. Environ. Sci. Tech. 7:671-679.

Donio M.B.S., Ronica F.A., Viji V.T., Velmurugan S., Jenifer J.S.C.A., Michaelbabu M., Dhar P., and Citarasu T. (2013) *Halomonas* sp. BS4, A biosurfactant producing halophilic bacterium isolated from solar salt works in India and their biomedical importance. SpringerPlus 2:149.

El-Sheshtawy H., and Doheim M. (2014) Selection of *Pseudomonas aeruginosa* for biosurfactant production and studies of its antimicrobial activity. Egypt. J. Petrol. 23:1-6.

Finnerty W.R. (1994) Biosurfactants in environmental biotechnology. Curr. Opinion Biotech. 5:291-295.

Fischer S. (1993) The role of macroeconomic factors in growth. J. Monetary. Econ. 32:485-512.

Geetha I., and Manonmani A. (2010) Surfactin: a novel mosquitocidal biosurfactant produced by *Bacillus subtilis* ssp. subtilis (VCRC B471) and influence of abiotic factors on its pupicidal efficacy. Lett. Appl. Microbiol. 51:406-412.

Grangemard I., Wallach J., Maget-Dana R., and Peypoux F. (2001) Lichenysin. Appl. Biochem. Biotech. 90:199-210.

Helmy Q., Kardena E., and Funamizu N. (2011) Strategies toward commercial scale of biosurfactant production as potential substitute for it's chemically counterparts. Int. J. Biotech. 12:66-86.

Henkel M., Müller M.M., Kügler J.H., Lovaglio R.B., Contiero J., Syldatk C., and Hausmann R. (2012) Rhamnolipids as biosurfactants from renewable resources: concepts for next-generation rhamnolipid production. Process Biochem. 47:1207-1219.

Herman D.C., Artiola J.F., and Miller R.M. (1995) Removal of cadmium, lead, and zinc from soil by a rhamnolipid biosurfactant. Envir. Sci. Tech. 29:2280-2285.

Horowitz S., and Currie J. (1990) Novel dispersants of silicon carbide and aluminum nitride. J. Disper. Sci. Technol. 11:637-659.

Iqbal S., Khalid Z.M., and Malik K.A. (1995) Enhanced biodegradation and emulsification of crude oil and hyperproduction of biosurfactants by a gamma ray-induced mutant of *Pseudomonas aeruginosa*. Lett. Appl. Microbiol. 21:176-9.

Ismail W., El Nayal A.M., Ramadan A.R., and Abotalib N. (2014) Sulfur source-mediated transcriptional regulation of the rhlABC genes involved in biosurfactants production by *Pseudomonas sp.* strain AK6U. Front. Microbiol. 5:423.

Jayme D.W., and Smith S.R. (2000) Media formulation options and manufacturing process controls to safeguard against introduction of animal origin contaminants in animal cell culture. Cytotechnology. 33:27-36.

Joshi S., Yadav S., Nerurkar A., and Desai A.J. (2007) Statistical optimization of medium components for the production of biosurfactant by *Bacillus licheniformis* K51. J. Microbiol. Biotechnol. 17:313.

Kaeppeli O., and Finnerty W. (1980) Characteristics of hexadecane partition by the growth medium of *Acinetobacter sp*. Biotechnol. Bioeng. 22:495-503.

Kapadia S.G., and Yagnik .B.N (2013) Current Trend and Potential for Microbial Biosurfactants. Asian J. Experiment. Biol. Sci. 4:1-8.

Kesting W., Tummuscheit M., Schacht H., and Schollmeyer E. (1996) Ecological washing of textiles with microbial surfactants, Interfaces, Surfactants and Colloids in Eng, Springer. pp. 125-130.

Kim P., Bai H., Bai D., Chae H., Chung S., Kim Y., Park R., and Chi Y.T. (2004) Purification and characterization of a lipopeptide produced by *Bacillus thuringiensis* CMB26. J. Appl. Microbiol. 97:942-949.

Kim P.I., Ryu J., Kim Y.H., and ChI Y.T. (2010) Production of biosurfactant lipopeptides Iturin A, fengycin and surfactin A from *Bacillus subtilis* CMB32 for control of *Colletotrichum gloeosporioides*. J. Microbiol. Biotechnol. 20:138-145.

Koch A.K., Reiser J., Käppeli O., and Fiechter A. (1988) Genetic construction of lactose-utilizing strains of *Pseudomonas aeruginosa* and their application in biosurfactant production. Nature Biotechnol. 6:1335-1339.

Koglin A., Doetsch V., and Bernhard F. (2010) Molecular engineering aspects for the production of new and modified biosurfactants. Adv. Exp. Med. Biol. 672:158-69.

Kretschmer M., Schellenberger W., Otto A., Kessler R., and Hofmann E. (1987) Fructose-2, 6-bisphosphatase and 6-phosphofructo-2-kinase are separable in yeast. Biochem. J. 246:755-759.

Kuiper I., Lagendijk E.L., Pickford R., Derrick J.P., Lamers G.E., Thomas Oates J.E., Lugtenberg B.J., and Bloemberg G.V. (2004) Characterization of two *Pseudomonas putida* lipopeptide biosurfactants, putisolvin I and II, which inhibit biofilm formation and break down existing biofilms. Mol. Microbiol. 51:97-113.

Kumar A.P., Janardhan A., Radha S., Viswanath B., and Narasimha G. (2015) Statistical approach to optimize production of biosurfactant by *Pseudomonas aeruginosa* 2297. 3 Biotech. 5:71-79.

Lang S., and Wullbrandt D. (1999) Rhamnose lipids–biosynthesis, microbial production and application potential. Appl. Microbiol. Biotechnol. 51:22-32.

Lee M., Kim M.K., Vancanneyt M., Swings J., Kim S.-H., Kang M.S., and Lee S.-T. (2005) *Tetragenococcus koreensis sp.* nov., a novel rhamnolipid-producing bacterium. Inter. J. Syst. Evol. Microbiol. 55:1409-1413.

Lin S.-C., Lin K.-G., Lo C.-C., and Lin Y.-M. (1998) Enhanced biosurfactant production by a *Bacillus licheniformis* mutant. Enzyme Microb. Technol. 23:267-273.

Maier R., and Soberon-Chavez G. (2000) *Pseudomonas aeruginosa* rhamnolipids: biosynthesis and potential applications. Appl. Microbiol. Biotech. 54:625-633.

Maneerat S., and Phetrong K. (2007) Isolation of biosurfactant-producing marine bacteria and characteristics of selected biosurfactant. Songklanakarin J. Sci. Technol. 29:781-791.

Manivasagan P., Sivasankar P., Venkatesan J., Sivakumar K., and Kim S.-K. (2014) Optimization, production and characterization of glycolipid biosurfactant from the marine actinobacterium, *Streptomyces sp.* MAB36. Bioproc. Biosyst. Eng. 37:783-797.

Mizumoto S., and Shoda M. (2007) Medium optimization of antifungal lipopeptide, iturin A, production by *Bacillus subtilis* in solid-state fermentation by response surface methodology. Appl. Microbiol. Biotech. 76:101-108.

Mukherjee S., Yang J.W., Hoffmann S., and List B. (2007) Asymmetric enamine catalysis. Chem. Rev. 107:5471-5569.

Mulligan C.N. (2005) Environmental applications for biosurfactants. Environ. Pollut. 133:183-198.

Mulligan C.N., Yong R.N., and Gibbs B.F. (1999a) On the use of biosurfactants for the removal of heavy metals from oil contaminated soil. Environ. Progress 18:50-54.

Mulligan C.N., Yong R.N., and Gibbs B.F. (2001) Heavy metal removal from sediments by biosurfactants. J. Hazard. Mater. 85:111-125.

Mulligan C.N., Yong R.N., Gibbs B.F., James S., and Bennett H. (1999b) Metal removal from contaminated soil and sediments by the biosurfactant surfactin. Environ. Sci. Technol. 33:3812-3820.

Nakano M.M., Corbell N., Besson J., and Zuber P. (1992) Isolation and characterization of sfp: a gene that functions in the production of the lipopeptide biosurfactant, surfactin, in *Bacillus subtilis*. Mol. Gen. Genet. 232:313-321.

Neu T.R. (1996) Significance of bacterial surface-active compounds in interaction of bacteria with interfaces. Microbiol. Rev. 60:151.

Nitschke M., and Costa S. (2007) Biosurfactants in food industry. Trends. Food Sci. Technol. 18:252-259.

Ochsner U.A., Reiser J., Fiechter A., and Witholt B. (1995) Production of *Pseudomonas aeruginosa* Rhamnolipid Biosurfactants in Heterologous Hosts. Appl. Env. Microbiol. 61:3503-3506.

Ohno A., Ano T., and Shoda M. (1995) Production of a lipopeptide antibiotic, surfactin, by recombinant *Bacillus subtilis* in solid state fermentation. Biotechnol. Bioeng. 47:209-214.

Olsen J.V., Blagoev B., Gnad F., Macek B., Kumar C., Mortensen P., and Mann M. (2006) Global, in vivo, and site-specific phosphorylation dynamics in signaling networks. Cell. 127:635-648.

Olvera C., Goldberg J.B., Sánchez R., and Soberón-Chávez G. (1999) The *Pseudomonas aeruginosa* algC gene product participates in rhamnolipid biosynthesis. FEMS microbiol. Lett. 179:85-90.

Ozyilmaz U., and Benlioglu K. (2013) Enhanced biological control of phytophthora blight of pepper by biosurfactant-producing *pseudomonas*. Plant Pathol. J. 29:418-26. DOI: 10.5423/ppj.oa.11.2012.0176.

Perneel M., Heyrman J., Adiobo A., De Maeyer K., Raaijmakers J., De Vos P., and Höfte M. (2007) Characterization of CMR5c and CMR12a, novel fluorescent *Pseudomonas* strains from the cocoyam rhizosphere with biocontrol activity. J. Appl. Microbiol. 103:1007-1020.

Persson A., and Molin G. (1987) Capacity for biosurfactant production of environmental *Pseudomonas* and *Vibrionaceae* growing on carbohydrates. Appl. Microbiol. Biotechnol. 26:439-442.

Peypoux F., Bonmatin J., and Wallach J. (1999) Recent trends in the biochemistry of surfactin. Appl. Microbiol. Biotechnol. 51:553-563.

Peypoux F., Guinand M., Michel G., Delcambe L., Das B.C., and Lederer E. (1978) Structure of iturine A, a peptidolipid antibiotic from *Bacillus subtilis*. Biochemistry 17:3992-3996.

Pornsunthorntawee O., Wongpanit P., Chavadej S., Abe M., and Rujiravanit R. (2008) Structural and physicochemical characterization of crude biosurfactant produced by *Pseudomonas aeruginosa* SP4 isolated from petroleum-contaminated soil. Bioresource Technol. 99:1589-1595.

Porob S., Nayak S., Fernandes A., Padmanabhan P., Patil B.A., Meena R.M., and Ramaiah N. (2013) PCR screening for the surfactin (sfp) gene in marine *Bacillus* strains and its molecular characterization from *Bacillus tequilensis* NIOS11. Turk. J. Biol. 37:212-221.

Qazi M.A., Subhan M., Nighat F., Ali M., and Ahmed S. (2013) Role of biosurfactant produced by *Fusarium sp.* BS-8 in enhanced oil recovery (EOR) through sand Pack column. Int. J. Biosci. Biochem. Bioinform. 3:6-10.

Rahman P.K., and Gakpe E. (2008) Production, characterisation and applications of biosurfactants-Review. Biotechnology.7(2)360-370.

Raza Z.A., Rehman A., Khan M.S., and Khalid Z.M. (2007) Improved production of biosurfactant by a *Pseudomonas aeruginosa* mutant using vegetable oil refinery wastes. Biodegradation 18:115-121. DOI: 10.1007/s10532-006-9047-9.

Romero D., de Vicente A., Rakotoaly R.H., Dufour S.E., Veening J.W., Arrebola E., Cazorla F.M., Kuipers O.P., Paquot M., and Perez-Garcia A. (2007) The iturin and fengycin families of lipopeptides are key factors in antagonism of *Bacillus subtilis* toward *Podosphaera fusca*. Mol. Plant Microb. Interact. 20:430-40. DOI: 10.1094/mpmi-20-4-0430.

Ron E.Z., and Rosenberg E. (2001) Natural roles of biosurfactants. Env. Microbiol. 3:229-236.

Rosenberg E., and Ron E. (1999) High-and low-molecular-mass microbial surfactants. Appl. Microbiol. Biotechnol. 52:154-162.

Rufino R., Rodrigues G., Campos-Takaki G., Sarubbo L., and Ferreira S. (2011) Application of a yeast biosurfactant in the removal of heavy metals and hydrophobic contaminant in a soil used as slurry barrier. Appl. Env. Soil Sci. 1-7.

Sachdev D.P., and Cameotra S.S. (2013) Biosurfactants in agriculture. Appl. Microbiol. Biotechnol. 97:1005-1016. DOI: 10.1007/s00253-012-4641-8.

Salihu A., Alam M.Z., AbdulKarim M.I., and Salleh H.M. (2012) Lipase production: an insight in the utilization of renewable agricultural residues. Resour. Conserv. Recycl. 58:36-44.

Samuels Y., Wang Z., Bardelli A., Silliman N., Ptak J., Szabo S., Yan H., Gazdar A., Powell S.M., and Riggins G.J. (2004) High frequency of mutations of the PIK3CA gene in human cancers. Science. 304:554-554.

Schneider J., Taraz K., Budzikiewicz H., Deleu M., Thonart P., and Jacques P. (1999) The structure of two fengycins from *Bacillus subtilis* S499. Zeitschrift. Fur. Naturforschung. C .54:859-866.

Sekhon K.K., Khanna S., and Cameotra S.S. (2011) Enhanced biosurfactant production through cloning of three genes and role of esterase in biosurfactant release. Microbial. cell factories. 10:1-10.

Sen R. (1997) Response surface optimization of the critical media components for the production of surfactin. J. Chem. Technol. Biot. 68:263-270.

Singh A., Van Hamme J.D., and Ward O.P. (2007) Surfactants in Microbiology and Biotechnology: Part 2. Application aspects. Biot. Advance. 25:99-121.

Singh A.K., and Cameotra S.S. (2013) Efficiency of lipopeptide biosurfactants in removal of petroleum hydrocarbons and heavy metals from contaminated soil. Env. Sci. Pollut. Res. 20:7367-7376.

Singh P., and Cameotra S.S. (2004) Potential applications of microbial surfactants in biomedical sciences. Trends Biotechnol. 22:142-146.

Sinha S.N., and Paul D. (2012) Detoxification of Heavy Metals by Biosurfactants. B. Environ Sci. Res. 1:1-3.

Tahzibi A., Kamal F., and Assadi M.M. (2004) Improved production of rhamnolipids by a *Pseudomonas aeruginosa* mutant. Iran. Biomed. J. 8:25-31.

Tan H., Champion J.T., Artiola J.F., Brusseau M.L., and Miller R.M. (1994) Complexation of cadmium by a rhamnolipid biosurfactant. Environ. Sci.Technol. 28:2402-2406.

Tatsi A., Zouboulis A., Matis K., and Samaras P. (2003) Coagulation–flocculation pretreatment of sanitary landfill leachates. Chemosphere 53:737-744.

Tomar S., Singh B., Lal M., Khan M., Hussain T., Sharma S., Kaushik S., and Kumar S. (2014) Screening of novel microorganisms for biosurfactant and biocontrol activity against *Phytophthora infestans*. J. Environ. Biol. 35(5) : 893-899.

Urum K., Pekdemir T., and Gopur M. (2003) Optimum conditions for washing of crude oil-contaminated soil with biosurfactant solutions. Process Saf. Environ. 81:203-209.

Van Bogaert I.N., Holvoet K., Roelants S.L., Li B., Lin Y.C., Van de Peer Y., and Soetaert W. (2013) The biosynthetic gene cluster for sophorolipids: a biotechnological interesting biosurfactant produced by *Starmerella bombicola*. Mol. Microbiol. 88:501-509.

Van Bogaert I.N., Saerens K., De Muynck C., Develter D., Soetaert W., and Vandamme E.J. (2007) Microbial production and application of sophorolipids. Appl. Microbiol. Biotechnol. 76:23-34.

Violeta O., Oana S., Matilda C., Maria C., Catalina V., Gheorghe C., and Petruta C. (2011) Production of biosurfactants and antifungal compounds by new strains of *Bacillus spp.* isolated from different sources. Roman. Biotech. Lett. 16:84-91.

Wang P., Hu X., Cook S., Begonia M., Lee K.S., and Hwang H.-M. (2008) Effect of culture conditions on the production of ligninolytic enzymes by white rot fungi *Phanerochaete chrysosporium* (ATCC 20696) and separation of its lignin peroxidase. World J. Microb.Biot. 24:2205-2212.

Yakimov M.M., Amro M.M., Bock M., Boseker K., Fredrickson H.L., Kessel D.G., and Timmis K.N. (1997) The potential of *Bacillus licheniformis* strains for in situ enhanced oil recovery. J. Petrol. Sci. Eng. 18:147-160.

Zeftawy E., and Monem M.A. (2006) Use of rhamnolipid to remove heavy metals from aqueous streams via micellar enhanced ultrafiltration, Concordia University.

Zhang L., Veres-Schalnat T.A., Somogyi A., Pemberton J.E., and Maier R.M. (2012) Fatty Acid Cosubstrates Provide β-Oxidation Precursors for Rhamnolipid Biosynthesis in *Pseudomonas aeruginosa*, as Evidenced by Isotope Tracing and Gene Expression Assays. Appl. Environ. Microbiol. 78:8611-8622.

Zheng C., He J., Wang Y., Wang M., and Huang Z. (2011) Hydrocarbon degradation and bioemulsifier production by thermophilic *Geobacillus pallidus* strains. Bioresource Technol. 102:9155-9161.

Zhu K., and Rock C.O. (2008) RhlA converts β-hydroxyacyl-acyl carrier protein intermediates in fatty acid synthesis to the β-hydroxydecanoyl-β-hydroxydecanoate component of rhamnolipids in *Pseudomonas aeruginosa.* J. Bacteriol. 190:3147-3154.

Zosim Z., Gutnick D., and Rosenberg E. (1982) Properties of hydrocarbon in water emulsions stabilized by *Acinetobacter* RAG 1 emulsan. Biotechnol. Bioeng. 24:281-292.

7

Hot Springs of Sikkim: Biotechnological and Sociological Importance

Sayak Das, Ishfaq Nabi Najar, Mingma Thundu Sherpa and Nagendra Thakur*

Department of Microbiology, School of Life Sciences, Sikkim University, 6th Mile, Samdur, Tadong, Gangtok – 737102, Sikkim, India.

**E-mail: nthakur.cus@gmail.com*

Abstract

Thermophiles mainly harbor in highly temperate environments such as hydrothermal vents, hot springs, fumaroles etc. The main importance of thermophiles is due to their thermostable proteins, enzymes and their adaptive mechanisms at higher temperature which are very useful for industrial applications. The main rationale behind this research was to isolate industrially important thermophilic bacteria from the Hot Springs of Sikkim which can be used as tools for genetic engineering and vector based research activities. Another major thrust area of the research is physicochemical quantification and analysis of chemical compounds like arsenic, fluoride content, silica, heavy metals or halogens in Hot springs or bathing ponds is required, as they can cause severe diseases or syndromes, which are detrimental to health, keeping in mind as these Hot Springs are Tourist destinations. Sikkim – a Northeast Indian state of pristine landscapes is a natural niche and habitat of many Hot springs. Hot Springs of Sikkim are regarded as holistic elements having importance of locale panacea. Dissolved minerals helps in curing many diseases, and this forms the basis of balneotherapy. The chemical analysis showed the presence of many naturally dissolved elements like sulfur, chloride and salts which have an antimicrobial effect and are regarded as potent biomedical therapeutic elements. The water being very scalding in nature acts as an antiseptic or inhibitor to many pathogens. Survey reports showed that the hot springs are used to completely cure off the osteo related diseases like arthritis, skin diseases like eczema, psoriasis, inflammations and Gastrointestinal diseases. Every year many tourists and patients throng in around the niche and bathe, take away and drink the *Tatopani* as a medicine.

Introduction

Life at extremes

Microbial World is a concealed world, whose existence is more authoritative and conundrum than the human species (the most intelligent and smart ones to evolve!!). Microorganisms metabolic functioning, adaptive prowess, rationale of retaliation to survive, and their exodus mechanism from the so called "immunity" of hosts are completely shrouded and is itself an enigma – unspecified and undefined till date and is completely veiled. They are omnipresent and ubiquitous i.e., existing all around us and within us. Only with the invention of microscope and advancement in instrumentation, we have been able to capture little information about them but still 98% remains unknown.

R. D. MacElroy in the year of 1974 coined the term "extremophiles" which described those microbes surviving and tolerating extreme conditions with respect to human life adaptability conditions (MacElroy 1974). Extreme conditions in the form of high temperature or minimum freezing temperatures, hyper saline conditions or hyper acidic or alkaline conditions, or hyper barophilic conditions etc. which were earlier thought to be free or rather impossible to sustain life, was then regarded as a habitat which sustains normal life metabolism. The protein folding and their enzymology is the main distinguishing factor which separates them from the rest (Madigan *et al.* 2003). Thus the most important factors for the growth of extremophilic microorganisms are high or low temperature, high or low pH, and high salinity (Stetter 1999; Brock 1978). The resulting environments, based on the elevated or low conditions, are then classified as thermophilic, psychrophilic, alkalophilic, acidophilic and halophilic, respectively (Brock 1978). The extremophiles are the first living thing on the earth, evolved during the primordial days of the earth when surface temperatures were quite hot, and thus has been called as "Universal Ancestor" (Stetter 1999).

From the anthropocentric point of view, the word "extreme" is only limited to human beings, as for some special microbes "our extreme" are their amicable and optimum conditions for growth. Microbiologists and researchers, initially discovered that at 60°C and above, at the Hot Springs of Yellowstone National Park, a bacteria named *Thermus aquaticus* grew optimally and later its study gave an insight into the "Extremophiles and their world" (Brock and Freeze 1969).

The aftermath saw a series of studies as scientists isolated microbes from extreme environments such as the hydrothermal vents (*Thermotomaculum hydrothermale*), fumaroles (*Clostridium subterminale*), geysers (*Anoxybacillus kamchatkensis)*, solfataric fields (*Thermogymnomonas acidicola*), acidic areas (*Picrophilus oshimae)*, severe alkaline regions (*Halanaerobium hydrogeniformans*), high saline areas (*Salinispirillum marinum*), glaciers (*Arctibacter pallidicorallinus*), Antarctica (*Polymorphobacter multimanifer*), deep trenches of Pacific Ocean (*Pseudomonas* sp.), spacecraft assemblage centre (*Deinococcus phoenicis*), geothermally heated marine sediments (*Pyrococcus furiosus*) etc. (Schleper *et al.* 1996; Madigan *et al.* 2003; Kevbrin *et al.* 2005; Itoh *et al.* 2007; Alexandrino *et al.* 2011; Izumi *et al.* 2012; Begemann *et al.* 2012; Shahinpei *et al.* 2014; Liu *et al.* 2014; Vaishampayan *et al.* 2014). Extremophiles are abundant in nature. The study of extremophiles can be linked to astrobiology and these microorganisms have influenced scientists to search for extraterrestrial life (Cavicchioli 2002; Stojanovic *et al.* 2008).

Thus, the study of this "extreme life" and its ability to metabolically function and adapt, in this fascinating world of extreme conditions, proves that life originated "in a primordial pool rich in amino acids and organic molecules." And this very reason, accounts for further investigations and researches to know about the evolution and the evolving pattern of higher hierarchal organisms from the most primitive microorganisms. Taking cue from this very thought, a Hot Spring of Sikkim has been studied and its microbial world evaluated.

It is important to study thermophiles of Hot Springs as their physiological constituents and metabolites are stable at higher temperatures. Thermo enzymes of these microbes have received special attention of scientists from all over the world since they are usually stable at high temperature, resistant to chemical reagents and extreme pH in comparison to their mesophilic homologues. These thermo enzymes have vast applications in industries and genetic engineering (Rothschild and Manicinelli, 2001).

For the first time, a microbiological investigation has been carried out in Hot Springs of Sikkim. Some of the industrially important bacteria have been screened out for further analysis and investigation for biotechnological importance will be done.

Hot Springs

Hot springs are any natural geothermal spring or a natural discharge of groundwater with elevated temperature with respect to the surrounding. When silica or other minerals encounters with the ascending pressurized hot water then they are carried on the earth's surface through the volcanic rocks and as the atmospheric temperature cools it down, layers of geyserites develops around the spring. Owing to the presence of the limestone, Calcium Carbonate saturated water forms Travertine on crystallization. Lucent visuals can be noticed around the hot springs as various types of geological components of physical abiotic and biotic matter like micro flora forms distinct stratums around the hot springs. Thermal springs have their definite and profound characteristics. The hot springs can be perennial or seasonal and their temperature can vary at various geographical locations. They have a wide range of temperature variation mainly, from atmospheric temperature to 100°C and above.

Hot springs can be distinguished with reference to the geology of the earth's water circulatory system and their parameter which governs the hydrological flux. Basically the underground hydro geological channel, internal heat wave intensity and the influence of the cooling phenomena by the percolated ground water results in thermal dilution. This thermal dilution creates a pressure gradient. The temperature of the water of the hot springs continuously boils with periodical agitation on confrontation with the thermal gases or vaporized pressure.

Importance

Thermal springs have been in use for religious and/or medicinal purposes before 2000 BC in India and for hundreds of years in Crete, Egypt, China, Japan, Turkey and many European and Middle-Eastern countries as shown by Archaeological marks. Many thermal springs developed into thriving centres of religion, culture and health, such as those at Bath in England, Vichy in France and Baden-Baden in Germany.

Although, medical sciences has reached its apex, still many people and ailed patients use the hot springs for their medical treatment. Scientifically, if observed, bathing at hot springs helps in efficient and rapid increase in the blood circulation and helps in better metabolic activities. It also provides relief from arthritis, muscular stress and strains, joint pains and also inflammatory skin diseases. Many cases of neuralgia, myalgia, rheumatiods, dermatosis, osteosis, Inflammatory Bowel Diseases, Metabolic Diseases due to indigestion and constipation and hyper conditions of osseous diseases has been completely cured by the usage of hot springs. Especially the dissolved minerals like sulphur, carbonates and other trace minerals are effective against high blood pressure related diseases and also can provide certain relief to cardiovascular ailments and arteriosclerosis (Korea 2005).

Microorganisms at Hot Springs

The various properties of these kinds of hot springs are also dependent upon the micro-biota present in it and thus it is interesting to study the micro-flora of Hot springs. Generally, microorganisms which can thrive under extreme temperatures are found in Hot springs. These microorganisms called thermophiles which can grow into the temperatures above 45°C. The ideal growth temperature of the thermophiles may vary from 45°C to above 100°C. Some extremophiles have been known for more than 40 years, but the detection and isolation of new thermophilic microorganisms has increased for the last decades or ever since thermophilic bacteria were first discovered in the 1960´s (Brock 1994). Many bacteria and archaea have been isolated from hot springs. *Aquificales* are the dominant group of thermophilic bacteria. Also *Thermotoga, Thermus (T. thermophilus), Bacillus, Clostridium, Synechococcus, Chloroflexus* etc., have been detected in many hot springs. In environments with temperature above 90°C *Archaea* are dominating (Reysenbach and Shock 2002). Other archaea, such as, *Methanopyrus, Pyrodictium, Thermoproteus, Methanothermus, Archaeoglobus, Thermofilum, Thermococcus, Sulfolobus* etc. have been isolated.

Thermophiles

Thermophiles are a branch of extremophiles which include the microbes residing at temperatures above 40°C. Thermophile derived its diction from the Greek origin "*thermotita*" meaning heat and "*philia*" meaning love (heat loving organism) (Brock 1978). Some of the most extreme thermophilic microbes are Strain 121, discovered with a maximum growth temperature of 121°C astonished the highest limit of temperature (Stetter 1996). The optimum temperatures of thermophilic prokaryotes vary and were classified (Aragno 1992) into several groups depending on their optimum temperature:

- Facultative thermophile: any thermophile whose temperature range extends largely in the mesophilic range (i.e. below 45°C).
- Thermotolerant: any organism with a temperature optimum d" 45°C but can grow at a temperature >45°C.
- Moderate thermophile: any organism with a temperature optimum between 45°C and 60°C.
- Strict thermophile: any organism with a temperature optimum between 60°C and 90°C.
- Extreme thermophile: any organism with temperature optimum 90°C.

Thermophiles can be found in all the three phylogenetically different Domains of life, Eukarya, Bacteria and Archaea (Stetter 1999). Thermophilic fungi (*Chaetomium thermophilum*), thermophilic algae (*Cyanidium caldarium*), thermophilic protozoa (*Naegleria fowleri*), thermophilic bacteria (*Thermus aquaticus*), thermophilic archaea (*Thermoplasma acidophilum*), hyperthermophiles (*Methanoccus jannaschii*) are some of the classic examples of thermophilic diversity.

The thermophiles can easily adapt and tolerate high temperatures, hence the physiology of the microorganisms prerequisites for modified genetic replication, transcription and translational machinery, thermostable proteins, lipids and other cellular components, which in general are thermolabile (Rothschild and Manicinelli, 2001). There are many intrinsic factors which enable thermophilic or thermotolerant microbes to grow optimally at high temperatures. Generally, it has been reported that on comparison with mesophilic bacteria, the membrane lipids of thermophilic bacteria contain higher percentage of saturated and straight chain fatty acids, which provides the essential, higher and optimum degree of fluidity required for membrane function at elevated temperatures.

Another important feature is the presence of numerous thermostable heat shock proteins, chaperones and histone like proteins that coil around the genetic material and render it very compact and protect it. The thermophiles show higher degree of stability due to the presence of polyamines, hydrophobic core, higher amount of hydrogen bonds and thermophilic amino acids and enzymes. These structural modifications play a crucial and significant role in maintaining the metabolites and normal functioning of the cell (Campbell and Pace 1968; Ladenstein and Antranikian 1998; Stetter 1996; Rothschild and Manicinelli 2001; Madigan and Marris 2007; Satyanarayana *et al.* 2005).

Biotechnological Applications of Thermophiles

The main importance of thermophiles is their proteins, enzymes, metabolism and their adaptive mechanism. The applicability of thermo stable enzymes as biocatalysts is gaining wide industrial and biotechnological interest.

The thermostable properties of these enzymes have been endorsed to many factors, such as: (a) amino acid composition, including decreases in thermolabile residues such as Asn and Cys; (b) hydrophobic interactions; (c) aromatic interactions, ion pairs, and increased salt bridge networks; (d) oligomerization and intersubunit interactions; (e) packing and reduction of solvent-exposed surface area; (f) metal binding; (g) substrate stabilization; and (h) a decrease in number and size of surface loops (Cowan 1996; Hough and Danson 1999). Thus, the application of thermophiles in biotechnology is a great boon for mankind. The DNA polymerase enzyme of *Thermus aquaticus* and *Thermococcus litoralis* is used for the polymerase chain reaction (PCR) which is one of the most commonly used reaction in biological sciences. As their enzymes are thermo stable at relatively high temperatures, it is extensively used in the PCR process which requires thermal cycles. The enzymes isolated from *Bacillus stearothermophilus* is used as detergents, hence the bio prospect of thermophilic enzymes are maximum at industrial point of view.

Thermophilic enzymes and its application

Thermal sensitivity has been the foremost problem to the widespread use of enzymes as far as industry is concerned. From the biotechnological standpoint the thermophilic microorganisms are the most attractive due to their ability to produce enzymes proficient to catalyze industrial relevant process at higher temperature than corresponding enzymes from mesophiles. Thus, the ability of thermophilic enzymes to suitably work at high temperature implies many advantages for their applications in industry. Extracellular-polymer-degradingenzymes (such as amylases, pullulanases, Cellulases,chitinases, xylanases, pectinases, isomerases, esterases, dehydrogenases and DNA-modifying enzymes with potential use in food, chemical and pharmaceutical industries and environmental biotechnology (Poonam and Dalel 1995; Cowan 1996; Ladenstein and Antranikian 1998; Mueller *et al.* 1998; Hough and Danson 1999; Galante and Formantici 2003; Bedford *et al.* 2003; Bernard *et al.* 2010).

Cellulases

Cellulose is the extensively used most organic polymer which is an exhaustible source of raw material for different products (Klemm *et al.* 2002). It was initially used for the bioconversion of biomass which gave way research in the industrial application of enzyme in various fields such as food, feed, textiles, and detergents and in the paper industry (Cowan 1996; Xia and Cen 1999; Galante and Formantici 2003). Thermophilic cellulases are used in the bio-stoning of denim garments replacing the use of palmic stones which were traditionally employed in the industry (Belghith *et al.* 2001). Cellulases have also been used in softening and in providing variation in the colour density of fibers (Galante and Formantici 2003). Cellulase in particularly ECIII and CBHI, are commonly used in detergents for cleaning textiles. In food industry, cellulases are used in extraction and clarification of fruits and vegetables juices, production of fruit nectars and purees, and in the extraction of Olive oil (Graham and Balnave 1995; Cowan 1996). Cellulases also used to improve the nutritive quality of forages (Bedford *et al.* 2003). In the pulp and paper industry cellulose and hemicellulases have been employed for mechanical pulping for modification of the coarse mechanical pulp and hand sheet strength properties (Franks *et al.* 1996).

Exoglucanases

β-1,4-Exoglucanases or cellobiohydrolases (CBHs) are of substantial importance to cellulolytic systems, facilitating the production of mostly cellobiose which can readily be converted to glucose by β-glucosidases. Exoglucanses are extensively used in wort production and other cellulolytic activities (Pajunen 1986; Bronnenmeier *et al.* 1991; Marcus *et al.* 2008). CBHs attach to the ends of cellulose microfibrils and then processively slide down the strands cleaving off cellobiose as they progress. *Clostridium thermocellum* produce most thermostable CBHs from cellulosomal complexes. The thermophilic bacterium *Thermotoga* sp. FjSS3-B can produce a thermostable CBH whose enzyme has maximal activity at 105°C and maintains a half-life of 70°C min at 108°C (Ruttersmith and Daniel 1991).

Hemicellulases

A mixture of vastly branched and complex polysaccharides forms Hemicellulose. These may include xylans, glucans, xyloglucans, callose, mannans, and glucomannans (Marcus *et al.* 2008). Mostly xylanases are endoenzymes and they catalyse the endo-hydrolysis of 1,4-α-D-xylosidic linkages (Dodd and Cann 2009). Xylanases have long been utilized in the, food, paper, and fine chemical industries and are well recognized as critical components in the deconstruction of lignocellulose for biofuels production (Aparicio *et al.,* 2007). They are also used in bioleaching of paper pulp (Giordano *et al.* 2006).

Amylases

Amylases are starch degrading enzyme that catalyses the breakdown of starch into sugar. The starch requires a combination of enzymes for its complete hydrolysis. These enzymes include α-amylases, Glucoamylases or α-amylases (Poonam and Dalel 1995). α-amylase is an endozyme which leads to the formation of linear and branch oligosaccharide, while the rest are exoenzyme and attacks the substrate from the non-reducing end, producing oligosaccharide/monosaccharide. Amylase is used in bread making and to break down complex sugar, such as starch into simple sugars. α- and β-amylases are important in brewing beer and liquor made from sugar derivative from starch (Poonam and Dalel 1995).

Proteases

Protease is the enzyme that converts the protein into amino acid and peptides. In the food, leather, pharmaceutical and textile industry, these are the major used enzymes (Cowan 1996; Ladenstein and Antranikian 1998; Mozersky *et al.* 2002). Serine alkaline protease is used as addition to detergents for laundering. Proteinases show high keratinolytic and elastolytic activities and are used for soaking in the leather industry. It has been found that most proteases from extremophiles belongs to the serine type and are stable at high temperature even in the presence of high concentration of detergents and denaturing agents (Cowan 1996).

Lipases

Lipases (triacylglycerol acylhydrolases, EC 3.1.1.3) catalyze the hydrolysis and the synthesis of esters formed from glycerol and long-chain fatty acids. Lipases are regarded as the most versatile enzymes of the industries which bring about a range of bioconversion reaction (Vulfson 1994), which includes hydrolysis, esterification, acidolysis (Panday *et al.* 1999; Kim *et al.* 2002). The esters produced play a vital role in the food industry as flavour and aroma constituents (Gandhi *et al.* 1995). The long chain methyl and ethyl esters of carboxylic acid moieties function as fuel for diesel engines and ester of long chain carboxylic acid and alcohol moieties have application as lubricants and additives in cosmetic formulations (Linko *et al.* 1994). Other applications includes are the removal of the pitch from pulp produced in the paper industry, for the hydrolysis of milk fat in the dairy industry, removal of non-cellulosic impurities from raw cotton before processing into dyes and finished product, drug formulations in the pharmaceuticals industry and in the removal of subcutaneous fat in the lather industry (Vulfson 1994; Linko *et al.* 1994; Kim *et al.* 1994; Gandhi *et al.* 1995; Wu *et al.* 1996; Luisa *et al.* 1997; Panday *et al.* 1999 Kim *et al.* 2002).

Esterases

The enzymes that catalyze the hydrolysis (and formation) of ester bonds are known as Esterases e.g. acetyl xylan esterases catalyze the removal of acetyl ester groups from C2 or C3 positions of D-xylopyranosyl residues (Biely 2003). Thermostable acetyl xylan esterases have been purified and characterized from the thermophilic anaerobic bacteria *Thermoanaerobacterium* sp. and *Clostridium thermocellum* (Correia *et al.* 2008). The phenolic acids released through the action of esterases are potential precursors to a variety of value-added products and may be exploited as a means to offset some of the costs currently limiting the feasibility of bio-refineries (Biely 2003).

Hot Springs in Sikkim

In 1973, the Geological Survey of India started the geothermal exploration in India and it was found that more than 350 hot springs with temperatures ranging from 30° to 98°C are known to occur throughout the country. The hot spring systems are considered to be the result of deep circulation of local ground waters as no recent volcanism or plutonic igneous activity has been recorded. Krishnaswami (1975) gave a detailed account of the hot spring occurrences and divided them into two groups on the basis of their tectonic setting: an orogenic group, represented by warm springs in the Himalayas and a non-orogenic group, represented by the warm springs in peninsular India. In orogenic group, many hot springs are located in Sikkim which is geographically located between 27°04'46" and 28°07'48" North latitude and 88°00'58" and 88°55'25" East longitude on the Eastern Himalayas.

Sikkim houses many natural Hot springs within its geography. It is a major tourist attractive state of India where the nature is in its juvenile form and a refresh season greets its visitors. In local languages these Hot Springs are called as *Tatopani* or *Tshachu*. *Tatopani* is a Nepali word where "*Tato*" means "Hot" and "*pani*" means "water" whereas *Tshachu* is a Tibetan word where "*Tsha*" means "Hot" and "*Chu*" means "water". Thus *Tatopani* refers to the Hot Springs. The Hot Springs of Sikkim are ancient and have a lot of aesthetic value since ages. It is believed that the water from these hot springs can cure several diseases and therefore, used as a social medicine, religious customs and practices. People living in vicinities of these hot springs not only use these hot springs for bathing, but also use water for drinking and cooking purpose.

There are numerous hot springs and respective bathing ponds at Sikkim; located at various locations like Polok, Reshi, Borong, Takrum, Yume Samdong, Yumthang, Zee, Shagyong Phedok and Tholung Kang of Sikkim.

There are many hot springs in Sikkim. But insufficient amount of research has lead to very few literatures on the hot springs of Sikkim. We have initiated a research into this and the following are our observations till date from the study of Hot Springs of Sikkim. Many industrially important microorganisms have been screened and isolated.

Polok Tatopani

It is also known as Ralang Tchu or Ralang Tsha chuu. Polok *Tatopani* is physically located at the base of Gangyab, West Sikkim by the banks of the river Rangeet. The river bifurcates South and West Sikkim. The Polok *Tatopani* is also referred to as Ralang *Tatopani* or Rabong *Tatopani*. There are two main natural ponds locally referred as "Paru Hang" and "Sumnima" which are Rai words referring to the ancestral legends of the Kirat Mythology. According to other legends these two ponds are referred as Holy Ponds where Hindu God "Lord Shiva" and Goddess "Parvati" had bathed (Das *et al.*, 2012). As per the survey, it was observed that the ancient Rai community had used this terminology to nomenclature the two ponds. But, unfortunately due to westernization, slowly these ancient name has been wiping off from the memory lanes of new civilization.

Pond A is located between 27°21.19' N latitude and 88°19.352' E longitude. The elevation of from the sea level is 884m. Pond B is located between 27°21.00' N latitude and 88°19.353' E longitude. The elevation of from the sea level is 925m. The trail to the *Tatopani* spot is from Polok, South Sikkim but the *Tatopani* and its source is located at Gangyab foothill, West Sikkim just away from the River Rangit.

Physical Analysis of the water samples

These are the parameters used to describe the physical state of the water. Parameters like Temperature, pH, D.O., Conductivity and T.D.S. of water was measured on site by the help of portable Orion benchtop. It was observed that Polok *Tatopani* had the highest temperature around 72.5°C as measured on April 2012. With respect to pH, it was observed that it ranged from 7.8 – 8.0. Polok *Tatopani* had the D.O. of 34.6% saturation. Polok *Tatopani* showed the highest conductivity of 859 to 860 μS/cm whereas highest dissolved solids of 398 to 400 mg/L.

Chemical Analysis of the water samples

The HiMedia Kit was used to determine the chemical properties of the water. It was observed that the chloride and fluoride concentration was maximum compared to other non metals. Among the various chemicals found at Polok *Tatopani*; both the ponds on average had 5.85 ppm Fluoride, 1.75 ppm Nitrate, 63 ppm Chloride, 40ppm $CaCO_3$ and 349 ppm alkalinity dissolved within.

Microbiology

The microbial load in the water, food and dairy products are enumerated by Total Plate Count Method in Plate Count Agar. It was found that, both the ponds at Polok *Tatopani* had 0.6×10^7 viable cells/mL or microbial load. A total of 11 thermophilic or thermotolerant bacterial isolates were isolated from this Hot spring water sample. The isolates were selected on the basis of their interesting pigmentation and shape of the colony formed at the Thermus Agar plate. Majority of the isolates formed pale yellow colored colonies, some white colored colonies and few produced yellow colored colony. Some colonies were as small as only 2 mm diameter size whereas some formed large colonies of 6 mm in diameter size. Most of the colonies were smooth and elevated. On Gram Staining, to understand the physical morphology of the bacteria isolated from *Tatopani*, it was observed that all the isolates were Gram negative and some were short rods whereas others were long and thick rods. According to Bergey's classification of bacteria, carbohydrate utilization is an important characteristic feature of bacteria through which differentiation and group classification can be done. Thus, carbohydrate fermentation test was carried out with 16 different carbohydrates and each of them were substituted and checked as a carbohydrate substrate for the different isolates. It was observed that none of the isolates were able to ferment the carbohydrates like arabinose, galactose, lactose, maltose, rhamnose, xylose, cellobiose and inositol. Except, one isolate all the isolates were able to ferment dextrose, fructose, sucrose, trehalose, Melibiose and sorbitol. Only some isolates were catalase positive whereas others were negative. Few of them showed protease activity, urease activity and amylase activity whereas others showed no enzymatic activity. None of the isolates showed gelatinase activity. All the isolates showed good growth at 60°C. Most of the isolates showed good growth when grown at 2.5% NaCl condition but interestingly all of them were able to survive 20% NaCl concentration. pH conditions are also important in deciphering the physiology of the thermophilic bacteria. Depending on the bacterial optimum pH conditions, they are classified as Acidophiles and Alkaliphiles. Acidophiles requires acidic conditions and Alkaliphiles requires basic conditions of pH. This same characteristic feature was investigated for the isolates isolated from the *Tatopani*. Most of the isolates preferred pH 8.0 as their optimum pH conditions (Das 2012).

Detection of coli – aerogenes contamination in water.

Water from these two hot springs are utilized for both drinking and bathing purpose and therefore, it is important to check if these hot springs have presence of fecal contamination. Water from these hot springs was checked for coli-aerogenes contamination on Violet Red Bile Agar. It was found that none of the sample showed any presence of coli-aerogenes contamination.

Borong Tatopani

Borong ***Tatopani*** is physically located at Lower Borong and the Ponds are situated at the banks of river Rangit, in West Sikkim. There are three ponds for bathing in Borong *Tatopani* but it depends on seasons. But the major bathing pond is located between the range of 27°21.944' N latitude and 88°19.668' E longitude. The elevation of from the sea level is 998 m.

Physical Analysis of the water samples

These are the parameters used to describe the physical state of the water. Parameters like Temperature, pH, D.O., Conductivity and T.D.S. of water was measured on site by the help of portable Orion benchtop. It was observed that the Borong *Tatopani* had the temperature of 60.3°C as measured on April 2012. Borong *Tatopani* had only showed conductivity of 240-242 μS/cm and T.D.S. count of 110-112 mg/L.

Chemical Analysis of the water samples

The HiMedia Kit was used to determine the chemical properties of the water. It was observed that the chloride and fluoride concentration was maximum compared to other non metals. Borong *Tatopani* on average had 3.96 ppm of Fluoride, 0.05 ppm Nitrate, 11 ppm Chloride, 3.96 ppm $CaCO_3$ and 75 ppm alkalinity dissolved within it.

Microbiological

It was found that the ponds at Borong *Tatopani* had 0.2 X 10^7 viable cells/mL or microbial load. A total of 5 thermophilic or thermotolerant bacterial isolates were isolated from this Hot spring water sample. The isolates were selected on the basis of their interesting pigmentation and shape of the colony formed at the Thermus Agar plate. Majority of the isolates formed pale yellow colored colonies, and white colored colonies. Some colonies were as small as only 1.5 mm diameter size whereas some formed large colonies of 4 mm in diameter size. Most of the colonies were smooth and elevated. On Gram Staining, to understand the physical morphology of the bacteria isolated from *Tatopani*, it was observed that all the isolates were Gram negative and some were short rods whereas others were long and thick rods. Carbohydrate fermentation test was carried out with 16 different carbohydrates and each of them were substituted and checked as a carbohydrate substrate for the different isolates. It was observed that all the isolates were able to ferment dextrose, fructose, sucrose, trehalose, Melibiose and sorbitol with few exceptions. Only some isolates were catalase positive whereas others were negative. Few of them showed protease activity, urease activity and amylase activity whereas others showed no enzymatic activity. None of the isolates showed gelatinase activity. All the isolates showed good growth at 60°C. Most of the isolates showed good growth when grown at 2.5% NaCl condition but interestingly all of them were able to survive 20% NaCl concentration. pH conditions are also important in deciphering the physiology of the thermophilic bacteria. Most of the isolates preferred pH 8.0 as their optimum pH conditions (Das, 2012).

Detection of *coli – aerogenes* contamination in water

It was found that none of the sample showed any presence of coli-aerogenes contamination.

Reshi Tatopani

It is also known as Phur Tsha Chu ("Phur" means bubble "Tsha" mean hot and "Chu" means water in Tibetan language) and is located approximately 25km from Jorethang to the east of Reshi (Tinkitam). Hot spring source is located near the bank of river Testa. Reshi hot spring is located at an altitude of 583 m above the mean sea level and the geographical location is 27°15.057' N, 88°18.148' E, situated on the eastern part of Reshi ,the border between south and west district of Sikkim. Separate time table for male and female hot spring goers were provided by Phur Tsha Chu committee (5-8 am for men, 8-11am men, 11-2 pm women and after 2 pm onwards till 5pm for men). The surrounded area of Reshi hot spring is covered with huge bolder an metamorphic rock which is suppose to carry by river action during rainy seasons and one can feel strong smell from hot spring which might be the smell of sulphur. Private hut is available there on ranted basis which is made up of bamboo and plastic roof for seasonal basis and the sanitary condition in and around the hut is poorly maintained. Reshi hot spring during December, second during March and last during April and we found that there is a variation of temperature (during December 52°C, March 47.3°C and April 39.2°C).

Physical Analysis of the water samples

It was observed that the Tatopani had the highest temperature around 43.2°C and minimum temperature of 40.3°C as measured on April 2012. With respect to pH, it was observed that there was no much variation between the Ponds as they ranged from 7.8 – 8.0. The Ponds lacked D.O as the values were almost zero.

Chemical Analysis of the water samples

It was observed that the chloride and fluoride concentration was maximum compared to other non metals. Among the various chemicals found at Reshi Tatopani; the pond on average had 3.85 ppm Fluoride, 0.5 ppm Nitrate, 80 ppm Chloride, 50 ppm $CaCO_3$ and 419 ppm alkalinity dissolved within it.

Microbiological

The microbial load in the water, food and dairy products are enumerated by Total Plate Count Method in Plate Count Agar. It was found that, both the ponds at Reshi Tatopani had 0.2 X 10^7 viable cells/mL or microbial load. A total of 7 thermophilic or thermotolerant bacterial isolates were isolated from this Hot spring water sample. The isolates were selected on the basis of their interesting pigmentation and shape of the colony formed at the Thermus Agar plate. Some of the isolates formed pale yellow colored colonies, few white colored colonies and majority produced yellow colored colony. Most of the colonies were smooth and elevated. Both gram positive and negative bacteria were observed and it was observed that some of the isolates were short rods whereas others were long and thick rods. Carbohydrate fermentation test was carried out with 16 different carbohydrates and each of them were substituted and checked as a carbohydrate substrate for the different isolates. It was observed that none of the isolates were able to utilize and ferment the carbohydrates like arabinose, galactose, lactose, maltose, rhamnose, xylose, cellobiose and inositol. Except, one isolate all the isolates were able to ferment dextrose, fructose, sucrose, trehalose. Few isolates were catalase positive whereas only two showed protease activity. None of the isolates showed gelatinase activity. All the isolates showed good growth at 60°C. The isolates showed good growth when grown at 2.5% NaCl condition. Most of the isolates preferred pH 8.0 as their optimum pH conditions (Sherpa, 2012).

Detection of coli – aerogenes contamination in water.

Water from these two hot springs are utilized for both drinking and bathing purpose and therefore, it is important to check if these hot springs have presence of fecal contamination. It was found that none of the sample showed any presence of coli-aerogenes contamination.

Yumthang Tatopani

Yumthang is a valley, a nibbling meadow, surrounded by the Himalayan Mountains in the North Sikkim. It is a flat valley readily called as "the valley of flowers". Adjoining the Singba forest, it has a very picturesque landscape and is lined with Rhododendrons and other orchids which bloom in different colors. It is renowned for its Hot Spring. Yumthang *Tatopani* is physically located on the base of the mountain across the river Lachung, a tributary of the river Teesta which flows past the valley and the town of Lachung. According to the GPS data the location of the *Yumthang Tatopani* lies between 27°47.565' N longitude and 88°42.525' E latitude. The elevation range of the location ranged showed 3619 m above the sea level.

Physical Analysis of the water samples

The main source of Yumthang *Tatopani* is outside the bathing pond and the source is connected to the bathing pond through a long pipe cemented outside. The temperature and pH was measured both on the source outside the bathing pond and within the bathing pond. It was observed that within the bathing pond temperature varies from 40-41°C and the temperature of the source outside possess 45°C. Also the temperature of the entrance point of the pipe into the pond was measured to be 43°C. The atmospheric temperature was around 13-14°C.

Chemical Analysis of the water samples

The HiMedia Kits were used to determine the chemical properties of the water. It was observed that Nitrite, Silica and Chloride concentrations are maximum as compared to other non- metals. Among the various chemicals found at Yumthang *Tatopani* has an average of 350 ppm Nitrite, 80 ppm Silica, 50 ppm Chloride, 15 ppm Sulfite, 10 ppm Alkalinity, 0.8 ppm Iron dissolved in it.

Microbiology

The microbial load enumerated by Total Plate Count Method was found to be around 1.4×10^4 viable cells/mL. 20 thermophilic or thermotolerant bacterial isolates were isolated on the basis of the general cellular morphology of isolates (pigmentation and shape of the colonies). Some isolates formed pale yellow colored colonies, whereas the rest formed cream white colored colonies. All of the colonies were circular in formation having entire margins, flat elevation and opaque density. The colonies formed by some were larger in size (4-5 mm) as compared to other isolates with only 2 mm as colony diameter. On Gram Staining, to understand the physical morphology of the bacteria isolated from the hot water spring, it was observed that all the isolates were Gram positive short rods, except some were of long rods. Spore staining was done to check the ability of the isolates to form spores under given circumstances. All the bacterial isolates showed the presence of spores when observed under compound microscope. It was found that all the isolates could tolerate a wide range of temperature from 30°C to 100°C as they showed growth within this range. Although the temperature of the *Tatopani* source and the bathing pond was in the range of 40 – 45°C, only few of the isolates shows maximum growth on 40°C. Most of the isolates showed maximum growth at temperatures 60°C and 50°C. pH conditions play an important role in deciphering the physiology of the thermophilic bacteria. The maximum growth of most of the isolates was observed at pH 7. Only some of the isolates showed maximum growth at pH 5. The other most important condition in deciphering the physiology of the thermophilic bacteria is the salt (NaCl) concentration or salinity. Against the concentrations of 2% NaCl which was used as control, the growth at various NaCl concentrations was checked. It was found that most of the isolates were able to grow maximum at 1% NaCl concentration and only few at 2.5% NaCl concentration. Carbohydrate fermentation test was carried out with nine different carbohydrates and each of them were substituted and checked as a carbohydrate substrate for the different isolates. Most of the isolates were able to utilize Dextrose, Lactose and Mannitol. All the isolates were able to utilize Fructose except two of the isolates. None of the isolates were able to utilize Galactose. All the bacterial isolates produced bubbles when treated with 3% Hydrogen peroxide. They showed catalase activity. Some of the isolates showed positive amylase and protease activity. None of the isolates were able to hydrolyze gelatin. All the bacterial isolates have shown negative result for Urease activity (Najar 2013).

Detection of coli – aerogenes contamination in water

As water from the Yumthang hot spring is utilized for both drinking and bathing purpose and therefore, it is important to check if these hot springs have presence of fecal contamination. Water from Yumthang *Tatopani* was checked for *coli-aerogenes* contamination on Violet Red Bile Agar. It was found that none of the sample showed any presence of *coli-aerogenes* contamination.

Dzongu Tatopani

Dzongu is the native place of Lepchas – the aborigines of Sikkim. The Dzongu Hot spring is located in the valley of Upper Dzongu, Sikkim. The geographical coordinates of the Main Source Point is 27°31'58.8"N – 88°28'10.1"E at an elevation of 1170 m above the sea level. 70 km away from Gangtok, the capital of Sikkim, at an altitude of approx. 900 m – 6100 m, lays Dzongu, which is a very peaceful and sacred region of North Sikkim. Dzongu is closely associated with three terms – Land of Lepcha, Natural Hub of medicinal plants and interaction of nature and culture. A huge pipe has been connected to this bore channel and it was connected to the Bathing House where, artificially two ponds have been constructed. Ponds are like modern pools for bathing purposes. Two separate Bathing ponds are present – one for males and the other for females in separate two rooms. The water is used only for bathing purpose.

Physical analysis of the water samples

The physical parameters which were used to describe the physical state of the water were temperature, pH, TDS, salinity and conductivity. These parameters were measured on site by the help of portable Orion bench top and PockeTester TRACER^{TM} (pH/EC/TDS). All the physical parameters were measured for all the water samples. It was observed that the temperature of the Main Source Point was 50.2°C but when it reached in the Bathing Pond it got reduced to around 45°C. The pH was measured to be in the range from 7.0-7.5 of all the water samples. The TDS of the Main Source Point was found to be 97 ppm and Bathing Pond had 120 ppm. The salinity and conductivity of the Main Source Point was 55 ppm and 208.3 µS/cm respectively; whereas the salinity and conductivity of the Bathing Pond was 104 ppm and 279.8 µS/cm.

Chemical analysis of the water samples

The HiMedia Kits were used to determine the chemical properties of the water. It was observed that Nitrate, Nitrite and Sulphite concentrations were maximum as compared to other non- metals. Fluoride was not detected. In the water samples, it was found that it contained 80 ppm Nitrate, 50 ppm Nitrite and 50 ppm Sulphite respectively. And the other chemical parameters were - 10 ppm Silica, 10 ppm Chloride, 10 ppm Orthophosphate, 10 ppm Alkalinity, 2 ppm Iron and the Total Hardness was found to be only 20 ppm in the water samples.

Microbiology

The microbial load in the water, food and dairy samples are usually enumerated by Total Plate Count Method in Plate Count Agar. It was observed that the Dzongu *Tatopani* had 6.4 X 10^6 viable cells/mL microbial loads. Fifty isolates were isolated from the Dzongu *Tatopani*. The isolates were selected on the basis of their pigmentation and shape of the colony formed at the Thermus Agar plate. Most of the isolates formed yellow colored colonies; some formed brown colored colonies, cream white colored and white colored colonies. The colonies were either circular or irregular, all had entire margins and their density of the colony was opaque, some had raised elevation whereas others were flat. It was observed that only few isolates were coccus in shape whereas all the others were rod in shape only differing in their length and diameter. It was observed that only few were motile and all the other isolates were non-motile. Both Gram positive and Gram negative bacterial isolates were obtained. It was observed most of the isolates were non-spore formers when observed under compound microscope. Most of the isolates showed maximum growth at pH 6.0-8.0. It was found that all the isolates were able to grow at 0% NaCl concentration and only few at 20% NaCl concentration. It was found that all the aerobic and anaerobic isolates could tolerate a wide range of temperature from 40°C to 70°C as they showed growth within this range. The temperature of the *Tatopani* source and the bathing pond was in the range of 46°C – 50°C. Carbohydrate fermentation test was carried out with 15 different carbohydrates and each of them were substituted and checked as a carbohydrate substrate for the different isolates. It was observed that most of the isolates were able to ferment Dextrose and Fructose whereas most of them were unable to ferment the following sugars – Arabinose, Cellobiose,

Galactose, Inositol, Lactose, Maltose, Melibiose, Rhamnose, Sorbitol, Sucrose and Xylose. Most of the aerobic isolates produced bubbles when treated with 3% Hydrogen peroxide and hence they were catalase positive whereas all the other aerobic and anaerobic isolates were catalase negative (Das, 2015).

Detection of coli-aerogenes contamination in the water of the Hot Spring

As water from the Dzongu Hot Spring is utilized for bathing purpose therefore, it is important to check if it has the presence of fecal contamination. The water samples were checked for *coli-aerogenes* contamination on Violet Red Bile Agar. It was found that none of the sample showed any growth on the test agar plate. Hence fecal contamination was absent from the Dzongu Hot Spring.

Balneotherapy

The scientific study of the therapeutic properties of naturally occurring mineral waters is known as Balneology. The application based study of the health benefits of water is referred to as Balneotherapy. In many European countries and South Asian countries like Japan, Balneology and hot springs therapy is regarded as an official licensed medical practice. In those regions hot springs are recognized as a preventative medicine. The alleged curative properties of hot spring baths are now on being deciphered in a scientific manner to find the possible explanations behind the cure of the various ailments.

The higher temperature of the water, minerals dissolved within it, mud, sulfur precipitates and the geographical positioning influencing the climate of the area, regulates the curative properties in balneology. The heated water influences the blood circulation and breath regulation. Due to the concentrated mineralized content, the absorption rate and discharge of ions are metabolically controlled through the permeable skin. The main minerals found in hot springs are like magnesia which provides anti – allergic effect on skin and bromine compounds which enhances the nervous system and it's functioning. The high salt concentration in the water can act as physiotherapy for various forms of joint pains, rheumatic associated pain, and muscle spasms due to stress and strain developed or treatment after fractures in osseous matter. Low humidity and high evaporation contributes in accelerating the metabolic activities of our system. These climatic properties have a considerable effect on the non – specific treatment of certain diseases based on stimulation of cells and tissues while acting against antigens and defunct stimuli response.

Chemical constituents and mineralogy of Polok, Borong and Reshi Tatopani

The applications of hot springs for balneotherapy depend upon its chemical constituents. According to our study, it was observed that Polok *Tatopani* contained following chemicals in the month of April 2012: Calcium (9 ppm); Magnesium (7 ppm); Sulfate (13.1 ppm); Chloride (60 ppm); Fluoride (5.9 ppm); Arsenic (0.004 ppm) and Boron <1 pp; Borong *Tatopani* contained following chemicals: Calcium (7 ppm); Magnesium (4 ppm); Sulfate (12.1 ppm); Chloride (56 ppm); Fluoride (5.4 ppm); Arsenic (0.004 ppm) and Boron <1ppm; whereas Reshi Tatopani contained the following constituents; Calcium (61 ppm); Magnesium (25 ppm); Sulfate (20 ppm); Chloride (103 ppm); Fluoride (1.6 ppm); Arsenic (0.006 ppm) and Boron <1 ppm.

Bathing at the *Tatopani,* without prior knowledge of bathing regime should be always avoided as it may lead to serious skin inflammation due to the scalding temperature of the water. Thus a particular regime or protocol has to be followed for the safety.

There are many famous Hot springs in the world where the balneotherapy is practiced and some of them have been already discussed. We compared our data with that with the Tecopa Hot spring, Wilbur Hot spring, Arkansas Hot spring and Patagonia Hot springs. Tecopa Hot spring has the average temperature of 45-50°C and is located at South East edge of Death Valley in Mojave desert, California. The Arkansas Hot spring has the average temperature of 58°C and is located at Arkansas. The Patagonia Hot spring is located at Patagonia. The average temperature of Sikkim's Hot springs where the study was carried out was: Polok *Tatopani* is 60°C, Borong *Tatopani* is 50°C and Reshi *Tatopani* is 40°C. Thus on comparison it can be said that the temperature of these *Tatopani* are hotter than the other discussed hot springs.

The Sulfur rich hot springs in many developed countries like France, Spain and Japan are used for the various treatments like skin inflammations and respiratory problems like asthma, bronchial infections etc. The element

Boron found in dissolved state in hot spring helps in building muscle mass and strengthening bones whereas bicarbonates of sodium, calcium and potassium helps in circulation and porosity enhancement of the skin membrane pores by opening the peripheral blood vessels and helps in improving the blood circulation. Thus bathing in hot springs having high bicarbonate concentration yields beneficial health properties. Bicarbonate water containing more than 250 ppm is regarded as a potential elixir in Spain. In European countries, hot springs containing the bicarbonate water is utilized for curing cardiovascular diseases, nervous system imbalance, hypertension, high blood pressure and mild atherosclerosis. Many of the stimulating benefits of hot springs water are temperature dependent. It was found that bathing in hot springs temporarily provided relief in chronic pain directly associated with inflammation, even in cases where inflammation has not been reduced.

The influx and efflux of minerals into the body depends on two factors: First, the ratio of body lipids and the amount of water permeable through the structure of the skin membrane and Second, the salts and minerals concentration of the pond. The minerals adsorption through skin via hot spring bath is in limited amount as it is concentration dependent of the dissolved minerals. Even minute amounts of therapeutic minerals adsorbed into the body via the skin have a significant therapeutic value.

Bathing at Hot spring is a type of thermal healing mechanism. The scalding temperature of the water serves as a shock stimuli initiator molecule when a person is exposed to a hot spring bath. Any increase in the abiotic conditions serves as a shock stimulus to the living system. This initiator molecule enhances mineral deposition in the subcutaneous cell tissue. The rhythmicity creates a stress – relax condition ultimately generating a localized physiotherapy in the strained zone of reduced blood circulatory area. Thus it provides instant relief to the patients.

The natural thermal water containing the dissolved minerals enhances and stimulates the immunity of the body indirectly. Again, through extensive sweating, which occurs on bath might purify the blood and helps in the elimination of toxic wastes through sweat secretion. Also, thermal temperature with dissolved minerals may acts as natural analgesic and pain killers. The hotness of the *Tatopani* sedates the nervous system and tones up the stressed muscles.

Possibly some of these beneficial properties, i.e., chemical dissolved in these hot springs are also present. We further checked what kind of patients visit these hot springs in Sikkim. Most of the visitors were mainly suffering from Chronic inflammatory rheumatism, Degenerative chronic rheumatism like severe arthritis in the joints, spine, hip, elbows, wrists, knees and hands and skin diseases like Eczemas, skin rashes, scabies, dermatosis and psoriasis. Even patients suffering from IBD, Gastrointestinal associated problems were observed.

Social significance

During the ancient periods, people were dependent on the Nature and its products. Traditional and indigenous medicinal practices were the major form of treatment in our primeval period. Diseases were then categorized or rather claimed to be a symbol of committed sin or gluttony. And it was the punishments of Him unto teach the moral basics of life. The only way to cure their ailments was to search and bath in the Holy River or in any mineral springs or in the natural hot water bodies. Only through their bath, one can get complete relief from their sufferings. Thus in order to get relief from their diseases or sins, through ages people have been bathing in many rivers like The Ganges of India, River Nile of African Continent, River Yangtze in China and even in European countries they used to bath in mineralised springs or hot water springs (Goodrich and Uysal 1994; Hembry 1990). People even used "to take away" the water and drank it as a blessed potion for its alleged curative properties. Thus it can be very truly placed that these hot water springs either due to religious customs or traditional customs, has been a part and parcel of life and was a describing parameter of society. It can be directly related to the culture. Thus, the methodology of touring a place with the intention of becoming medically fit, transformed into a new industry called Medical Tourism. The faith and belief in the Nature and its product as a healing device has been the evolutionary cause of medical tourism. With the advent of Medical tourism, it has been a major economic force and even has become a backbone for developed countries to promote and financially uplift the eco-socio status of the nation.

Today, in this 21st century where medical science has reached its zenith, still many common people and ailed patients use the hot springs for their medical treatment. Scientifically, if observed, bathing at hot springs helps in efficient and rapid increase in the blood circulation and helps in better metabolic activity. It also provides relief from arthritis, muscular stress and strains, joint pains and also inflammatory skin diseases. Many cases of neuralgia, myalgia, rheumatiods, dermatosis, osteosis, Inflammatory Bowel Diseases, Metabolic Diseases due to indigestion and constipation and hyper conditions of osseous diseases has been completely cured by the usage of hot springs. Especially the dissolved minerals like sulphur, carbonates and other trace minerals are effective against high blood pressure related diseases and also can provide certain relief to cardiovascular ailments and arteriosclerosis (Korea 2005).

But even the cases of spread of infectious diseases mainly the Legionnaire's disease outbreak is a certain possibility which has had been proved to be fatal at many hot water spring bath ponds. *Legionella* causing pneumonia is a certain disease which can surface on bathing in the recirculated water through the water particles. The worst case was a Legionnaire's disease outbreak at a spa in Hyuga, Miyazaki Prefecture, in 2002 which led to seven deaths and possibly infected about 300 people.

The temperature of rocks within the earth increases with increase in the depth and if water passes through these rocks it also get heated, i.e., because of geothermal heat and when it comes out, creation of hot springs takes place where the temperature of water is usually higher than body temperature. In non-volcanic areas this is the usual process for creation of hot springs whereas in volcanic areas it leads to formation of geysers. These hot springs usually have high mineral content like Calcium, Lithium, Sulfur, Fluorides, Carbonic acid; Hydrogen sulfide etc. as heated water have more capacity to dissolve solids. Due to high content of these kinds of salts and minerals and warm temperature of water, hot springs are quite often serve as centers for social medicine, religious and faith healing centers and thus popular destinations for tourists (Gitelson and Crompton 1984; Sheldon and Fox 1988; Crompton 1992; Connell 2005; Lee and King 2008; Das 2012; Sherpa 2012). Japan has more than 3000 hot springs which are believed to cure or at least lessen neuralgia, myalgia, rheumatism, dermatosis, high blood pressure and arteriosclerosis and therefore act as centers for medical tourism (Nakata 2008).

As mentioned earlier that these ponds are not only centers for social medicine but also for religious beliefs and tourism. More than 20,000 people visit these three hot springs annually, however no studies have been conducted to find out which group of people in terms of their nationality and religion visit these hot springs and for which purpose they visit these hot springs. Also, many people's livelihood is dependent on these hot springs and therefore it requires attention to find out the ways to uplift the social status of these groups of people. It will be also interesting to scientifically check if these hot springs have capacities to cure diseases and if yes, how these hot springs can be really developed as social centers for medical tourism and faith healing centers.

The higher temperature of the water, minerals dissolved within it, mud, sulfur precipitates and the geographical positioning influencing the climate of the area, regulates the curative properties in balneology. The heated water influences the blood circulation and breath regulation. Due to the concentrated mineralized content, the absorption rate and discharge of ions are metabolically controlled through the permeable skin. The main minerals found in hot springs are like magnesia which provides anti – allergic effect on skin and bromine compounds which enhances the nervous system and it's functioning.

The mud is rich in organic compounds and the microbial composition. The high salt concentration in the water can act as physiotherapy for various forms of joint pains, rheumatic associated pain, and muscle spasms due to stress and strain developed or treatment after fractures in osseous matter. Low humidity and high evaporation contributes in accelerating the metabolic activities of our system. These climatic properties have a considerable effect on the non – specific treatment of certain diseases based on stimulation of cells and tissues while acting against antigens and defunct stimuli response (Korea 2005). The balneology is widely accepted and is obligatory component of medical practice and preventive medicine in European and South East Asian countries.

According to the present survey, 54% of the respondents were males and 46% were females who visited these hot springs. It was observed that among the three survey sites, at *Reshi* there were more number of female visitors as compared to the other two hot springs i.e., *Polok* and *Borong* hot springs. This may be due to comfortable location of *Reshi* hot spring. Secondly, the temperature of the *Reshi* hot spring water was nominal and could be easily tolerated. The temperature of the water of Reshi *Tatopani,* in the month of December was 57.1°C, whereas in March it was 48.9°C and in April it was around 39.2°C. The atmospheric temperature of the Reshi in December was 27.5°C, in March 35.4°C and in the month of April it was around 33.8°C. Thirdly, a very sacred and religious natural cave is present at the site, *Khado Sangphug* (meaning "The sacred cave of

Dakinis"), which adds to the religious quotient of the place increasing its healing powers (Das 2012).

Since many people visit these hot springs due to religious reasons, therefore, their religious backgrounds were also studied. It was found that 47.33% of the respondents at these hot springs were Hindus whereas 40.33% were Buddhist. Around 13% were from other religions that came as tourists to these hot springs. There were more Hindu visitors at Reshi and Borong compared to Polok as they were visited by many tourists, most of them Hindus and also the people inhabiting the vicinity of the Hot springs were Hindus.

Since these hot springs are popular tourist hot spots, the nationality of these visitors was also studied. It was found that the majority of respondents were from Nepal (47.33%). The strong religious belief and faith on hot spring as sources of social medicine encouraged many patients and weak people to visit these hot springs. These patients were carried by people of the Sherpa tribes on a bamboo carrier. Some of them have been coming here from ages and for some it is a traditional custom to bathe at these hot springs and get relieved of their diseases and sins. Among the 40% Indian respondents majority of the people were from Sikkim and others were from Darjeeling, Kurseong, Kalimpong, Mirik – North hills of West Bengal.

As per the profession wise distribution we found that 31.33% respondents were farmers, 17.33% were house wives and11.33% were women vendors both from India and Nepal. There were also 10.66% students from schools who came to enjoy the hot water bath. According to one of the respondent's statement this is a "natural hot spa" and also some followed it as a traditional custom because it was practiced by their ancestors. When asked the students about their source of knowledge, most of the respondents said that it was from their family members and now they themselves have started to follow their elders. Thus, in a way, the traditional knowledge was passed from generation to generation.

As mentioned earlier that the strong religious beliefs and faith on hot spring as sources of social medicine encouraged many people to visit these hot springs. We further checked if the education level of respondents also played a role for them to visit these hot springs. We found that the majority of the respondents had very low level of education, i.e., majority of the respondents were having less that high school education or they had no formal education. Most of the visitors from Nepal were uneducated in comparison to their Indian counterparts where majority of them (at least 33%) having at least high school education. These data indicate that majority of people having poor education or strong social and religious faith on these hot springs visited these places. Further we checked the various reasons for which the visitors come to these places. We found that 47.33% came to get cured from their respective diseases, 31.33% respondents were indirect tourists as they came only to bathe and some of them came just to visit these hot springs and not to take bath. 22% of the respondents came due to their religious beliefs and traditional customs which they have been following from ages, learned from their ancestors.

Among the diseased patients there were 36.98% respondents suffering from various forms of arthritis like osteoarthritis, rheumatoid arthritis, gout, psoriatic arthritis, osteoporosis and severe cases of fibromyalgia has also been found. 30.14% respondents were suffering from severe joint pain which were on the way to become arthritic. 19.18% respondents had bowel problems and were suffering from indigestion, constipation, low enzyme activity and poor digestion. Scleroderma and scabies patients were the majority ones among the 9.59% skin disease patients. There were 4.11% respondents who were suffering from paralysis. A paralysis patient visits the Polok *Tatopani* and has a regime of bathing in it for three months continuously for 5 years.

Other patients bathe in hot water for a period of two hours at a stretch followed by a break. People even used to cook their meals from the hot water of the spring. At Polok *Tatopani* many people drink water directly coming from the source which was seeping through the manually made bamboo pipe. Bacteriological tests were carried out on this water used as a drinking purpose and to our observation it was found that it was free from fecal contamination.

Apart from the Hot spring water, another material is used for medical treatment and i.e., *Khar*. Basically it is a kind of sulfur ball obtained from Hot springs. *Khar* is used as an antiseptic agent which is used in the cases of skin burn, skin rashes, skin allergies and other inflammatory forms of skin diseases.

According to the survey, it was observed that around 69% of the respondents visited the *Tatopani* due to its beneficial properties and traditional belief system. Apart from curing various diseases, it also helped in muscular stress relief and hyper blood circulation which leads to excessive perspiration.

Among the Polok *Tatopani*, Borong *Tatopani* and Reshi *Tatopani*, most spring goers have a usual practice of "Taking the waters". The people take water mostly from Polok source as it is very hot compared to the others and also the source is very clear dripping from the mountain. In the case of Borong and Reshi, people collect water early in the morning or late at night, after human activity has ceased and the solfataric mud has completely settled down. According to the respondents, this water is daily consumed in small amounts by the patients suffering from Gastro intestinal diseases and several cases of bowel problems and their constipation have been successively cured.

Undoubtedly, the *Tatopani* water is an effective and alternative medicine for the commoners. Their frequency and regular visit to the sites helped to come to this hypothesis. According to the literature studied, it found that in primeval period many patients or relatives related to diseased individuals travelled far across many places in search of hot water springs or river bodies or minerals springs for the assertive remedial and therapeutic properties (Goodrich and Uysal 1994; Hembry 1990). Hot water springs have been a part of medical tourism and favorite choice or destination for vacation and relaxing (Gitelson and Crompton 1984; Sheldon and Fox 1988; Crompton 1992; Connell 2005; Lee and King 2008).

Science and the advent of pharmacy have discovered many important drugs and medicines, yet individuals utilize the *Tatopani* for its beneficial purposes. In Sikkim, it is cohesively attached with traditions and community belief systems. While cleansing or dipping inside the *Tatopani* ponds, it has been observed that unlike other hot springs of various sources all around the world, it helps in improving blood circulation, bails out various body aches and pains, soothes the stressed and strained muscles, cures osseous matter related problems and is a good therapeutic agent for skin diseases (Korea 2005).

Reshi Tatopani showed the highest concentration of Calcium and Magnesium among others and that may be due to the geographic location and the type of rocks in contact with the source water. But, this definitely aids the body indirectly, helps in other problems like relief from muscular pains and other aches of the body. Thus, the therapeutic activity at 40°C of the Tatopani might play a crucial role in a beneficiary role. The fluoride content of these areas is higher and thus they can be helpful in chemotherapeutic treatment of fungal diseases and other skin inflammations.

Conclusion

Every nation has their cultural identity and behavioral pattern, which gets reflected in their traditional belief systems and practices. *Tatopani* or hot springs and its curative properties have been a part of traditional knowledge, it is like a cult for the local inhabitants of Sikkim and as well as people from the neighboring areas. *Tatopani* elixir is an antediluvian theory which has been an ancient source of traditional medicine in Sikkim. The credulity of *Tatopani* is all across the neighboring states and overseas.

The second most important part of these niches is their microbial content which is a colossal treasure. The search for this huge potential has begun and hoped that many novel and industrial microorganisms can be explored and future biotechnological industries can be built through it as thermozymes are of an immense importance. These are the future of next generation need. Hence, Sikkim in all aspects is a natural hub and through this type of study can be a leader in biotech hub of India.

These hot springs are rich in minerals like sulfur, fluoride, chloride, potassium, magnesium and boron. The high content of these chemicals or minerals as well the high temperature of these hot springs are responsible for its medicinal values and therefore acts as the centers for medical tourism. Sikkim has an outright advantage to implement and grow its balneotherapy industry. However, the mineralogy has to be completely deciphered to understand the other potential applications of the *Tatopani*.

The present study is preliminary in nature which provides the base line data and leaves tremendous scope for researchers and scientists from various fields to explore the various potential applications of these hot springs for medicinal values.

References

Aparicio M.P.G., Ballesteros M., Manzanares P., Ballesteros I., Gonzalez A. and Negro M.J. (2007) "Xylanase contribution to the efficiency of cellulose enzymatic hydrolysis of barley straw", Applied Biochemistry and Biotechnology. 137(140): 353–365.

Aragno M. (1992) "Aerobic, chemolithoautotrophic, thermophilic bacteria", In: J. K. Kristjansson (ed.) Thermophilic Bacteria, Boca Baton: CRC Press. pp. 77-103.

Barnard D., Casanueva, A., Tuffin M. and Cowan D. (2010) "Extremophiles in biofuel synthesis", Environmental Technology. 31(8-9): 871–888.

Bedford M. R., Morgan A. J., Fowler T., Clarkson K. A., Ward M. A., Collier K. D. and Larenas E. A. (2003) "Enzyme feed additive and animal feed including it", US Pat 6562340.

Belghith, H., Ellouz, C. S. and Gargouri, A. (2001) "Biostoning of denims by *Penicilllin occitanis*", Journal of Biotechnology. 89(6): 257-262.

Biely P. (2003) "Xylanolytic enzymes", In: Whitaker J.R., Voragen A.G.J. and Wong D. W.S. (Eds.), "Handbook of Food Enzymology", New York.

Brock T. (1994) "Yellowstone's hidden biological sources": *Life at high temperatures*: http://www.bact.wisc.edu/themicrobialworld/LAHT/b3

Brock T.D. (1978) Thermophilic microorganisms and life at high temperature, 1st edition, Springer-Verlag, New York. pp. 23-55.

Bronnenmeier K., Rucknagel K.P. and Staudenbauer W.L. (1991) "Purification and properties of a novel type of exo-1, 4-â-glucanase (Avicelase II) from the cellulolytic thermophile *Clostridium stercorarium*", Eur. J. Biochem. 200: 379–385.

Campbell L.L. and Pace B. (1968) "Physiology of growth at high temperature", Journal of Applied Bacteriology. 31: 24-35.

Connell J. (2005) "Medical tourism: sea, sun, sand and surgery", Tourism Management. 27(6): 1093-1100.

Correia M.A., Prates J.A., Bras J., Fontes C.M., Newman J.A., Lewis R.J., Gilbert H.J. and Flint J.E. (2008) "Crystal structure of a cellulosomal family 3 carbohydrate esterase from *Clostridium thermocellum* provides insights into the mechanism of substrate recognition", Journal of Molecular Biology. 379: 64-72.

Cowan D. (1996) "Industrial enzyme technology", Trends Biotechnol, 14(6): 177-178.

Crompton J.L. (1992) "Structure of Vacation Destination Choice Sets." Annals of Tourism Research. 19: 420-34.

Das S. (2012) "Physicochemical Analysis, Isolation and Characterization of The Thermophilic Microorganisms from The Hot Water Springs of Sikkim (Polok *Tatopani*, and Borong *Tatopani*)." Master's Thesis, Sikkim Universit, Gangtok, India.

Das S. (2015) "Physicochemical and Microbiological Analysis of Dzongu *Tatopani* – The Hot Spring of Sikkim," M.Phil. Thesis, Sikkim University, Gangtok, India.

Dodd D. and Cann I.K.O. (2009) "Enzymatic deconstruction of xylan for biofuel production", G.C.B. Bioenergy. 1: 2-17.

Franks N.E., Bazewicz S.E. and Holm H.C. (1996) "Use of monocomponent cellulose for removing inks, coating, and toners from printed paper", US Pat 5525193.

Galante Y.M. and Formantici C. (2003) "Enzyme Applications in detergents and in manufacturing industries", Current Organic Chemistry. 7: 1399-1422.

Gandhi N., Sawant S. and Joshi J. (1995) "Studies on the lipozyme catalyzed synthesis of butyl laurate", Biotechnol. Bioeng 46: 1-12.

Giordano A., Andreotti G, Tramice A. and Trincone K.M. (2006) "Marine glycosyl hydrolases in the hydrolysis and synthesis of oligosaccharides", J. Biotechnol. 1: 511–530.

Gitelson R. J., and Crompton J. L. (1984) "Insights into the Repeat Vacation Phenomenon", Annals of Tourism Research. 11: 199-217.

Goodrich J.N. and Uysal M. (1994) "Health tourism: A new positioning strategy for tourist destinations." Journal of International Consumer Marketing. 6(3): 227-238.

Graham H. and Balnave D. (1995) "Dietary enzymes for increasing energy availability", In: Wallace R.J. and Chesson A. (Eds.) Biotechnology in animal feeds and animal feedings, Weinheim, Germany, VHC pp. 296–309.

Hembry P.M. (1990) The English Spa. A Social History. Athlone Press, London pp. 1560-1815.

Hough W.D. and Danson J.M. (1999) "Extremozymes", Current Opinion in Chemical Biology 3: 39–46.

Kim H., Sung M., Kim M. and Oh T. (1994) "Occurrence of thermostable lipase in thermophilic *Bacillus sp.* Strain 398", Biosci. Biotech.Biochem 58: 961-962.

Kim I., Kim H., Lee K., Chung S. and Ko S. (2002) "Lipase-catalyzed acidolysis of perilla oil with caprylic acid to produce structured lipids", Journal of the American Oil Chemists' Society. 79: 363-367.

Klemm D., Schmauder H. P. and Heinze T. (2002) Biopolymers, ed. Vandamme, E., de Beats, S. and Steinb-chel, A., Weinheim, Wiley-VCH. 6: 290-292.

Ladenstein R. and Antranikian G. (1998) "Proteins from hyperthermophiles: stability and enzymatic catalysis close to the boiling point of water", Advances in Biochemical Engineering/Biotechnology 61: 37-85.

Linko Y., Lamsa M., Huhatala A. and Linko P. (1994) "Lipase catalysed transesteriûcation of rape seed oil snf-2-ethyl-1-hexanol", Journal of the American Oil Chemists' Society 71: 1411-1414.

Luisa M.R., Schmidt C., Wahl S., Sprauer A. and Schmid R. (1997) "Thermo-alkalophilic lipase of *Bacillus thermocatenulatus*. Large scale production, puriûcation and properties: aggregation behaviour and its eûect on activity". J. Biotechnol, 56: 89-102.

Madigan M.T. and Marris B.L. (2007) "Extremophiles", Scientific American. 276(4): 82-87.

Mahamuni K. and Kulkarni H. (2011) "Hydrogeological action research for spring recharge and development and hill-top lake restoration in parts of Southern District, State of Sikkim, India. Rural Management and Development Department, Govt. of Sikkim", Advanced Center for Water Resources Development and Management (ACWADAM Report), Pune, India.

Marcus S.E., Verhertbruggen Y., Herve C., Ordaz-Ortiz J.J., Farkas V., Pedersen H.L., Willats W.G. and Knox J.P. (2008) "Pectic homo galacturonan masks abundant sets of xyloglucan epitopes in plant cell walls", BMC Plant Biology. 8: 60.

Mozersky S., Marmer W., Dale A.O. (2002) "Vigorous proteolysis: Relining in the presence of an alkaline protease and bating (Post Liming) with an extremophile protease", JALCA. 97: 150-155.

Mueller R., Antranikian G., Maloney S. and Sharp R. (1998) "Thermophilic degradation of environmental pollutants", Advances in Biochemical Engineering / Biotechnology 61: 155-169.

Najar I.N. (2013) "*Microbiological and physicochemical study of Yumthang Hot Spring (Tatopani) – Sikkim*", M. Phil. Thesis, Sikkim University, Gangtok, India.

Nakata H. (2008) "Japan's Hot Springs Part of Social, Geologic, Historic Fabric." The Japan Times Online Issue: January 28.

Pajunen E. (1986) "Optimal use of β-glucanases in wort production", In: EBC-Symposium on Wort Production, Monograph XI, Maffliers, France. pp. 137-148.

Panday A., Selvakumar P., Soccol C.R. and Nigam P. (1999) "Solid state fermentation for the production of industrial enzymes", Curr. Sci. 77: 149-162.

Korea P. (2005) Hot Springs: the greatest health and leisure activity for winter, pp. 32-35.

Poonam N. and Dalel S. (1995) "Enzyme and microbial systems involved in starch processing", Enzyme Microbiology Technology. 17: 770-778.

Reysenbach A.L. and Shock E. (2002) "Merging Genomes with Geochemistry in Hydrothermal Ecosystems", Science. 296: 1077-1082.

Rothschild L.J. and Manicinelli R.L. (2001) "Life in extreme environments", Nature. 409: 1092-1101.

Ruttersmith L.D. and Daniel R.M. (1991) "Thermostable cellobiohydrolase from the thermophilic eubacterium *Thermotoga sp.* strain FjSS3-B1", Biochemical Journal. 277: 887–890.

Satyanarayana T., Raghukumar C. and Shivaji S. (2005) "Extremophilic microbes: Diversity and perspectives", Curr. Science. 89(1): 78-90.

Sheldon P. and Fox M. (1988) "The Role of Food service in Vacation Choice and Experience: A Cross-Cultural Analysis." Journal of Travel Research, 27(3): 9-15.

Sherpa M.T. (2012) "Physicochemical and Microbiological Analysis of Reshi (*Tatopani*)" Master's Thesis, Sikkim University, Unpublished.

Stetter K. (1999) "Extremophiles and their adaptation to hot environments", FEBS Letters. 452: 22–25.

Stetter K.O. (1996) "Hyperthermophilic prokaryotes", FEMS Microbiology Reviews, 18: 149-158.

Subba J.R. (2009) "Indigenous Knowledge on Bio-Resources Management for Livelihood of the People of Sikkim," Indian Journal of Traditional Knowledge, 8(1): 56-64.

Vulfson E. (1994) "Industrial applications of lipases", In: Lipases: Their Structure, Biochemistry, and Application, ed. Wooley, P. and Peterson, S.B, Cambridge, UK: Cambridge University Press. pp. 271-288.

Wu X., Jaaskelainen S. and Linko Y. (1996) "An investigation of crude lipase for hydrolysis, esteriûcation, and transesteriûcation", Enzyme and Microbial Technology 19: 226-231.

Xia L. and Cen P. (1999) "Cellulase production by solid state fermentation on lignocellulosic waste from the xylose industry", Process Biochem. 34: 909-912.

8

Medicinal Plants and Their Conservation: An Indian Perspective

Mithilesh Singh[a]* and Shailendra Pradhan[b]
[a]G.B. Pant Institute of Himalayan Environment and Development, Sikkim Unit, Pangthang, Gangtok, Sikkim-737101, India
[b]Department of Dravyaguna, Uttarakhand Ayurved University, Gurukul Campus, Haridwar, Uttarakhand-249404, India
*Email: singmithilesh@gmail.com

Abstract

Since the beginning of the civilization, medicinal plants have been used to prevent and treat various ailments. At present, most of the pharmaceutical/ nutraceutical industries highly depend on wild medicinal plants for the supply of raw materials for extraction of bioactive compounds. As medicinal plants receive increased scientific and commercial attention, there is increasing pressure on the wild medicinal plant populations. Illegal and indiscriminate collection of medicinal plants, lack of proper cultivation practices and destruction of plant habitats have placed many medicinal species at risk of extinction. Disappearance of medicinal plants has come to the forefront as one of the important issues that needs immediate attention. There is an urgent need to look into possibilities of conserving high value medicinal plants. In this context, biotechnological interventions namely, *in vitro* plant propagation, short-to-medium term storage, cryopreservation and *in vitro* metabolite production play a significant role. The present chapter discusses state-of-the-art of medicinal plant conservation through biotechnological interventions, stressing the most recent advances, limitations and prospects. It is aimed to provide an experience backdrop to researchers who intend to work on medicinal plant conservation.

Keywords : Medicinal Plant, Conservation, Tissue culture, Secondary metabolite

Introduction

Medicinal plants have played a crucial role in combating multiple and complex health problems since time immemorial. At present, about 80% population of the developing countries relies predominantly on plants for their basic health related needs (Vines 2004). According to World Health Organization, 11% of the 252 drugs which are considered as indispensable for human health are mainly derived from ûowering plants. In spite of the considerable progress in synthetic chemistry, nearly 40% drugs of modern pharmacopeias are derived from plant. The global demand of medicinal plants for herbal medicine, nutraceuticals and cosmetics is not only large, but increasing exponentially in both developing and developed countries due to the fact that they are natural products, generally have no side effects and sometimes only source of healthcare available to rural people (Acharya and Acharya 2009). At present, only 10% of medicinal plant species are cultivated, and the remaining 90% of the species are harvested from wild populations (Julsing *et al.* 2007). Many pharmaceuticals are produced solely from massive quantities of whole plant parts. Excessive and unplanned destructive collection of plant material can diminish local plant populations and erode genetic diversity. Besides, habitat destruction, population growth, global warming and urbanization also cause great loss of medicinal plant diversity. Confronted by such unprecedented disappearance of species, conservation of medicinal plants assumes paramount urgency.

In situ conservation practices through biosphere reserves, sanctuaries and national parks can preserve medicinal plants, but this strategy alone do not ensure sufficient supply of medicinal plants for commercial use. Moreover, the conventional methods of plant propagation are very lengthy and time consuming. The long cultivation periods between planting and harvesting make selection of high yielding strains difficult which in turn results into high cost of the drugs. Besides, the wild populations are circumvented with the problems of disease, drought, environmental fluctuations, low rate of fruit set, poor seed yield/germination/viability, and in most cases the clonal uniformity cannot be maintained through seeds due to high heterozygosity in populations. Therefore, there is an urgent need of alternative and complimentary methods for medicinal plants conservation. In this context, advantages of biotechnological techniques have been widely recognized for preservation of elite germplasm and for pilot scale production of high value, low volume medicinal plants. The technique is evolving and may provide better systems of plant propagation and conservation when compared to conventional methods. In the following sections, various techniques for medicinal plant conservation along with various challenges being faced are discussed. We attempted to provide the detailed in-depth information on status and prospects of Indian medicinal plant conservation through biotechnological interventions.

Scenario of Indian Medicinal Plants

India, ranked sixth among the 12 mega biodiversity centers of the world, has been considered as a treasure house of valuable plants (Kala and Sajwan 2007). Of the 34 globally identified biodiversity hotspots, India harbours four hotspots, i.e., Himalaya, Indo-Burma, Western Ghats and Sri Lanka and Sundaland. This is mainly because of diverse edaphic, climatic and topographic conditions that are met within India. The biogeographic position of India is very unique that has resulted in a wide range of ecosystems and habitat such as forests, grasslands, wetlands, deserts, coastal and marine ecosystems. India is rich in all the three levels of biodiversity viz. species diversity, genetic diversity, and habitat diversity (Mukherjee and Wahile 2006). India has 426 biomes representing different habitat diversities that gave rise to one of the richest centers for plant genetic resources in the world (Pushpangadan and Narayanan Nair 2001).

In India, around 45,500 plant species are present that represent about 11% of the World's known biodiversity, and thus ranks tenth in the world and fourth in Asia (Khandekar and Srivastava 2014). Concerning the total number of plant species, 18,665 flowering plants are present, which represent approximately 8% of the world known flowering plants and belong to 2984 genera and 247 families (Verma *et al.* 2012). Out of 18,665 flowering plants, the Indian classic systems of medicines like Ayurveda, Siddha, Homeopath and Unani make use of about 8000 plants in various formulations. However, the survey and documentation of medicinal plants in the country is itself incomplete and so far only 3000 plants have been scientifically documented (Patwardhan *et al.* 2004).

International market of medicinal plant is over $ 70 billion per year, which is growing at the rate of 7% per year, and according to an estimate of

WHO, the demand for medicinal plants is expected to grow to $ 5 trillion by the year 2050 (Verma *et al.* 2012). In India, 7800 existing drug manufacturing units consume about 2000 tons of medicinal herbs annually, and about 2000 drugs in use are of plant origin. Moreover, at present, India is exporting herbal materials and herbal medicine to the tune of approximately US $ 128 million (Scindia 2010). China and India are two great producers of medicinal plants having more than 40% of global biodiversity. China, besides fulfilling its domestic demand, is earning $5 billion per year from herbal trade. However, despite of being a hub of medicinal plants, India's contribution to the International market is comparatively very low.

Need for medicinal plant conservation

The medicinal plant resources are getting depleted at an alarming rate. Today, 90% of the medicinal plants consumed domestically and for export are collected from the wild, and only 70 out of around 700 species in the trade are obtained purely from cultivated sources (Soumya 2012). Ever-increasing demand of herbal products has put the valuable plant resources under great stress and brought many medicinal plants at the verge of extinction. In addition to this, other threats to the medicinal plants are deforestation, destructive harvesting because of the use of plant parts like root, stem, bark, wood and whole plant in case of herb, extensive industrialization, forest fire and climate change. It is estimated that in India about 246 plants species are threatened, a bulk of which are medicinal plants (IUCN 2011). Of these, seven species are already extinct and 44 are critically endangered (IUCN 2011). Thus there is an urgent need to conserve the wild populations of medicinal plant diversity.

Strategies for medicinal plant conservation

Conservation of medicinal plants has become a subject of primary importance considering a constant addition of a rare or endangered plant species in Red List of Threatened Species. There are mainly two strategies for medicinal plants conservation *i.e. in situ* conservation and *ex situ* conservation. *In situ* conservation focuses on maintenance of plant species in their natural environment where they have developed their unique characteristics through the protection of the environment itself. On the other hand *ex situ* approach involves plant conservation outside the native habitat and is generally used to protect populations in danger of extinction. *In situ* technique is idealistically the most appropriate method of plant conservation that allows plants to interact and co-evolve with other component of ecosystems. However, this conservation method has several limitations. This technique of plant conservation is time consuming because of the slow growth and multiplication process (Vinoth and Ravindhran 2013). Additionally, *in situ* conservation is costly to maintain and is highly vulnerable to natural calamities like forest fires, extreme climate conditions and destruction by diseases and disasters.

Ex situ conservation techniques can be used to complement *in situ* methods and, in some junctures, may be the only option for some species (Maunder *et al.* 1998; Ramsay *et al.* 2000). The foremost drawback of this method is that the *ex situ* conserved species may epitomize a small range of genetic variation than that occurs in the natural conditions. Approaches to *ex situ* conservation include methods like seed storage, field genebanks and botanical gardens. Among the different *ex situ* conservation techniques, seed storage is the most appropriate technique for long-term conservation of plant genetic resources. However, conventional seed storage approaches are not very satisfactory for a large number of important tree species which produce recalcitrant seeds that do not survive dehydration and lose viability in a short period of time (Vinoth and Ravindhran 2013). Field gene bank is another technique to conserve plant material but they have risk of destruction by natural calamities, pests and diseases. Hence, *in vitro* conservation through biotechnological interventions is the safest and efficient alternative for medicinal plant conservation. Biotechnology offers an opportunity to exploit plant cell, tissue or organ by growing them *in vitro* to get large number of plants and desired compounds. There are different biotechnological approaches that can be used for conservation of the critical genotype of medicinal plants. This include: (a) *in vitro* propagation, (b) short to medium term storage, (c) cryopreservation and (d) *in vitro* metabolite production which are discussed below.

In vitro propagation

In vitro propagation or micropropagation is a method of producing large number of progeny plants from different somatic cells, tissues or organs of plants under controlled *in vitro* conditions in a relatively short time compared to conventional propagation. This implies that *in vitro* plant production based on the verity that different plant parts are capable of regeneration into whole plants under favorable *in vitro* conditions. This characteristic of plant cells in which they retain a latent capacity to produce a whole plant is called totipotency (Verdeil *et al.* 2007). *In vitro* propagation is an efficient technique for plant conservation with several distinct advantages. Some of these are as follows: (1) plant material collection may occur at anytime independent of flowering period because seed is not always required for propagation, (2) disease free plants can be produced through meristem culture technique, (3) clonal plants can be produced where conservation of elite genotypes is required, (4) large scale multiplication of plants may occur at any time where stocks are required using various propagation techniques, (5) germination of difficult or immature seed or embryo may be enabled for breeding programs, and (6) delivery across the border may be safer, in terms of germplasm health status using *in vitro* cultures. Some more advantages of *in vitro* techniques include the fact that storage space requirements are greatly reduced compared with field storage. Tissue culture facilities can be established at any geographical site and cultures are not affected by environmental fluctuations such as temperature variation, insect, pests and pathogen (Bhojwani and Denis 1999; Shibli *et al.* 2006). Moreover, *in vitro* culture techniques are very useful for the species with reproductive problem and / or extremely reduced population.

In vitro propagation involves the following basic steps- Stage 0: the preparative stage, involve germplasm collection, Stage 1: the establishment stage, include the production of sterilized cultures, Stage 2: the multiplication stage, during this stage number of propagules is increased, Stage 3: the plantlet production- production of seedlings of sufficient size and quality for transfer to *ex vitro* conditions, and stage 4: establishment under *ex vitro* conditions, involving the acclimatization of plantlets to greenhouse conditions (Lynch 1999). This technique involves the use of different plant parts such as shoot tip, axillary bud, leaf disc, zygotic embryo etc. Following two *in vitro* techniques *viz.* organogenesis and somatic embryogenesis, a number of medicinal plants have been propagated, of which some examples from the work in India are given in Table 1.

In vitro propagation have a vital role within *ex situ* conservation strategies, including for endangered species, particularly where it is important to conserve specific genotypes or where normal propagules, such as recalcitrant seeds may not be suitable for long-term storage. Although standard *in vitro* propagation methods are, in general accessible, endangered species may have specific growth requirements and, thus, may need modified techniques for *in vitro* culture. Besides, the limited amount of plant material available from rare and endangered species poses major challenges in the application of *in vitro* techniques (Sarasan *et al.* 2006).

Table 1 : *In vitro* regeneration protocol standardized in some important medicinal plants of India.

Plant	Family	Explant used	References
Aconitum heterophyllum	Ranunculaceae	Shoot tips, leaf, petioles, seeds	Giri *et al.* 1993; Solanki and Siwach 2012
Adhatoda beddomei	Acanthaceae	Nodes	Sudha and Seeni 1994
Adhatoda vasica	Acanthaceae	Nodal segments, shoot tips, axillary buds	Rahman *et al.* 2004; Nath and Buragohain 2005; Soni *et al.* 2012
Aloe vera	Liliaceae	Shoot apex, inflorescence axis	Cavallini *et al.* 1993; Aggarwal and Barna 2004; Rathore *et al.* 2011
Angelica glauca	Apiaceae	Seedling explants (epicotyls, hypocotyls, and cotyledonary nodes)	Pandey *et al.* 2011
Arnebia euchroma	Boraginaceae	Intact leaves	Malik *et al.* 2010
Artemisia annua	Asteraceae	Nodal stem segment, leaf explants	Gulati *et al.* 1996; Ganesan and Paulsamy 2011
Atropa belladonna	Solanaceae	Various parts of plant, leaves, root segments, leaf midribs, petiols	Chaturvedi *et al.* 1982; Asha rani and Prasad 2013
Azadirachta indica	Meliaceae	Nodal stem segment	Sharma *et al.* 2002; Chaturvedi *et al.* 2004
Bacopa monniera	Scrophulariaceae	Leaf, internodes, nodal, stem segments	Ali *et al.*1996; Tiwari *et al.* 2000; 2001; 2006
Catharanthus roseus	Apocynacea	Shoot tip, axillary buds	Bajaj *et al.*1988; Kumar *et al.* 2013
Chlorophytum borivilianum	Liliaceae	Shoot bases	Purohit *et al.* 1994; Jain *et al.* 1997; Dave *et al.* 2003
Coleus forskohlii	Lamiaceae	Stem tips	Neelam *et al.* 1991; Dube *et al.* 2011

Contd.

Commiphora wightii	Burseraceae	Nodal explants, Stem cuttings	Barve and Mehta 1993; Kant *et al.* 2010
Coptis teeta	Ranunculaceae	Hypocotyls, axillary buds	Tandon and Rathore 1992; Tandon *et al.* 2007
Curculigo orchioides	Amaryllidaceae	Rhizomes, shoot buds	Neelam *et al.* 2007
Curcuma domestica	Zingiberacae	Shoot apexs, Rhizome bud	Balachandran *et al.* 1990; Salvi *et al.* 2002
Dactylorhiza hatagirea	Orchidaceae	Green pods, shoot buds, tuber and leaf segments	Vij *et al.* 1995; Dinesh Giri and Tamta 2012
Ephedra gerardiana	Ephedraceae	Nodal segments, internodes	Sharma *et al.* 2012
Gentiana kurroo	Gentianaceae	Shoot tip, nodes, leaves,petioles, roots	Sharma *et al.* 1993; Sharma *et al.* 2014
Glycyrrhiza glabra	Fabaceae	Shoot buds, Shoot tips and nodal stem segments	Shah and Dalal 1982;Thengane *et al.* 1998; Arya *et al.* 2009
Holostemma annulare	Asclepiadaceae	Shoot tips, nodes	Sudha *et al.* 1998
Hyoscyamus niger	Solanaceae	Hypocotyls, cotyledons,stem explants	Uranbey 2005
Kaempferia galanga	Zingiberaceae	Rhizomes	Fatima *et al.* 2000; Parida *et al.* 2010
Nardostachys jatamansi	Valerianaceae	Petioles	Mathur 1992
Picrorhiza kurroa	Scorphulariaceae	Leaf discs, cotyledonary nodes, shoot tips	Chandra *et al.* 2004; Sood and Chauhan 2009
Podophyllum hexandrum	Podophyllaceae	Zygotic embryos	Arumugam and Bhojwani 1990
Rauwolfia micrantha	Apocynaceae	Shoot tips, nodes, root segments	Sudha and Seeni 1996; Sudha and Seeni 2006
Rauwolfia serpentina	Apocynaceae	Shoot tips, nodes	Bhattacharyya *et al.* 2008
Rheum emodi	Polygonaceae	Shoot tips	Lal and Ahuja 1993
Saussurea obvallata	Asteraceae	Roots, hypocotyls, cotyledons, leaves	Joshi and Dhar 2003
Steculia urens	Sterculiaceae	*In vitro* germs	Purohit and Dave 1996
Swertia chirayita	Gentianaceae	Shoot tips, leaves,	Balaraju *et al.* 2009;

Contd.

		roots	Balaraju *et al.* 2011; Kumar and Chandra 2014
Taxus wallichiana	Taxaceae	Zygotic embryos	Datta *et al.* 2006; Datta and Jha 2008
Trichopus zeylanicus	Dioscoreaceae	Shoot tips of axenic seedlings	Krishnan *et al.* 1995
Valeriana wallichiana	Valerianaceae	Shoot tips, nodes	Mathur *et al.* 1988
Woodfordia fruticosa	Lythraceae	Shoot tips, node	Krishnan and seeni 1994
Zingiber officinale	Zingiberacae	Excised shoot buds, meristem culture	Nadgauda *et al.* 1978;Bhagyalakshmi and singh 1988

Short to medium term storage

This technique allows the storage of plant material from few months to 2–3 years by delaying periods of subculture. Short to medium term storage of *in vitro* materials can be achieved by following several strategies viz. (i) by decreasing mineral concentration, (ii) by reducing sugar concentration, (iii) changing type and/or concentration of plant growth regulators, and (iv) by incorporating some osmotically active compounds such as mannitol and sorbital, and by the use of growth retardants (Engelmann 2011). Along with this, changing the culture condition such as reducing the temperature, combined or not, with a decrease in light intensity or by keeping cultures in complete darkness also reduce growth rate and thus, favorable for *in vitro* plant conservation. Generally, low temperatures (0-5 °C) are used for conservation of cold tolerant species and high temperatures (15–20 °C) are used for tropical species which are cold-sensitive. The most popular combination of physical and chemical factors involves decrease of temperature, reduction of strength of nutrient, concentration of carbon source and the use of low light intensity (Holobiuc *et al.* 2007). In addition to this, other factors such as the type of explants, their physiological state, as well as culture vessels also influence the efficiency of slow growth storage, (Engelmann 1991). Sometimes normal *in vitro* culture conditions are also used for medium-term storage when dealing with species that have a natural slow growing habit. Instead, paraffin, mineral oil or liquid medium may be used to reduce the growth rate of the explants. *In vitro* cultures desiccation and/or encapsulation are other possible alternative of medium term storage (Engelmann 2012). Synthetic seeds, which are produced by encapsulating plant propagules such as shoot buds or somatic embryos in an artificial medium, facilitate medium-term conservation of plant species (Devi *et al.* 1998). In India, several plant species have been conserved *in vitro* at NBPGR, New Delhi by culturing them at room temperature (Mandal 2000).

The benefits of slow growth techniques are that they use the same fundamental amenities that are used for plant propagation and that the storage system are based on modifying the conditions previously established for multiplication. However, these techniques do not lessen the main problem related with the high costs of labour and space requirements of any micropropagation system, in addition to the risks of somaclonal variation for some species (Blakesley *et al.* 1996).

Cryopreservation

To avoid the genetic variations that may occur in long tissue cultures storage, cryopreservation has been developed. Plant cryopreservation involves the storage of plant tissues at very low temperature (-196 ºC) in such a way that the viability of stored tissues is retained following re-warming (Hamilton *et al.* 2009). Any plant material like cells, protoplasts, shoot apices, somatic embryos, seed or excised zygotic embryos can be used for cryopreservation. At present, it is the only available method to ensure the safe and cost efficient long-term conservation of plant. Cryopreservation is generally applied to species with recalcitrant seeds that cannot be preserved by any other means or storage of specific cultivars of vegetatively propagated plants (Kaczmarczyk *et al.* 2011). Another justification to use cryostorage is to conserve endangered plant species, particularly where seeds may be extremely insufficient or of lower quality and/or the species is threatened with upcoming extinction (Paunescu 2009). This preservation technique is currently at the forefront of conservation activities.

Cryopreservation of medicinal plants has several distinct advantages such as (i) at ultra low temperature metabolic processes and cell division of plant material cease and plant material can thus be stored without change or modification for an indefinite period of time, (ii) it decrease the risk of contamination, cost of preservation and cost of labour, (iii) very small space is needed for conservation of large number of plant materials. Moreover, this technique of plant conservation requires a very limited maintenance (Engelmann 2004; Rao 2004). Cryopreservation is today used worldwide as a safe backup of valuable medicinal plants. There has been great interest and progress in cryopreservation of Indian medicinal plants, and the number of species has increased substantially in the past two decades (Table 2).

Table 2: Medicinal plants of India for which cryopreservation protocol have been standardized.

Plant	Family	Preserved material	% Response after storage	Reference
Celastrus paniculatus	Celastracea	Zygotic embryos	65	Radha *et al.* 2010a
Crateva nurvala	Capparidaceae	Shoot tips	56.6	Sanayaima *et al.* 2006
Dendrobium chrysanthum	Orchidaceae	Protocorms	63.2	Mohanty *et al.* 2013
Dioscorea bulbifera	Dioscoreaceae	Embryogenic callus	67.8	Mandal and Sharma 2009
Dioscorea wallichi	Dioscoreaceae	Shoot tips	37	Mandal *et al.* 1996
Holostemma annulare	Asclepiadaceae	Shoot tips	54.2	Decruse and Seeni 2002; Decruse *et al.* 2004
Myristica malabarica	Myristicaceae	Zygotic embryos	60	Radha *et al.* 2006
Nothapodytes nimmoniana	Icacinaceae	Zygotic embryos	60	Radha *et al.* 2010b
Picrorhiza kurrora	Scorphulariaceae	Shoot tips	35	Sharma and Sharma 2003
Rauvolfia serpentin	Apocynaceae	Nodal segments	66	Avik and Sabita 2008
Zingiber officinale	Zingiberaceae	*In vitro* shoot tips	80	Yamuna *et al.* 2007

There are mainly two types of cryopreservation protocols: classical cryopreservation procedures, in which cooling is performed in the presence of ice; and the procedures based on vitrification, in which cooling normally takes place without ice formation (Engelmann 2000; Gonzalez-Arnao *et al.* 2008). Classical freezing procedures involve cryoprotection by using different cryoprotective solutions combined or not with pregrowth of material and followed by slow cooling (0.5–2.0 °C/min) to a determined prefreezing temperature (usually around "40 °C), rapid immersion of samples in liquid nitrogen, storage, rapid thawing and recovery. They are generally operationally complex, since they require the use of complicated and expensive programmable freezers (Cruz-Cruz *et al.* 2013). By contrast, the vitrification-based procedures involve cell dehydration prior to cooling by exposure of samples to highly concentrated cryoprotective media (usually called plant vitrification solutions, PVS) and/or by air desiccation. Vitrification as such is a physical procedure, defined as the transition of the liquid phase to an amorphous glassy solid at the glass transition temperature (Fahy *et al*, 1984). Therefore, the freeze-induced dehydration step specific of conventional procedures is eliminated and the slow freezing system is replaced by a rapid or ultra-rapid cooling process, characteristic of the vitrification-based protocols.

Organized plant tissues such as twigs, dormant buds, apical and axillary buds are generally preferred in cryopreservation of vegetatively propagated plants. Shoot tips (containing the apical meristem) are the most commonly used plant material for cryostorage. It is mainly due to the fact that the apical meristem is made up of small unvacuolated cells with small vascular system. The organised structure of apical meristems generally show direct organogenesis after re-warming, thus reduce the risk of somaclonal variation. Though unorganised callus tissue can also be cryostored, but the risk of genetic variations is higher when using callus. Besides above, cell cultures, somatic embryos, pollen or plant buds as well as recalcitrant and orthodox seeds can be used as explants in plant cryopreservation (Reed 2008).

Inspiring results in medicinal plants have been published in recent years which present extensive list of plant species whose embryos and or embryonic axes have been successfully cryopreserved (Engelmann *et al.* 1995). It is important to stress, however, that cryopreservation is not seen as a substitute for conventional *ex situ* approaches (Withers and Engelmann 1997).

In vitro secondary metabolite production

Medicinal plants virtues are mainly due to the presence of a wide array of secondary metabolites, which are used directly or indirectly for pharmaceutical production. Secondary metabolites do not participate in the vital metabolic processes of the living cells, but are involved in the interaction of the producing plant with its environment. Each plant species has its own specific set of secondary metabolites. The spectrum of chemical structures synthesized by the plant kingdom is broader than that of perhaps any other group of organisms (Rao and Ravishankar 2002).

Currently many secondary metabolites of medicinal values are solely obtained from the naturally grown whole plants. The continued overexploitation of plants for medicinal purpose has dragged certain species to fall in danger of getting threatened or extinct. Clearly the development of alternative and complimentary methods to whole plant extraction is an issue of great socioeconomic importance. In this perspective, plant cell culture technology shows promise for the large-scale production of high-value secondary metabolites of industrial importance. The production of pharmaceuticals by plant tissue culture offers a number of advantages such as:

- This technique of metabolite production is more reliable, simpler, and more predictable.
- Extraction of the metabolites from *in vitro* cultures can be rapid and efficient, as compared to extraction from complex whole plants.
- Interfering compounds that occur in the field-grown plant can be avoided in cell cultures.
- Tissue and cell cultures can yield a source of defined standard phytochemicals in large volumes.
- It is independent of geographical and seasonal variations and environmental factors.

All these factors make plant cell cultures an attractive alternative option for producing important and novel metabolites all the year round without destroying the source plants and thus improve the conservation status of medicinal plants.

Limitation and prospects of medicinal plant conservation

Several governmental and non-governmental agencies are focusing their attention on medicinal plant conservation. However, medicinal plant sector in India is not developed as it has been expected. The reason is multifarious and the key factors identified include:

1. Lack of proper documentation and data base of medicinal plants.
2. Most of the important medicinal plants are found in very extreme regions like at very high altitude region of Himalayan from where collection of seeds and plant material is very difficult.
3. Lack of proper estimation of quantity for sustainable harvesting of medicinal plants known to found in India.
4. Lack of proper initiative by developing infrastructure facilities for marketing, quality control of crude drugs, sustainable harvesting mechanism etc.
5. Lack of proper regulatory mechanism (in some states) to control the collection of medicinal plants from wild habitat.
6. Lack of awareness among the common people about the prospects of medicinal plants in India.
7. Lack of initiatives for commercial cultivation of medicinal plants.

Overcoming the above constrains in medicinal plant sector looks to be herculean task but is not impossible if proper approach is made in right directions. Authors are giving here some suggestions for medicinal plant conservation:

1. The Universities and the Research institute patronizing research work on medicinal plant should be encouraged to channelize their research towards conservation of medicinal plants.
2. A compulsory course on conservation of medicinal plant should be introduced at undergraduate level.
3. The government should be asked to give high priority to biotechnology based conservation of plant-genetic resources in their plan.
4. A close linkage between universities and research institution should be established for concerted efforts in conservation of medicinal plants.
5. Government funding for research and development in the area of medicinal plant conservation must be enhanced, particularly for the rare and endangered medicinal plant.
6. Some more Institutes focusing on medicinal plant conservation should be open.

7. Local people awareness towards medicinal plants importance and conservation should be increased by seminars and interactions,
8. More opportunity should be developed in medicinal plant research sector.

At the present situation, young scientists and academicians are more interested towards molecular biology work and interest towards basic science discipline like taxonomy, plant conservation, plant tissue culture is decreasing. Main reasons of this detraction are lack of good teacher who can develop interest for these subjects. In order to boost the medicinal plant research in the country, collective efforts of botanist, biotechnologist and science academies are needed to rejuvenate the subject through planned research and intensive training programme.

Conclusion

India, considered as a store house of medicinal plants, is losing its medicinal plant diversity at a very high rate due to indiscriminate and destructive collection of plant materials, population growth and habitat destruction. It is feared that unless immediate actions are taken most of the important medicinal plants will be lost or endangered. To prevent extinction of important medicinal plants of India, stride by scientist and government has been made but the situation is still out of control. The various conservation efforts need to be geared up to combat the existing challenges. Biotechnological techniques of plant conservation need to be harnessed to its full potential. Large scale production of medicinal plants through tissue culture can reduce the existing ever-growing burden on wild population of medicinal plants and thus will improve the conservation status of important medicinal plants.

References

Acharya K.P. and Acharya R. (2009) Ethnobotanical study of medicinal plants used by Tharu Community of Parroha V.D.C., Rupandehi District, Nepal. Sci. World 7: 80-84.

Aggarwal D. and Barna K.S. (2004) Tissue culture propagation of elite plant of *Aloe vera* Linn. J. Plant Biochem. Biotech. 13: 77–79.

Ali G., Purohit M., Mughal M.H., Iqbal M. and Srivastava P.S. (1996) A rapid protocol for micropropagation of *Bacopa monniera* (L) Wettst., a medicinal plant. Plant Tissue Cult. Biotech. 2: 208-211.

Arumugam N. and Bhojwani S.S. (1990) Somatic embryogenesis in tissue cultures of *Podophyllum hexandrum*. Canadian J. Bot. 68: 487-491.

Arya S., Rathi N. and Arya I.D. (2009) Micropropagation protocol for *Glycyrrhiza glabra* L. Phytomorphology 59: 71-76.

Asha Rani N.S. and Prasad M.P. (2013) Studies on the organogenesis of *Atropa belladonna* in *in-vitro* conditions. Int. J. Biotechnol. Bioeng. Res. 4: 457-464.

Avik R. and Sabita B. (2008) Cryopreservation of *in vitro* grown nodal segments of *Rauvolfia serpentina* by pvs2 vitrification. CryoLetters 29: 321-328.

Bajaj Y.P.S., Furmanowa M. and Olszowska (1988) Biotechnology of the micropropagation of medicinal and aromatic plants. In: Bajaj Y.P.S. (ed.) Biotechnology in agriculture and forestry, Vol. 4, Medicinal and Aromatic Plants I, Springer-Verlag, Berlin, pp. 60-103.

Balachandran S.M., Bhat S.R. and Chandel K.P.S. (1990) *In vitro* clonal multiplication of turmeric (*Curcuma* spp.) and ginger (*Zingiber officinale* Rosc.). Plant Cell Rep. 8: 521-524.

Balaraju K., Agasthan P. and Lguacinauthu S. (2009) Micropropagation of *Swertia chirata* Buch-Hames. Ex Wall: a critically endangered medicinal herb. Acta Physiol. Plant 31: 487-494.

Balaraju K., Saravanan S., Agastian P. and Ignacimuthu S. (2011) A rapid system for micropropagation of *Swertia chirata* Buch.- Ham. ex Wall.: an endangered medicinal herb via direct somatic embryogenesis. Acta Physiol. Plant. 33: 1123–1133.

Barve D.M., and Mehta A.R. (1993) Clonal propagation of mature elite trees of *Commiphora wightii,* Plant Cell Tissue Organ Cult. 35: 237-244.

Bhagyalakshmi and Singh N.S. (1988) Meristem culture and micropropagation of a variety of ginger (*Zingiber officinale* Rosc.) with a high yield of oleoresin. J. Hort. Sci. 63: 321-327.

Bhattacharyya R., Arif M., Gaur A.K. and Rao P.B. (2008) *Rauwolfia serpentina*: protocol optimization for *in vitro* propagation. Afr. J. Biotechnol. 7(23): 4265-4268.

Bhojwani S.S. and Dennis T.T. (1999) *In vitro* conservation of plant genetic resources. Botanica 49: 47–52.

Blakesley D., Pask N., Henshaw G.G. and Fay M.F. (1996) Biotechnology and the conservation of forest genetic resources: *In vitro* strategies and cryopreservation. Plant Growth Regul. 20: 11–16.

Cavallini A., Natali L., Cionini G., Sassoli O. and Castorena Sanchez I. (1993) *In vitro* culture of *Aloe barbadensis* Mill.: quantitative DNA variations in regenerated plants. Plant Sci. 91: 223- 229.

Chandra B., Palni L.M.S. and Nandi S.K. (2004) Micropropagation of *Picrorhiza kurroa* Royle ex Benth., an endangered alpine herb, using cotyledonary node and shoot tip explants. Phytomorphology 54: 303-316.

Chaturvedi H.C., Sharma M. and Prasad R.N. (1982) Morphogenesis, micropropagation and germplasm preservation of some economic medicinal plants by tissue culture. In: Plant tissue culture, (ed. Fugiwara, A), Maruzen, Tokyo, Japan, pp. 687-688.

Chaturvedi R., Razdan M.K. and Bhojwani S.S. (2004) *In vitro* clonal propagation of an adult tree of neem (*Azadirachta indica* A. Juss.) by forced axillary branching. Plant Sci. 166: 501-506.

Cruz-Cruz C.A., González-Arnao M.T. and Engelmann (2013) Biotechnology and conservation of plant biodiversity. Resources 2: 73-95.

Datta M.M. and Jha S. (2008) Plant regeneration through somatic embryogenesis in *Taxus wallichiana.* J. Plant Biochem. Biotechnol. 17: 37-44.

Datta M.M., Majumdar A. and Jha S. (2006) Organogenesis and plant regeneration in *Taxus wallichiana* (Zucc.). Plant Cell Rep. 25: 11-18.

Dave A., Bilochi G. and Purohit S.D. (2003) Scaling-up production and field performance of micropropagated medicinal herb 'Safed Musli' (*Chlorophytum borivilianum*), *In vitro* Cell Dev. Biol-Plant 39: 419-424.

Decruse S.W. and Seeni S. (2002) Ammonium nitrate in the culture medium influences regeneration potential of cryopreserved shoot tips of *Holostemma annulare*. CryoLetters 23: 55-60.

Decruse S.W., Seeni S. and Nair G.M. (2004) Preparative procedure and culture media effect on the success of cryostorage of *Holostemma annulare* shoot tips. Plant Cell Tissue Organ Cult. 76: 179-182.

Devi J., Ray B.K., Chetia S. and Deka P.C. (1998) Regeneration of low temperature stored encapsulated protocorms of orchids. J. Orchid Soc. India 12: 39-41.

Giri D. and Tamta S (2012) Propagation and conservation of *Dactylorhiza hatagirea* (D. Don), an endangered alpine orchid. African J. Biotechnol. 11(62): 12586-12594.

Dube P., Gangopadhyay M., Dewanjee S. and Ali M.N. (2011) Establishment of a rapid multiplication of *Coleus forskohlii* Briq. and *in vitro* conservation by reduced growth. Indian J. Biotechnol. 10: 228-231.

Engelmann F. (1991) *In vitro* conservation of tropical plant germplasm-A review. Euphytica 57: 227–243.

Engelmann F. (2000) Importance of cryopreservation for the conservation of plant genetic resources. In: Cryopreservation of tropical plant germplasm: Current research progress and application, (eds. Engelmann F. and Takagi H.), IPGRI, Rome, Italy, pp. 8-20.

Engelmann F. (2004) Plant cryopreservation: Progress and prospects. *In vitro* Cell Dev. Biol. Plant 40: 427–433.

Engelmann F. (2011) Use of biotechnologies for the conservation of plant biodiversity. *In vitro* Cell Dev. Biol. Plant 47: 5-16.

Engelmann F. (2012) Germplasm collection, storage and conservation. In: Plant biotechnology and agriculture, (eds. Altman A, Hasegawa PM), Academic Press, Oxford, UK, pp. 255–268.

Engelmann F., Assy-Bah B., Bagniol S., Dumet D. and Michaux-Ferriere N. (1995) Cryopreservation of Date Palm, Oil Palm and Coconut. In: Biotecnology in agriculture and forestry, (ed. Bajaj Y.P.S.), Springer-Verlag: Berlin, Germany, pp. 148–167.

Fahy G.M., MacFarlane D.R., Angell C.A. and Meryman H.T. (1984) Vitrification as an approach to cryopreservation. Cryobiology 21(4): 407-426.

Fatima S., Kumar S. and Mishra Y. (2000) *In vitro* plantlet production system for *Kaempferia galanga*, a rare Indian medicinal herb. Plant Cell Tissue Organ Cult. 63: 193-197.

Ganesan C.M. and Paulsamy S. (2011) Standardized protocol for the *in vitro* culture of *Artemisia annua* L. – A medicinal plant at high altitudes of Nilgiris, the Western Ghats. J. Res. Biol. 3: 173-178.

Giri A., Ahuja P.S. and Kumar A. (1993) Somatic embryogenesis and plant regeneration from callus cultures of *Aconitum heterophyllum* Wall. Plant Cell Tissue Organ Cult. 32: 213-218.

Gonzalez-Arnao M.T., Panta A., Roca W.M., Escobar R.H. and Engelmann F. (2008) Development and large scale application of cryopreservation techniques for shoot and somatic embryo cultures of tropical crops. Plant Cell Tissue Organ Cult. 92: 1-13.

Gulati A., Bharel S., Abdin M.Z., Jain S.K. and Srivastava P.S. (1996) *In vitro* micropropagation and flowering in *Artemisia annua.* J. Plant Biochem. Biotechnol. 5: 31-32.

Hamilton K.N., Ashmore S.E. and Pritchard H.W. (2009) Thermal analysis and cryopreservation of seeds of Australian wild Citrus species (Rutaceae): *Citrus australasica, C. inodora* and *C. garrawayi.* CryoLetters 30: 268-279.

Holobiuc I., Paunescu A. and Blindu R. (2007) *Ex situ* conservation using *in vitro* methods in some Caryophyllaceae plant species from the Red List of vascular plants in Romania. Rom. J. Biol. Plant Biol. 49: 3-16.

IUCN (2011) The IUCN red list of threatened species. http://www.iucnredlist.org.

Jain S., Ramawat K.G. and Sonie K.C. (1997) Somatic embryogenesis in *Chlorophytum borivilianum-* A rare medicinal plant of Arawali Hills. In: Biotechnological applications of plant tissue and cell culture, (eds. Ravishankar G.A., Venkataraman L.V.), Oxford & IBH Publishing Co. New Delhi, pp. 199-203.

Joshi M. and Dhar U. (2003) *In vitro* propagation of *Saussurea obvallata* (DC) Edgew-an endangered ethno-religious medicinal herb of Himalaya. Plant Cell Rep. 21: 933-939.

Julsing K.M., Quax J.W. and Kayser O. (2007) The engineering of medicinal plants: prospects and limitations of medicinal plant biotechnology. In: Medicinal plant biotechnology, (eds. Oliver K. and Wim J. Quax), WILEY-VCH Verlag GmbH and Co. KGaA, Weinheim.

Kaczmarczyk A., Rokka V. M. and Keller E.R.J. (2011) Potato shoot tip cryopreservation, a review. Potato Res. 54: 45-79.

Kala C.P. and Sajwan B.S. (2007) Revitalizing Indian systems of herbal medicines by the national medicinal plant board through institutional networking and capacity building. Curr. Sci. 93: 797-806.

Kant T., Tomar U.K., Prajapati S. and Parmar A.K. (2010) *In vitro* propagation as a viable conservation strategy for *Commiphora wightii*, an endangered medicinally important desert tree, India. Conserv. Evidence 7: 94-99.

Khandekar V. and Srivastava A. (2014) Ecosystem biodiversity of India. In: Biodiversity-The dynamic balance of the planet, (ed. Grillo S), Intech Publisher, pp 233-274.

Krishnan P.N., Seeni S. (1994) Rapid micropropagation of *Woodforaia fruticosa* (L.) Kurz (Lythraceae), a medicinal plant. Plant Cell Rep. 14: 55-58.

Krishnan P.N., Sudha C.G. and Seeni S. (1995) Rapid propagation through shoot-tip culture of *Trichopus zeylanicus* Gaertn, a rare ethnomedicinal plant. Plant Cell Rep. 14:708–711

Kumar A., Prakash K., Sinha R.K. and Kumar N. (2013) *In vitro* plant propagation of *Catharanthus roseus* and assessment of genetic fidelity of micropropagated plants by RAPD marker assay. App. Biochem. Biotechnol. 169: 894-900.

Kumar V. and Chandra S. (2014) High frequency somatic embryogenesis and synthetic seed production of the endangered species *Swertia chirayita*. Biologia. 69: 186-192.

Lal N. and Ahuja P.S. (1993) Assessment of liquid culture procedures for *in vitro* propagation of *Rheum emodi*. Plant Cell Tissue Organ Cult. 34: 223-226.

Lynch P.T. (1999) Tissue culture techniques in *in vitro* plant conservation. In: Plant conservation biotechnology, (ed. Bensen EE), Taylor and Francis, London, pp 41-62.

Malik S., Sharma S., Sharma M. and Ahuja P.S. (2010) Direct shoot regeneration from intact leaves of *Arnebia euchroma* (Royle) Johnston using thidiazuron. Cell Biol. International 34: 537-542.

Mandal B.B. (2000) Cryopreservation research in India: Current status and future perspectives. In: Cryopreservation of tropical plant germplasm-current research progress and applications, (eds. Engelmann F. and Takagi H.), Japan International Research Center for Agriculture: Tsukuba, Japan; International Plant Genetic Resources Institute: Rome, Italy, pp 282–286.

Mandal B.B., Chandel K.P.S. and Dwivedi S. (1996) Cryopreservation of Yams (*Dioscorea* spp.) shoot apices by encapsulation-dehydration. Cryoletters 17: 165-174.

Mandal B.B. and Sharma D.S. (2009) Embryogenic cultures of *Dioscorea bulbifera* L. by encapsulation-dehydration. Cryoletters 30: 440-448.

Mathur J. (1992) *In vitro* morphogenesis in *Nardostachys jatamansi* DC: Shoot regeneration from callus derived roots. Ann. Bot. 70(5):419-422.

Mathur J., Ahuja P.S., Mathur A., Kukreja A.K. and Shah N.C. (1988) *In vitro* propagation of *Valeriana wallichii*. Planta Med. 54: 82-83.

Maunder M., Higgens S. and Culham A. (1998) Neither common nor garden: the garden as a refuge for threatened plant species. Curtis's Bot. Mag. 15: 124-132.

Mohanty P., Das M.C., Kumaria S. and Tandon P. (2013) Cryopreservation of pharmaceutically important orchid *Dendrobium chrysanthum* Wall. ex Lindl. using vitrification based method. Acta Physiol. Plant 35: 1373-1379.

Mukherjee P.K. and Wahile A. (2006) Integrated approaches towards drug development from Ayurveda and other Indian system of medicines. J. Ethnopharmacol. 103 (1): 25–35.

Nadgauda R.S., Mascarenhas A.F., Hendre R.R. and Jagannathan (1978) Rapid multiplication of turmeric (*Curcuma longa* Linn.) plants by tissue culture. Indian J. Exp. Biol. 16: 120-122.

Nath S. and Buragohain A.K. (2005) Micropropagation of *Adhatoda vasica* Nees–A woody medicinal plant by shoot tip culture. Indian J. Biotechnol. 4: 396-399.

Neelam S., Chandel K.P.S. and Srivastava V.K. (1991) *In vitro* propagation of *Coleus* forskohlii Briq., a threatened medicinal plant. Plant Cell Rep. 10: 67-70.

Neelam S., Vimala D.S. and Ruchira P. (2007) *In vitro* propagation of a threatened, anticarcinogenic herb *Curculigo orchioides* Gaertn. J. Plant Biochem. Biotechnol. 16: 63-65.

Pandey M., Dhar U., Samant S.S., Shirgurkar M.R. and Thengane S.R. (2011) Recurrent somatic embryogenesis and plant regeneration in *Angelica glauca* Edgew., a critically endangered medicinal plant of Western Himalaya. J. Horticult. Soc. Biotechnol. 86: 493-498.

Parida R., Mohanty S., Kuanar A. and Nayak S. (2010) Rapid multiplication and *in vitro* production of leaf biomass in *Kaempferia galanga* through tissue culture. Electr. J. Biotechnol. 13: 1-8.

Patwardhan B., Ashok D.B., Vaidya and Chorghade M. (2004). Ayurveda and natural products drugs discovery. Curr. Sci. 86(6): 789-799.

Paunescu A. (2009) Biotechnology for endangered plant conservation: A critical overview. Rom. Biotech. Letters 14(1): 4095-4103.

Purohit S.D. and Dave A. (1996) Micropropagation of *Sterculia urens* Roxb. An endangered tree species. Plant Cell Rep. 15: 704-706.

Purohit S.D., Dave A. and Kukda G. (1994) Micropropagation of Safed musli (*Chlorophytum borivilianum*), a rare Indian medicinal herb. Plant Cell Tissue Organ Cult. 39: 93-96.

Pushpangadan P. and Narayanan Nair K. (2001) Future of systematics and biodiversity research in India. Curr. Sci. 80 (5): 631–638.

Radha R.K., William Decruse S., Amy M.V. and Krishnan P.N. (2010a) Zygotic embryo cryopreservation of *Celastrus paniculatus*. In: Golden Jubilee National Symposium on Plant Diversity Utilization and Management, Department of Botany, University of Kerala, Kariavattom, India.

Radha R.K., William Decruse S. and Krishnan P.N. (2010b) Cryopreservation of excised embryonic axes of *Nothapodytes nimmoniana* (Graham) Mebberly, a vulnerable tree species of the Western Ghats. Indian J. Biotechnol. 9: 435-437.

Radha R.K., William Decruse S., Seeni S. and Ganeshan S. (2006) Cryopreservation of embryonic axes of recalcitrant seed species *Myristica malabarica* Lam., a rare medicinal plant of the Western Ghats. In: National Seminar on Plant Resource of Western Ghats. Indian Institute of Science, Bangalore, Karnataka, India.

Rahman S.M.M., Sen P.K., Afroz F. and Sultana K. (2004) *In vitro* propagation of *Adhatoda vasica* from shoot tip. Mol. Biol. Biotechnol. J. 2: 33-35.

Ramsay M.M., Jacskon A.D. and Porley R.D. (2000) A Pilot Study for *ex Situ* Conservation of UK Bryophytes. In: Proceedings of EuroGard 2000—II European Botanic Garden Congress, Las Palmas de Gran Canaria, Spain, 10–15 April 2000; Botanic Garden Conservation International: Richmond, UK, 2000; pp. 52–57.

Rao K.N. (2004) Plant genetic resources: Advancing conservation and use through biotechnology. African J. Biotechnol. 3: 136-145.

Rao R.S. and Ravishankar G.A. (2002) Plant cell cultures: Chemical factories of secondary metabolites. Biotechnol. Adv. 20: 101-153.

Rathore M.S., Chikara J. and Shekhawat N.S. (2011) Plantlet regeneration from callus cultures of selected genotype of *Aloe vera* L.—An ancient plant for modern herbal industries. Appl. Biochem. Biotechnol. 163: 860-868.

Reed B.M. (2008). Plant Cryopreservation: A practical guide. Springer, New York.

Salvi N.D., George L. and Eapen S. (2002) Micropropagation and field evaluation of micropropagated plants of turmeric. Plant Cell Tissue Organ Cult. 68: 143-151.

Sanayaima R.K., Kaur A., Agrawal A. and Babbar S.B. (2006) Cryopreservation of *in vitro*-grown shoot tips of *Crateva nurvala* Buch. Ham, an important medicinal tree. Cryoletters 27(6): 375-86.

Sarasan V., Cripps R., Ramsay M.M., Atherton C., McMichen M., Prendergast G. and Rowntree J.K. (2006) Conservation *in vitro* of threatened plants—Progress in the past decade. *In vitro* Cell Dev. Biol. Plant 42: 206–214.

Shah R.R. and Dalal K.J. (1982) *Glycyrrhiza glabra* (Liquorice): From test tube to field. In: Plant Tissue Culture, (ed. Fujiwara A.), Tokyo, pp. 685-686.

Sharma A., Kaur R. and Sharma N. (2014) *In vitro* morphogenic response of different explants of *Gentiana kurroo* Royle from Western Himalayas—an endangered medicinal plant. Physiol. Mol. Biol. Plant 20: 249-256.

Sharma A.K., Sharma M. and Chaturvedi H.C. (2002) Conservation of phytodiversity of *Azadirachta indica* A. Juss. through *in vitro* strategies. In: Role of Plant tissue culture in biodiversity conservation and economic development, (eds. Nandi S.K., Palni L.M.S. and Kumar A.), Gyanodaya Prakashan, Nainital, India, pp. 513-520.

Sharma N., Chandel K.P.S. and Paul A. (1993) *In vitro* propagation of *Gentiana Kurroo*-an indigenous threatened plant of medicinal importance. Plant Cell Tissue Organ Cult. 34: 307-309.

Sharma N. and Sharma B. (2003) Cryopreservation of shoot tips of *Picrorrhiza kurroa* Royle ex Benth., an indigenous endangered medicinal plant through vitrification. CryoLetters 24: 181-190.

Sharma V., Gupta S.K. and Dhiman M. (2012) Regeneration of plants from nodal and internodal segment cultures of *Ephedra gerardiana* using thidiazuron. Plant Tissue Cult. Biotech. 22: 153-161.

Shibli R.A., Shatnawi M.A., Subaih W.S. and Ajlouni M.M. (2006) *In vitro* conservation and cryopreservation of plant genetic resources: a review. World J. Agricult. Sci. 2: 372–382.

Fatima S., Kumar S. and Mishra Y. (2000) *In vitro* plantlet production system for *Kaempferia galanga*, a rare indian medicinal herb. Plant Cell Tissue Organ Cult. 63(3):193-197.

Solanki P. and Siwach P. (2012) Optimization of conditions for *in vitro* seed germination and shoot multiplication of *Aconitum heterophyllum* Wall. Int. J. Med. Arom. Plant 2: 481-487.

Soni P., Bahadur A.N. and Kanungo V.K. (2012) Study of *in-vitro* micropropagation of a medicinally important plant *Adhatoda Vasica* Nees by auxillary-bud culture. Indian J. Innovations Dev. 1: 199-202.

Sood H. and Chauhan R.S. (2009) Development of a low cost micropropagation technology for an endangered medicinal herb (*Picrorhiza kurroa*) of north-western Himalayas. J. Plant Sci. 4: 21-31.

Soumya K.V. (2012) *Nothapodytes nimmoniana*- A unique medicinal plant in red list. Res. Rev. Biomed. Biotechnol. 3: 1-4.

Sudha C.G., Krishnan P.N. and Pushpangadan P. (1998) *In vitro* propagation of *Holostemma annulare* (Roxb.) Schum, a rare medicinal plant. *In vitro* Cell Dev. Biol. Plant 33: 57-63.

Sudha C.G. and Seeni S. (1994) *In vitro* multiplication and field establishment of *Adhatoda beddomei* C.B. Clarke, a rare medicinal plant. Plant Cell Rep. 13: 203-207.

Sudha C.G. and Seeni S. (1996) *In vitro* propagation of *Rauvolfia micrantha*, a rare medicinal plant. Plant Cell Tissue Organ Cult. 44: 243-248.

Sudha C.G. and Seeni S. (2006) Spontaneous somatic embryogenesis on *in vitro* root segment cultures of *Rauvolfia micrantha* Hook. F.-A rare medicinal plant. *In vitro* Cell Dev. Biol.-Plant 42: 119-123.

Tandon P. and Rathore T.S. (1992) Regeneration of plantlets from hypocotyl derived callus of *Coptis teeta*. Plant Cell Tissue Organ Cult. 28: 115-117.

Tandon P., Rathore T.S. and Kumaria S. (2007) Micropropagation of *Coptis teeta* Wall.– Threatened medicinal plant of Arunachal Pradesh, India. Indian J. Biotechnol. 6: 280-282.

Thengane S.R., Kulkarni D.K. and Krishnamurthy K.V. (1998) Micropropagation of licorice (*Glycyrrhiza glabra* L.) through shoot tip and nodal cultures. *In vitro* Cell Dev. Biol.-Plant 34: 331-334.

Tiwari V., Tiwari K.N. and Singh B.D. (2000) Suitability of liquid cultures for *in vitro* multiplication of *Bacopa monniera* (L.) Wettst. Phytomorphology 50: 337-342.

Tiwari V., Tiwari K.N. and Singh B.D. (2001) Comparative studies of cytokinins on *in vitro* propagation of *Bacopa monniera*. Plant Cell Tissue Organ Cult. 66: 9-16.

Tiwari V., Tiwari K.N. and Singh B.D. (2006) Shoot bud regeneration from different explants of *Bacopa monniera* (L.) Wettst. by trimethoprim and bavisitin. Plant Cell Rep. 25: 629-635.

Uranbey S. (2005) Thidiazuron induced adventitious shoot regeneration in *Hyoscyamus niger*. Biol. Plant 49: 427-430.

Verdeil J-L., Alemanno L., Niemenak N. and Tranbarger T.J. (2007) Pluripotent versus totipotent plant stem cells: dependence versus autonomy. Trends Plant Sci. 12: 245–252.

Verma P., Mathur A.K., Jain S.P. and Mathur A. (2012) *In vitro* conservation of twenty-three overexploited medicinal plants belonging to the Indian sub-continent. Sci. World J. 1-10.

Vij S.P., Pathak P. and Mahant K.C. (1995) Green pod culture of a therapeutically important species *Dactylorhiza hatagirea* (D. Don) Soo. J. Orchid Soc. India 9: 7-12.

Vines G. (2004) Herbal harvests with a future: towards a sustainable source for medicinal plants. Plant life International. www.plantlife.org.uk.

Vinoth A. and Ravindhran R. (2013) *In vitro* propagation-a potential method for plant conservation. Int. J. Computing Algorithm 2: 268-272.

Withers L.A. and Engelmann F. (1997) *In vitro* conservation of plant genetic resources. In: Biotechnology in agriculture, (ed. Altman A), Marcel Dekker Inc. New York, pp 57-88.

Yamuna G., Sumathi V., Geetha S.P., Praveen K, Swapna N. and Nirmal B.K. (2007) Cryopreservation of *in vitro* grown shoot of Ginger (*Zingiber officinale* Rosc). CryoLetters.28: 241-252.

9

Algal Biofuel: A Sustainable Green Energy

Laxuman Sharma[1] and Akhilesh Kumar Singh[2*]
[1]Department of Horticulture, Sikkim University, 6th Mile, Gangtok 737102 India
[2]Amity Institute of Biotechnology, Amity University Uttar Pradesh, Lucknow Campus 226028, India
**E-mail: aksingh2@lko.amity.edu*

Abstract

Modern world is experiencing energy insecurity because of extreme population pressure and depleting energy resources. Fossil fuels though have highly applauded for contributing to today's development, are non renewable energy and debated for not sustainable. This has posed serious challenges to the scientific fraternity for finding the suitable alternatives. A cheap, eco-friendly, sustainable or renewable type of energy is visualized for development with added advantage of mitigating green house gases. Biofuel produced from different feedstock has been researched, trialed and commercialized with many advantages and disadvantages. Biofuel production from the oil producing crops like maize, soyabean, jatropha *etc.* growing in arable land are not considered ideal because it is not climate neutral and pose competition for food production. Biofuel production using non-food feedstock such as wood processing residues, forest residues, agricultural residues and non-edible constituents of beet, sugarcane, corn *etc.*, which do not directly involved in competition with human food chain, are also not feasible owing to requirement of costly and sophisticated technologies. Biofuel production using algal biomass as feedstock has attracted the researchers and policy makers having several advantages of rapid growth rate, high space efficiency, high productivity, sulphur free fuel, non toxic and biodegradability. Several strains of algal species like *Scenedesmus* sp., *Scenedesmus obliquus, Chlorella vulgaris*, *Chlamydomonas* sp., *Chlorococcum littorale* etc. are reported to produce high biomass with high lipid content, making them ideal candidate for biodiesel production. Bioethanol production from species like *Spirogyra, Chlorococcum humicola and Chlorococcum infusionum* have also been reported. In addition to the production of the oil, these species have high CO_2 sequestration ability and can be used for waste water treatment. Despite constant efforts for producing biofuels from microalgae, the economic production is still questionable. The cutting edge technology for improving cultivation techniques, growing methods, species selection and bio-engineering are still to be applied holistically for cost effective production and commercialization of algal biofuel in near future, as algal fuel is only alternative to the fossil fuels.

Keywords: Algae, Biodiesel, Bioenergy, Bioethanol, Biofuel, Biogas.

Introduction

The rapid civilization and ever expanding economy have posed mammoth scale energy demand worldwide. Fossil fuels, produced from carbon reserves of millions of years has now been debated for being non renewable and its increased used has threatened closed carbon cycle. Burning of fossils fuels is tremendously contributing to increased green house gases and climate change. The sustenance of environment, economy and the pace of development can only be achieved by shifting to the renewable source of energy. An alternative energy now sought for exploitation of biological agents to transform source of energy, such as sunlight and biomass, into fuels such as biodiesel, bioethanol etc. Biomass referred to entire cellular dry weight or organic substances manufactured by an organism generally from sunlight in conjugation with carbon dioxide. Biologically synthesized fuels or solid, liquid or gaseous fuels derived from organic matter are typically termed as 'Biofuels'. Now, biofuel have received much attention as alternative green renewable energy over conventional fossil fuel owing to severe environmental concerns, increasing crude oil price, apprehension about energy security and diminishing non-renewable fossil fuels reserves (Chhetri *et al.* 2008). Interestingly, biofuels are categorically a green renewable energy of the choice with positive characteristics *viz.* biodegradability, non-toxicity and lower releases of greenhouse gases. Currently, the exploitation of liquid biofuels *viz.* biodiesel, bioethanol and jet fuel has been intensified enormously particularly in the transport sector (Tabatabaei *et al.* 2011).

Biofuels and their Categories

Biofuels have evolved into an innovative area of research globally. These are broadly classified into three categories: (i) Natural biofuels, (ii) Primary biofuels and (iii) Secondary biofuels (Table 1). The natural biofuels are usually manufactured from organic sources, such as landfill gas, animal waste and vegetables. Primary biofuels involve fuel-woods, which are exploited primarily for heating, cooking, electricity production or brick furnace. On the other hand, the secondary biofuels composed of biodiesel and bioethanol that are manufactured by processing of biomass (Nigam and Singh 2011). Based on features *viz.* feedstock, processing technology and/ or development levels, the secondary biofuels are further categorized into three sub-classes, *i.e.* first, second and third generation biofuels (Table 1, Dragone *et al.* 2010). The first generation biofuels (bioethanol), which predominantly manufactured from food crops including oil seeds are restricted in their capability to attain their goals for biofuel synthesis owing to alarming socio-economic consequences (competition of biodiesel feedstock with food commodities), climate change mitigation and requirement of more arable agricultural lands (Chhetri *et al.* 2008; Alam *et.al.* 2010). These issues have intensified the attention towards manufacturing/ producing second generation biofuels by exploiting non-food feedstocks such as wood processing residues (e.g. saw dust), forest harvesting residues, agricultural residues (wood chips, straws, sheared branches, leaves etc.) and non-edible constituents of beet, sugarcane, corn etc., which do not directly involved in competition with human food chain (Alam *et al.* 2010). However, the productions of second generation biofuels are not feasible owing to requirement of costly and sophisticated technologies including special enzymes (Lee *et al.* 2014; Espinosa-Gonzalez *et al.* 2014). Land based higher plants are also seen as alternative biofuels, but they are debated for not being climate neutral and potentially compete with net food production (Scharfemann and Larance 2008).

Table 1 : Classification of biofuels with their production

Categories of Biofuels				
Natural Biofuels	**Primary Biofuels**	**Secondary Biofuels**		
		1st Generation biofuel	**2nd Generation biofuel**	**3rd Generation biofuel**
Manufactured from sources *viz.* vegetable, animal waste and landfill gas	Produced from sources such as firewood, forest, plants and wood chips	Bioethanol manufactured from used cooking oil, wheat, corn, beet, sugarcane, potato, oil seeds such as coconut, sunflower etc., used animal fat	Bioethanol and biodiesel manufactured from straw, grass, wood, jatropha, miscanthus and cassava.	Biodiesel manufactured from microalgae and microbes

Source: Alam *et. al.* 2012

Overcoming the shortfalls of first and second generation, the third generation biofuels offers pronounced feasibility (Dragone *et al.* 2010; Nigam and Singh 2011; Brennan and Owendue 2010; Li *et al.* 2008). It is anticipated that the feasibility towards biodiesel synthesis by exploiting microalgae is 15 to 300 times greater compared to conventional/ traditional crops on area basis (Dragone *et al.* 2010). Apart from this, microalgae have a very short harvesting cycle over conventional crop plants that facilitate continuous or multiple harvesting in conjunction with significant enhanced yields (Dragone *et al.* 2010; Scheuk *et al.* 2008). In general, the microalgae do not require good quality agricultural land and exhibit more productivity compared to land based plants, which is evident from the fact that several species have very short doubling times of just few hours and produce an enormous quantities of triacylglycerides (Reith *et al.* 2012). The production of biodiesel from algal biomass offers the added advantage of low cost CO_2 mitigation (Wang *et.al*. 2008).

Algae: Potential feedstock for biodiesel production

Algae exhibit autotrophic, heterotrophic, mixotrophic and photoheterotrophic mode of nutrition (types of metabolisms) and are proficient in metabolic shift as a result of alterations in the environmental conditions.

The autotrophic algal species are photosynthetic in nature and exploits substances such as salts, carbon dioxide as well as an energy source in the form of light for their growth and developments. However the algae exhibiting heterotrophic mode of nutrition, being non-photosynthetic exploit organic materials and nutrients as source of their energy, growth and developments (Brennan and Owende 2010). Mixotrophic growth of algae involve photosynthesis as the key source of energy, however both carbon dioxide and organic substances are important. Furthermore, subclass of mixotrophy called amphitrophy involves organisms, which are capable to grow either heterotrophically or autotrophically depending on the intensity of light accessible and amount of organic substance present in the environment. In photoheterotrophycally (also termed as photoorganitrophy, photometabolism, and photoassimilation), the light is needed to exploit organic substance as carbon substrate.

Microalgae are eukaryotic (green algae, red algae, diatoms and others) or prokaryotic (blue-green algae/ cyanobacteria) photosynthetic microscopic algae that generally grow in ponds/ water bodies typically ranging from unicellular to colonies and filaments composed of a few hundred cells (Benemann *et al.* 1996). Interestingly, Lee *et al.* (2014) reported that microalgal species not only possess high lipid contents as biofuel feedstock but also exhibit faster growth compared to macroalgae. In the mid-1980s, investigations were commenced to exploit the microalgae as feedstock for the synthesis of biofuel (Cohen *et al.* 1995).

Biodiesel is an amalgamation of monoalkyl esters of long chain fatty acids that can be acquired from biomass and diverse renewable lipid feed stocks. It can be directly utilized in different diesel engines (Clark and Deswarte 2008; Demirbas 2009). Amongst 50,000 of microalgae species, about 30,000 were studied and analyzed so far for the production of biofuel (Fionnuala *et al.* 2013). They are apposite and potential feedstock for the manufacture of biofuel because of their high carbon dioxide fixation, high photosynthetic growth rates, quite high lipid content, little land usage and capability to exploit waste streams/ municipal wastewater as culture medium (Mata *et al.* 2010). Furthermore, lipid contents from microalgae can be extracted as a feedstock for the production of biodiesel, whereas other products can be exploited to synthesize biogas and ethanol. The manufacturing of biofuel involves a complex process such as growth, harvest, extraction and conversion process that are outlined in Table 2. Various microalgae

Table 2 : Summary of biofuel production processes from microalgae

Overview of various stages from microalgae to biofuel production					
Stage 1 (Growth)	**Stage 2 (Harvest)**	**Stage 3 (Extraction)**		**Stage 4 (Conversion)**	
Cultivation of microalgae in the presence of light, nutrients, water and carbon dioxide	Algal biomass settlement and their removal physically (cells separation from the growth medium)	Cell disruption followed by lipid/ oil extraction extraction biomass exploiting chemical methods	Transformation of algal lipid/ oil for biodiesel synthesis by transesterification process (algal species with rich in lipid content are highly suitable for biodiesel production)	Starch hydrolysis, fermentation and distillation for bioethanol production (algae rich in carbohydrate contents are more appropriate for bioethanol production)	Combinations are also feasible, where the lipid/oil content is used for the production of biodiesel and residual biomass is subjected to fermentation for biogas production.

Source : Alam *et. al*. 2012

species can be triggered to produce significant amounts of lipid content. The average content of lipid ranges from 1 to 70% dry cell weight (dcw) is reported but under particular environment certain species can accumulate up to 90% of dcw (Chisti 2007; Li *et al.* 2008). *Scenedesmus* sp. strain JPCC GA0024 has been reported to accumulate 70% (dcw) lipid content using artificial seawater under photoautotrophic condition (Matsunaga *et al.* 2009). *Scenedesmus obliquus* CNW-N was on the other hand, reported to accumulate lipid content up to 38.9% of dcw, when grown in supplementation with carbon dioxide (Ho *et al.* 2010). Under photo-autotrophic condition, *Chlorella vulgaris* ESP-31 was found to synthesis lipid up to 55.9% of dcw (Yeh *et al.* 2011). Whey permeate was found to be a prospective feedstock for lipid synthesis, where *Chlorella protothecoides* was found to manufacture up to 49.9% (dcw) lipid content when grown in such whey permeate (Espinosa-Gonzalez *et al.* 2014). Domestic wastewater was exploited as a nutrient for growth and lipid accumulation up to 32.2% (dcw) by *Scenedesmus* sp. ZTY1 (Zhang *et al.* 2014). Zhang *et al.* (2013) reported 60.6% (dcw) lipid content in *Eustigmatos* cf. *polyphem* under photoautotrophic condition. Molasses and Corn steep liquor were also found to be the potential feedstock for lipid synthesis (Mohammad Mirzaie *et al.* 2015). In order to enhance the lipid content and growth of microalgae, many approaches have been exploited such as optimization of the composition of medium (carbon substrate, salts, phosphorus, vitamins, and nitrogen), physical factors (light intensity, temperature and pH) and mode of nutrition (phototrophic, mixotrophic, heterotrophic and photoheterotrophic growth). Table 3 summarizes several significant outcomes on the production of lipid contents by various microalgae species.

Table 3: Accumulation of lipid/oil content (as biodiesel feedstock) by diverse microalgae species

Microalgae species (Feedstock)	Substrate/ Condition	Lipid/Oil content (% dcw)	Reference
Scenedesmus sp. strain JPCC GA0024	Photoautotrophic condition + Artificial seawater	73.0	Matsunaga *et al.* (2009)
Scenedesmus obliquus CNW-N	Carbon dioxide	38.9	Ho *et al.* (2010)
Chlorella vulgaris ESP-31	Photoautotrophic condition	55.9	Yeh *et al.* (2011)
Chlorella sp.	Raw digested liquid swine manure rich in volatile fatty acids	10.9	Hu *et al.* (2012)
Chlorella sorokiniana	Glucose	31.5	Li *et al.* (2013)
Eustigmatos cf. *polyphem*	Photoautotrophic condition	60.6	Zhang *et al.* (2013)
Scenedesmus obtusiusculus	Carbon dioxide	55.7	Toledo-Cervantes *et al.* (2013)
Scenedesmus sp. ZTY1	Domestic wastewater	32.2	Zhang *et al.* (2014a)
Chlorella protothecoides	Whey permeate	49.9	Espinosa-Gonzalez *et al.* (2014)
Chlorella sp. HQ	Photoautotrophic condition	63.0	Zhang *et al.* (2014b)
Leptolyngbya sp. ISTCY101	Sodium bicarbonate	16-21	Singh *et al.* (2014)
Chlorella ellipsoidea	Photoautotrophic condition	36.3	Wang *et al.* (2014)

Contd.

Chlamydomonas sp. JSC4	Photoautotrophic condition	43.1	Ho *et al.* (2015)
Chlorococcum littorale	Carbon dioxide	47.8	Ota *et al.* (2015)
Chlamydomonas reinhardtii CC1010	Glucose	61.0	Karpagam *et al.* (2015)
Chlorella vulgaris	Molasses + Corn steep liquor	32.8	Mohammad Mirzaie *et al.* (2015)
Scenedesmus sp. CCNM 1077	Glucose	25	Pancha *et al.* (2015)

Conversion of Algal Lipid into Biodiesel : Transesterification

Microalgae generally produce neutral (e.g. triglycerides) and polar lipids (e.g. phospholipids). Triglycerides are the main materials in the production of biodiesel, but are not apposite for most of available diesel engines (as fuel) owing to prominently viscous nature that hampers good ûow in addition to combustion. These algal lipids/ oils are subjected to modification for utilization in diesel engines through a chemical process called 'transesterification', where the addition of alcohol is carry out in a base mixture to transform the feedstock lipid (triglycerides) into shorter hydrocarbon chains for the synthesis of an alternative biofuel in diesel engines. The fatty acid methyl ester produced as a result of reaction of algal lipid/oil with methyl alcohol is commercialized under the trade name of 'Biodiesel' (Gallagher 2011). The transesterification reaction for biodiesel production from lipid/ oil is very simple and involves equilibrium reaction:

$$\underset{\text{(Lipid/ Oil)}}{\text{1 Triglyceride}} + \underset{\text{(Alcohol)}}{\text{3 Methanol}} \xrightleftharpoons{\text{Catalyst}} \text{1Glycerol} + \underset{\text{(Biodiesel)}}{\text{3 Methyl Esters}}$$

In the above reaction, one molecule of lipid (triglyceride)/ organic oil reacts with three molecules of methanol in the presence of catalyst and alkali (such as potassium hydroxide), which results into the formation of three molecule of methyl esters of fatty acids, i.e. biodiesel along with one molecule of glycerol (Christi 2007; Demirbas 2007). Furthermore, this reaction involves excess of methanol as one of the reactant in order to drive the reaction towards the biodiesel formation. Transesterification process is catalyzed by alkalis, acids (Fukuda *et al.* 2001; Meher *et al.* 2006) as well as lipase enzymes (Sharma *et al.* 2001). It has been reported that transesterification catalyzed by alkali is approximately 4000 times faster compared to the reaction catalyzed by acid (Fukuda *et al.* 2001). However, the rate of reaction is very slow with respect to the transformation of triglycerides into methyl esters in the presence of the acid catalysts. This is evident from the fact that microalga *Scenedesmus* sp. was exploited for biodiesel synthesis by acid and alkali transesterification process, where 55.07 $\pm$ 2.18% and 48.41 $\pm$ 0.21% were recorded based on lipid by weight of biodiesel production in the presence of sodium hydroxide (as alkaline catalyst) and sulfuric acid (acid catalyst), respectively (Kim *et al.* 2014). On the other hand, compared to alkalies and acid catalysts, lipases as biocatalyst exhibit different benefits such as regioselectivity, substrate specificity, enantioselectivity, versatility as well as high catalytic activity at ambient pressure and temperature (Knezevic *et al.* 2004).

Benefits of algal biodiesel over petroleum diesel

Algal biodiesel depicts many noteworthy properties not exhibited by petroleum diesel. Algal biodiesel practically do not contain sulfur or particulate substance that actually contributes towards air pollution in the form of black smoke and sour odor (Gallagher 2011). It possesses better lubricity compared to petroleum diesel (that helps in decreasing friction of moving parts). Interestingly, algal biodiesel is a carbon-neutral fuel, i.e. it fixes approximately as much carbon dioxide during algal growth as it set free upon combustion of fuel (Gallagher 2011). On the basis of life-cycle investigation, biodiesel utilization can decrease atmospheric concentration of carbon dioxide (and thus, reducing the global warming), decrease engine maintenance and improve quality of air. Algal biodiesel is effortlessly mingled with petroleum diesel to produce a superior fuel with far better performance (Gallagher 2011). Furthermore, the biodiesel exhibits energy density analogous to conventional petroleum diesel but high heating value of biodiesel fluctuate and reliant on the source of biomass. However, a value for petroleum diesel is 42.7 MJ/kg. Interestingly, biodiesel synthesize through seed oils (*viz.* soybean and rapeseed) and algal biomass yields 37 MJ/kg and 41 MJ/kg, respectively.

Algae: Feedstock for Bioethanol and biogas production

In addition to biodiesel, algae are also appropriate feedstock for bioethanol and biogas synthesis owing to potential source of natural sugars including other carbohydrates. (Murphy *et al.* 2013). Several algae that produce starch under condition of stress can be utilize for bioethanol synthesis through the process of fermentation. On the other hand, algae having low lipid content are particularly appropriate for the production of biogas through anaerobic fermentation (Murphy *et al*. 2013).

Bioethanol production

A critical review of literature depicted that there are many reports accessible on the production of bioethanol from algal species, where they are capable to accumulate huge quantity of carbohydrates as reserve materials. Currently, algal biomass receiving considerable attention as suitable feedstock for biodiesel production worldwide due to low hemicelluloses and lignin content as compared to lignocellulosic biomass (Chen *et al.* 2013). Presently, efforts have been directed towards the bioethanol production by means of fermentation utilizing algae as the potential feedstocks (Nguyen and Vu 2012; Chaudhary *et al*. 2014). There are diverse form of microalgae including macroalgae *viz. Chlorococcum* sp., *Prymnesium parvum*, *Gelidium amansii*, *Laminaria* sp., *Gracilaria* sp., *Spirogyra* sp. and *Sargassum* sp. that have been exploited for the production of bioethanol (Eshaq *et al.* 2011; Rajkumar *et al.* 2014). Generally, these algae need nutrients, carbon dioxide and light to yield huge amount of polysaccharides *viz.* starch and cellulose that can be extracted to fermentable sugars by hydrolysis and further fermentation to bioethanol followed by separation by means of distillation. Table 4 depicts algae as prospective feedstock for the production of bioethanol.

Table 4: Algal species as potential feedstock for the production of bioethanol

Microalgae species (Feedstock)	Conditions	Bioethanol	Reference
Spirogyra	Alkaline pre-treatment, synthetic media growth, Saccharification of biomass by *Aspergillus niger*, fermentation by *Saccharomyces cerevisiae*	80 g ethanol/kg algae	Eshaq *et al.* (2010)
Chlorococcum humicola	Acid pre-treatment, temp. 160°C, *Saccharomyces cerevisiae*	520 g ethanol/kg microalgae	Harun and Danquah (2011)
Chlorococcum infusionum	Alkaline pre-treatment, temp. 120°C, *Saccharomyces cerevisiae*	260 g ethanol/kg algae	Harun *et al.* (2011)

Source : Behera *et al.* 2015

Biogas production

Biogas constitute a combination of diverse gases (such as carbon dioxide, methane, and traces of hydrogen sulûde) manufactured owing to action of anaerobic microorganism on organic content existing in the eco-friendly feedstock *viz.* marine algae (Behera *et al.* 2015). Fundamentally four key steps are involved in the anaerobic transformation of biodegradable feedstock to biogas. The first step involves the exploitation of enzyme (produced by certain obligate anaerobic microorganism such as *Streptococci* and *Clostridia*) for the hydrolysis of high molecular weight materials such as carbohydrates, proteins and lipids including insoluble organic material into soluble organic material. These soluble organic materials act as substrates for the enzymes released by the acidogenic bacteria that resulting into the formation of volatile fatty acids and alcohols and thus, constituting second step termed as acidogenesis. Furthermore, these volatile fatty acids and alcohols are transformed to hydrogen and acetic acid by exploiting acetogenic bacteria with the acetogenesis process that ultimately converted to methane as well as carbon dioxide with the action of methanogens (Cantrell *et al.* 2008; Vergara-Fernandez *et al.* 2008; Brennan and Owende 2010; Romagnoli *et al.* 2011). Certain algal species such as *Scenedesmus obliquus* and *Chlamydomonas reinhardtii* act as green feedstocks for biogas production with 287 and 587 ml biogas/g of volatile solids, respectively (Mussgnug *et al.* 2010). Review of literature revealed that little investigations have been carried out, where microalgae exhibiting the influence of diverse pretreatment such as ultrasound, microwave and thermal for the enhanced formation of biogas (Gonzalez-Fernandez *et al.* 2012a, b; Passos *et al.* 2013). Table 5 shows the production of biogas from algal biomass as feedstock.

Table 5 : Algal biomass as feedstock for the production of biogas

Algal biomass (Feedstock)	Conditions	Biogas	Reference
Chlamydomonas reinhardtii	Drying as the pre-treatment, batch	587 ml/g of volatile solids	Mussgnug et al. (2010)
Scenedesmus obliquus	fermentation, temperature 38°C	287 ml/g of volatile solids	
Blue algae	pH-6.8, microcystin biodegradation	189.89 ml/g of volatile solids	Yuan et al. (2011)

Contd.

Ulva sp.	Batch reactor,	191 ml/g of VS	Vanegas and
Laminaria digitata	co-digestion with bovine slurry, temperature 35°C	246 ml/g of volatile solids	Bartlett (2013)
Saccorhiza polyschides		255 ml/g of volatile solids	
Saccharina latissima		235 ml/g of volatile solids	

Source : Behera *et al.* 2015

Economics of producing microalgal biofuel

The production expenditure of algal oil/lipid contents (as feedstock) reliant on several factors: (i) Lipid/ oil content; (ii) Production of algal biomass from the cultivation system; (iii) Price of recovering lipid/oil from algal biomass and (iv) Scale-up of production systems. At present, algal biofuel production is still not economical compared to petroleum fuels and this fact is supported by various researchers. Chisti (2007) (as one of the researcher) conducted a study on the manufacturing cost of algae lipid/oil content from a photobioreactor with an annual production capacity of 10,000 tons/year, where a production cost of \$2.80/liter (\$10.50/gallon) of algal oil/lipid were estimated considering the lipid/oil content of the algal biomass to be about 30% (but excluding taxes, marketing expenditure of biodiesel, costs of transforming algal lipid/oil into biodiesel and distribution. The petroleum diesel cost in Virginia at the same time, was \$3.80 to \$4.50/gallon. In the near future, whether lipid/oil from algal biomass could be a cost-effective source for biofuel is still vastly reliant on the cost of fossil fuel oil (petroleum oil). Chisti (2007) estimate the price of algal lipid/oil by exploiting the following equation (to correlate the feasibility of algal lipid/oil as competitive substitute for petroleum diesel):

$$C_{\text{Algal oil}} = 25.9 \times 10^{-3} C_{\text{Petroleum oil}}$$

Where: $C_{\text{Algal oil}}$ is the cost of microalgal oil /lipid in \$/gallon

$C_{\text{Petroleum oil}}$ is the cost of crude petroleum in \$/barrel

This equation presumes that algal oil/lipid has approximately 80% of the caloric energy value of crude petroleum oil. For instance, algal oil/lipid should cost not more than \$2.59/gallon in order to be viable with petroleum diesel when the petroleum oil exhibiting cost at \$100/barrel.

The above facts established that the triumphant and feasible production of an algal biofuels warrant the optimal amalgamation of technological innovations linked with economic viability and integrated scale-up for profitable production and marketing. Currently, microalgae as sustainable renewable biomass feedstocks for biofuel production is in its infancy irrespective of the enormous prospective of algal feedstocks to substitute substantial amounts of petroleum based fuels. Over the past several decades, biofuel has been exploited as an eco-friendly alternative substitute for fossil fuel (petroleum fuel) but so far there is no commercially feasible system accessible for industrial scale production. The economic viability for the production of algal biofuels is hugely reliant on the oil/ lipid feedstock price in conjunction with reducing the maintenance and operational expenditure. Therefore, algal productivity is the key production cost determining factor (in addition to maintenance and operational costs) and consequently, intensifying research should be concentrated on many characteristics of algal biology, which have the highest effect on rate of growth including lipid accumulation. With a ultimate objective to commercializing algal biofuel, the research and development efforts can be classified into several areas: (i) Exploration of an efficient oil-extraction method; (ii) Enhancement in the cultivation rate of algae; (iii) Enhancement in oil/lipid content of accessible strains or screening and selecting novel strains with elevated oil content; (iv) Production of co-products in addition to oil/lipid content; (v) Introduction of robust algal growing systems both in open-air or closed environments and (vi) Exploitation of algae in bioremediation.

Conclusion

At present, algal biofuel emerged as sustainable, renewable and eco-friendly substitute to non eco-friendly conventional petroleum based fuel. One notable bottleneck linked with microalgae green feedstocks for biofuel production is the low production of algal biomass, which is an obstacle for industrial production of biofuel. Synthesis of lipid/oil in microalgae is triggered under stress conditions such as nitrogen starvation/ limitation and thereby, limiting the algal growth and lipid production. Various investigations conducted by researchers for developing techniques to improve density and oil/lipid content, revealed that there is still plentiful unrealized prospective. With intensifying interest and research towards algal fuel, cultivation of algae should be capable to attain histrionic improvements in density, oil/lipid accumulation and growth rates. This will be achieved by improve cultivation techniques, growing methods, species selection and bio-engineering and therefore, paving the cost effective and commercialization path for algal biofuel in near future. Production of algal biomass can be visualized for huge CO_2^- sequestration and production of biofuel. In combating the challenges posed by green house gases and the energy scarcity, algal biomass is the only alternatives that link energy and climate.

References

Alam F., Date A., Rasjidin R., Mobin S., Moria H. and Baqui A. (2012) Biofuel from algae- Is it a viable alternative? Procedia Engineer. 49: 221-227.

Behera S., Singh R., Arora R., Sharma N.K., Shukla M. and Kumar S. (2015) Scope of algae as third generation biofuels. Front Bioeng. Biotechnol. (Ahead of Print).

Benemann J.R., Oswald W.J. (1996) Systems and economic analysis of microalgae ponds for conversion of CO_2 to biomass. University of California: Berkeley, CA, USA, 1996.

Brennan L., Owende P. (2010) Biofuels from microalgae - A review of technologies for production, processing, and extractions of biofuels and co-products. Renew. Sustain. Energy Rev. 14: 557-577.

Bruton T., Lyons H., Lerat Y., Stanley M. and Rasmussen, M.B. (2009) A Review of the Potential of Marine Algae as a Source of Biofuel in Ireland. Sustainable Energy Ireland: Dublin, Ireland.

Cantrell K.B., Ducey T., Ro K.S. and Hunt P.G. (2008) Livestock waste to bioenergy generation opportunities. Bioresour. Technol. 99: 7941-7953.

Chaudhary L., Pradhan P., Soni N., Singh P. and Tiwari A. (2014) Algae as a feedstock for bioethanol production: new entrance in biofuel world. Int. J. Chem. Technol. Res. 6: 1381-1389.

Chen C.Y., Zhao X.Q., Yen H.W., Ho S.H., Cheng, C.L., Bai F., et al. (2013) Microalgae-based carbohydrates for biofuel production. Biochem. Eng. J. 78: 1-10.

Chhetri A.B., Tango M.S., Budge S.M., Watts K.C. and Islam M. R. (2008) Non-Edible Plant Oils as New Sources for Biodiesel Production. Int. J. Mol. Sci. 9:169-180.

Chisti Y. (2007) Biodiesel from microalgae. Biotechnol. Adv. 25: 294-306.

Clark J. and Deswarte F. (2008) Introduction to Chemicals from Biomass. Wiley series in renewable resources. Hoboken: John Wiley & Sons.

Cohen Z., Norman H.A. and Heimer, Y.M. (1995) Microalgae as a source of x-3 fatty acids. World Rev. Nutr. Diet. 77: 1-31.

Demirbas A. (2007) Importance of Biodiesel as Transportation Fuel. Energy Policy. 35: 4661-4670.

Demirbas, A. (2009) Progress and recent trends in biodiesel fuels. Energy Convers. Manag. 50: 14-34.

Dragone G., Fernandes B., Vicente A.A. and Teixeira J.A. (2010) Third generation biofuels from microalgae. in: Mendez-Vilas A (ed.) Current Research, Technology and Education Topics in Applied Microbiology and Microbial Biotechnology. Formatex, pp.1355-1366.

Eshaq F.S., Ali M.N and Mohd M.K. (2010) *Spirogyra* biomass a renewable source for biofuel (bioethanol) production. Int. J. Eng. Sci. Technol. 2: 7045-7054.

Eshaq F.S., Ali M.N. and Mohd M.K. (2011) Production of bioethanol from next generation feed-stock alga *Spirogyra* species. Int. J. Eng. Sci. Technol. 3: 1749-1755.

Espinosa-Gonzalez I., Parashar A. and Bressler D.C. (2014) Heterotrophic growth and lipid accumulation of *Chlorella protothecoides* in whey permeate, a dairy by-product stream, for biofuel production. Bioresour. Technol. 155: 170-176.

Fukuda H., Kondo A. and Noda H. (2001) Biodiesel fuel production by transesterification of oils. J. Biosci. Bioeng. 92: 405-416.

Gallagher B.J. (2011) The economics of producing biodiesel from algae. Renew. Energy 36: 1158-162.

Gonzalez-Fernandez C., Sialve B., Bernet N. and Steyer J.P. (2012a) Comparison of ultrasound and thermal pretreatment of *Scenedesmus* biomass on methane production. Bioresour. Technol. 110: 610-616.

Gonzalez-Fernandez C., Sialve B., Bernet N. and Steyer J.P. (2012b) Thermal pretreatment to improve methane production of *Scenedesmus* biomass. Biomass Energy 40: 105-111.

Harun R. and Danquah M.K. (2011) Influence of acid pre-treatment on microalgal biomass for bioethanol production. Process Biochem. 46: 304-309.

Harun R., Jason W.S.Y., Cherrington T. and Danquah M.K. (2011) Exploring alkaline pretreatment of micrcalgal biomass for bioethanol production. Appl. Energy 88: 3464-3467.

Ho S.H., Chen W.M. and Chang J.S. (2010) *Scenedesmus obliquus* CNW-N as a potential candidate for CO_2 mitigation and biodiesel production. Bioresour. Technol. 101: 8725-8730.

Ho S.H., Nakanishi A., Ye X, Chang J.S, Hara K. and Hasunuma T. (2014) Optimizing biodiesel production in marine *Chlamydomonas* sp. JSC4 through metabolic profiling and an innovative salinity-gradient strategy. Biotechnol. Biofuels. 7: 97.

Ho S.H., Nakanishi A., Ye X., Chang J.S, Chen C.Y, Hasunuma T. and Kondo A. (2015) Dynamic metabolic profiling of the marine microalga *Chlamydomonas* sp. JSC4 and enhancing its oil production by optimizing light intensity. Biotechnol Biofuels. (Ahead of Print).

Hu B., Min M., Zhou W., Du Z., Mohr M., Chen P., Zhu J., Cheng Y., Liu Y. and Ruan R. (2012) Enhanced mixotrophic growth of microalga *Chlorella* sp. on pretreated swine manure for simultaneous biofuel feedstock production and nutrient removal. Bioresour Technol. 126: 71-79.

Jiang L., Luo S., Fan X., Yang Z. and Guo, R. (2011) Biomass and lipid production of marine microalgae using municipal wastewater and high concentration of CO_2. Appl. Energ. 88: 3336-3341.

Karpagam R., Preeti R., Ashokkumar B. and Varalakshmi P. (2015) Enhancement of lipid production and fatty acid profiling in *Chlamydomonas reinhardtii*, CC1010 for biodiesel production. Ecotoxicol. Environ. Saf. (Ahead of Print).

Kim G.V., Choi W.Y., Kang D.H., Lee S.Y. and Lee H.Y. (2014) Enhancement of biodiesel production from marine alga, *Scenedesmus* sp. through in situ transesteriûcation process associated with acidic catalyst. BioMed. Res. Int. (Ahead of Print).

Knezevic Z.D., Siler-Marinkovic S.S. and Mojovic L.V. (2004) Immobilized lipases as practical catalysts. Acta Period. Technol. 35: 151-164.

Lee K., Eisterhold M.L., Rindi F., Palanisami S. and Nam P.K. (2014) Isolation and screening of microalgae from natural habitats in the Midwestern United States of America for biomass and biodiesel sources. J. Nat. Sci. Biol. Med. 5: 333-339.

Li T., Zheng Y., Yu L. and Chen S. (2013) High productivity cultivation of a heat-resistant microalga *Chlorella sorokiniana* for biofuel production. Bioresour Technol. 131: 60-67.

Li Y., Horsman M., Wu N., Lan C.Q. and Dubois-Calero N. (2008) Biofuels from microalgae. Biotechnol. Prog. 24: 815-820.

Li Y., Wang B., Wu N. and Lan CQ. (2008) Effects of nitrogen sources on cell growth and lipid production of *Neochloris oleoabundans*. Appl. Microbiol. Biotechnol. 81: 629-636.

Mata T.M., Martins A.A. and Caetano N.S. (2010) Microalgae for biodiesel production and other applications: a review. Renew. Sustain. Energy Rev. 14: 217-232.

Matsunaga T., Matsumoto M., Maeda Y., Sugiyama H., Sato R. and Tanaka T. (2009) Characterization of marine microalga, *Scenedesmus* sp. strain JPCC GA0024 toward biofuel production. Biotechnol Lett. 31(9):1367-72.

Meher L.C., Vidya Sagar D. and Naik S.N. (2006) Technical aspects of biodiesel production by transesterification-a review. Renew. Sustain. Energy Rev.10: 248-68.

Mohammad Mirzaie M.A., Kalbasi M., Mousavi S.M. and Ghobadian B. (2015) Investigation of Mixotrophic, Heterotrophic and Autotrophic Growth of *Chlorella vulgaris* Under Agricultural Waste Medium. Prep. Biochem. Biotechnol. (Ahead of Print).

Murphy F., Devlin G., Deverell R., and McDonnell K (2013) Biofuel Production in Ireland-An Approach to 2020 Targets with a Focus on Algal Biomass. Energies 6: 6391-6412.

Mussgnug J.H., Klassen V., Schluter A. and Kruse O. (2010) Microalgae as substrates for fermentative biogas production in a combined bioreûnery concept. J. Biotechnol. 150: 51-56.

Nguyen T.H.M. and Vu V.H. (2012) Bioethanol production from marine algae biomass: prospect and troubles. J. Viet. Env. 3: 25-29.

Nigam P.S. and Singh A. (2011) Production of liquid biofuels from renewable resources, Prog. Ener. Combustion Sc. 37: 52-68.

Ota M., Takenaka M., Sato Y., Smith R.L. and Inomata H. (2015) Variation of photoautotrophic fatty acid production from a highly CO_2 tolerant alga, *Chlorococcum littorale*, with inorganic carbon over narrow ranges of pH. Biotechnol Prog. (Ahead of Print).

Pancha I., Chokshi K. and Mishra S. (2015) Enhanced biofuel production potential with nutritional stress amelioration through optimization of carbon source and light intensity in *Scenedesmus* sp. CCNM 1077. Bioresour Technol. 179: 565-572.

Passos F., Sole M., Garcia J. and Ferrer I. (2013) Biogas production from microalgae grown in wastewater: effect of microwave pretreatment. Appl. Ener. 108: 168-175.

Pittman J.K., Dean A.P. and Osundeko O. (2011) The potential of sustainable algal biofuel production using wastewater resources. Bioresour. Tech. 102: 17-25.

Rajkumar R., Yaakob Z. and Takriff M.S. (2014) Potential of the micro and macro algae for biofuel production: a brief review. Bioresour. 9: 1606-1633.

Reith J.H., van Hal J.W. and Lenstra, W.J. (2012) Large-Scale Carbon Recycling via Cultivation and Biorefinery of Seaweeds for Production of Biobased Chemicals and Fuels. in: Proceedings of the Conference on Carbon Dioxide as Feedstock for Chemistry and Polymers. Essen, Germany, 10-11 October 2012.

Romagnoli F., Blumberga D. and Gigli E. (2011) Biogas from marine macroalgae: a new environmental technology-life cycle inventory for a further LCA. Environ. Climate Technol. 4: 97-108.

Scharfemann J.P. W. and Larance W.F. (2008) How green are biofuels? Science 319: 43-44.

Schenk P., Thomas-Hall S., Stephens E., Marx U., Mussgnug J., Posten C., Kruse O. and Hankamer B. (2008) Second generation biofuels: high efficiency microalgae for biodiesel production, BioEnergy Res. 1: 20-43.

Sharma R., Chisti Y. and Banerjee U.C. (2001) Production, purification, characterization, and applications of lipases. Biotechnol. Adv. 19: 627-62.

Sheehan J., Dunahay T., Benemann J. and Roessler P. (1998) A look back at the U.S. Department of Energy's aquatic species program: biodiesel from algae. NREL/TP-580-24190, National Renewable Energy Laboratory, USA.

Singh J., Tripathi R. and Thakur I.S. (2014) Characterization of endolithic cyanobacterial strain, *Leptolyngbya* sp. ISTCY101, for prospective recycling of CO and biodiesel production. Bioresour Technol. 166: 345-52.

Spolaore P., Joannis-Cassan C., Duran E. and Isambert A. (2006) Commercial applications of microalgae. J. Biosc. Bioengineer. 101: 87-96.

Tabatabaei M., Tohidfar M., Jouzani G.S., Safarnejad M. and Pazouki M. (2011) Biodiesel production from genetically engineered microalgae: future of bioenergy in Iran. Renew. Sustain. Energy Rev. 15: 1918-1927.

Toledo-Cervantes A., Morales M., Novelo E. and Revah, S. (2013) Carbon dioxide fixation and lipid storage by *Scenedesmus obtusiusculus*. Bioresour Technol. 130: 652-658.

Vergara-Fernandez A., Vargas G., Alarcon N. and Velasco A. (2008) Evaluation of marine algae as a source of biogas in a two-stage anaerobic reactor system. Biomass Bioenergy 32: 338-344.

Wang B., Li Y., Wu N. and Lan, Q. (2008) CO_2 mitigation using microalgae. Appl. Microbiol. Biotechnol. 79: 707-718.

Wang Y., Liu Z. and Qin S. (2013) Effects of iron on fatty acid and astaxanthin accumulation in mixotrophic *Chromochloris zofingiensis*. Biotechnol Lett. 35: 351-357.

Yeh K.L. and Chang J.S. (2011) Nitrogen starvation strategies and photobioreactor design for enhancing lipid content and lipid production of a newly isolated microalga *Chlorella vulgaris* ESP-31: implications for biofuels. Biotechnol. J. 6(11):1358-66.

Zhang J., Wan L., Xia S., Li A. and Zhang C. (2013) Morphological and spectrometric analyses of lipids accumulation in a novel oleaginous microalga, *Eustigmatos* cf. *polyphem* (Eustigmatophyceae). Bioprocess Biosyst. Eng. 36: 1125-1130.

Zhang Q., Wang T. and Hong Y. (2014b) Investigation of initial pH effects on growth of an oleaginous microalgae *Chlorella* sp. HQ for lipid production and nutrient uptake. Water Sci. Technol. 70:712-719.

Zhang T.Y., Wu Y.H. and Hu H.Y. (2014a) Domestic wastewater treatment and biofuel production by using microalga *Scenedesmus* sp. ZTY1. Water Sci. Technol. 69(12):2492-2496.

10

Nanobiotechnology Perspectives: Role of Nanobiotechnology in Medical and Dental Health

Prabhanjana K.V.[1], Manoharan P.S.[2], Soundravally R.[3] and Agieshkumar B.[1,4*]

[1]School of Chemical and Biotechnology, SASTRA University, Thanjavur-613401, India

[2]Indira Gandhi Institute of Dental Sciences, Puducherry - 605 006, India

[3]Department of Biochemistry, JIPMER, Puducherry - 605 006, India

[4]Central Inter-Disciplinary Research Facility (CIDRF), A Unit of Sri Balaji Educational and Charitable Public Trust, Puducherry -607403, India

**Email: agiesh.b@gmail.com*

Abstract

Nanotechnology is an emerging field that offers various potential applications in the areas of agriculture, health and medicine. Recent advances in the design and synthesis of nano-materials aids for developing sophisticated nano-devices for clinical research and practice. Applications of nano materials in medicine includes disease diagnosis, early prognosis, detection of pathogens, tumor detection, tissue engineering, target specific drug delivery, development of biomaterials for oral health and understanding of cellular mechanisms in live cells using molecular imaging. The present chapter summarizes the current trends and advances in nanomedicine in three major diseases: cancer, cardiovascular diseases, and diabetes. The review presents various nanomaterials developed for these diseases in terms of diagnosis and treatment. Besides, the chapter presents a brief review on significance of nanotechnology in the field of dentistry and translational health research.

Introduction

In the recent years synthesis of particles at nano range received enormous applications in the field ofhealth and medicine, computing, catalysis and other engineering fields. Nanobiotechnology is an interdisciplinary of nanotechnology, whichhas a great impact on biological and biochemical resourcesand thus holds great promise of boosting pharmaceuticals and health care sectors (Kawadkar *et al.* 2011). Nanotechnology uses nano-sized tools for the improvement of the quality of life. For example the nano technology enabled delivery ofagriculture chemicals like fertilizers, pesticides, herbicides, plant growth regulators, *etc.*, field sensing systems to monitor the environmental stresses and crop condition, improving plant traits against diseases involves the principles ofnanotechnology for the drastic improvement of agriculture (Chen and Yada 2011). In the field of medicine which is the main focus area of the present chapter nanoparticles are used as carriers for traditional and modern drugs, peptides, antibodies and vaccines (Yeole *et al.* 2013). Several therapeutically significant nanoparticles are being synthesized that are biodegradable and biocompatible to the physiological environment. Thus nanobiotechnology offers sophisticated methods for the diagnosis, prevention and treatment of the disease and to understand the mechanisms underlying pathophysiology ofthe disease. In this context, the present chapter summarizes various developments in nanoresearch (*in vitro*/ *in vivo* models) that has potential clinical applications with special emphasis on diagnosis and treatment methods for three major diseases: cancer, cardiovascular diseases and diabetes. Also the review gives an introduction of how nanotechnology is applied in dental research and clinical practice.

Nanobiotechnology and Medicine

The contribution of nanotechnology in the field of medicine is huge. The technology supplies nanoparticles that are efficient with therapeutic and diagnostic properties (Sachiko and Keiji 2013). Nanoparticles are reported to have potential role in different branches of medicine including new diagnostic tools, imaging techniques, pharmaceutics, implant, new drug delivery systems and tissue engineering methods for treating various diseases such as cancer, cardiovascular diseases,diabetes, bone implants, dental problems, musculoskeletal disorders, neurodegenerative diseases,bacterial and viral infectious diseases *etc.* (Jooyeon *et al.* 2015; Kim *et al.* 2007; Lei *et al.* 2014; Masahiro *et al.* 2012; Xu*et al.* 2011; Fenniri 2006; Rutledge 2007; Stoimenov *et al.* 2002; Abraham *et al.* 2008). A brief outline of nanotechnology based diagnostics and therapeutics is given below along with a preview of various applications of nanoparticles in medicine (Fig. 1)

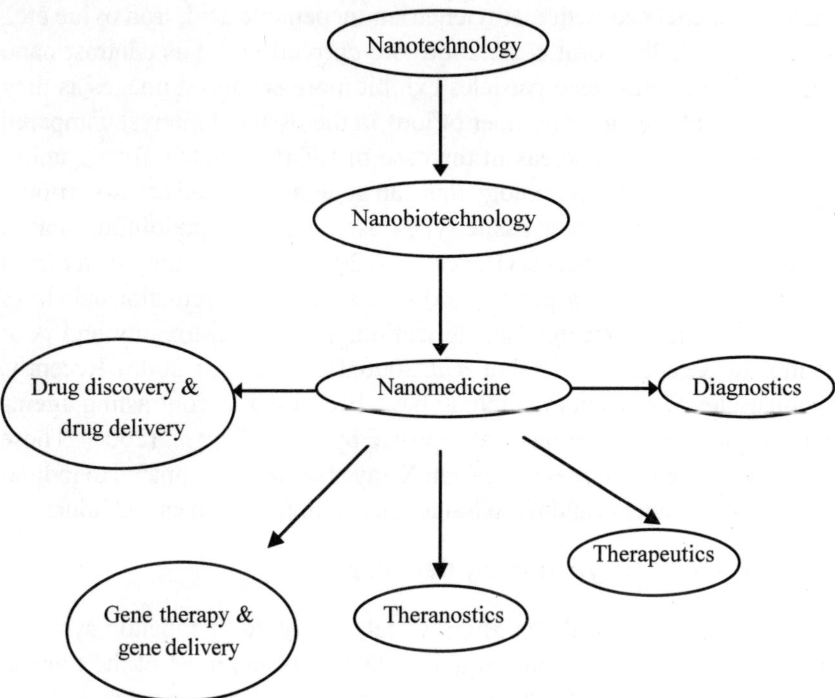

Fig. 1 : Various applications of nanobiotechnology in health industry

Nanotechnology in disease diagnostics-brief overview

Nanoparticle associated imaging techniques were reported to have a more pronounced effects in terms of precision. Nanodiagnostics technique is a conventional imaging technique where nano contrast agents are used to detect and characterize early stages which can develop into a disease (Khalid and Yazeed 2014).The technique uses different class of molecules like superparamagnetic, fluorescent, paramagnetic and radioactive materials. Usually nanoparticles coupled with target specific substances so that the particles bind to the specific region for more precise imaging (Yang *et al.* 2015). Based on the type of nanoparticle used, different imaging techniques like positron emission tomography (PET), computed tomography (CT), Magnetic resonance imaging (MRI), photo acoustic imaging and fluorescence imaging obtained.For example in the case of MRI (Magnetic Resonance Imaging) which involves the usage of powerful magneticfields and radiofrequency waves to produce images of internal structures,

Gadoliniumchelated with diethylenetriaminepentetic acid, iron oxide etc., (Chen *et al.* 2008; Corot *et al.* 2006) are currently used as contrast nano agents. These nano-sized particles exhibit more enhanced images as they could concentrate larger number of ions in the tissue of interest compared to macro particles. Whereas in the case of CT (Computed Tomography) X-ray based imaging technology that can generate detailed cross-sectional images of many different tissue types uses iodine or gadolinium-based molecules as contrast agents (Rosen *et al.* 2011) however, they suffer from limitations such as a nonspecific biodistribution, short circulation half-lives (<10min), requirement for catheterization, slight renal toxicity and poor contrastin larger patients (Rabin *et al.* 2006; Hainfeld *et al.* 2006). Recently, nanoparicles like bismuth sulphide have been used as contrasting agents with polymeric preparations was reported by Hainfeld *et al.* (2006). These particles showed a 5-times increasein X-ray absorption compared to iodine-based contrast agents andmuch longer circulation half- lives (>2 hours)

Nanobiotechnology in drug delivery

Conventional methods offers widely accepted therapeutic systems. However it suffers from undesirable side effects and most of the time an inadequate delivery of the therapeutic component to the target site. These limitations could be overcome by incorporating drug components with a nanocarrier which improves the delivery process. Usually thermo carriers are specially coated with hydrophilic polymers such as PEGinorder to protect the drug molecules from endosomal degradation (Andresen *et al.* 2005; Chellat *et al.* 2005). These kind of nanocarriers are called stealth nanocarriers which results in a durable circulation and effective drug delivery. Nanotechnology also offers a tissue specific drug delivery by coupling the nanoparticles with target specific molecules. This leads to target specific quantitative binding (Jaracz *et al.* 2005; Sapra and Allen 2003). As an example TRAIL ligand coupled with human serum albumin is used in the case of lung cancer treatment (Seong *et al.* 2015; Roby *et al.* 2006; Omid and Robert 2009)

Thus nanobased delivery systems offer the following potential advantages

- Enhanced delivery of poorly water soluble drugs
- Tissue specific targeted delivery

- A method for transfer of drug by a process called transcytosis, of through epithelial and endothelial barriers
- Release of large macromolecule drugs to intracellular sites of action
- Combination therapy involves co-delivery of two or more drugs.
- Confirmation of drug delivery sites by combining therapeutic agents with imaging system (theranostics)

Nanomedicine in cancer therapy

Cancer is one of the main causes of mortality worldwide. The main constraints are lack of biomarkers for an early prediction and a complete cure. Nanotechnology provides a wide range of new tools such as earlier diagnostics, improved imaging methods, more efficient targeted therapies (Biana *et al.* 2010). Diagnosis and treatment for cancer remains difficult because normally cancer biomarkers are present in low concentration and make it difficult to detect at early stages of the disease. This can be overcome by a targeted delivery of specific nanoparticles directly into the affected tissues, which may provide a significant signal for the detection of biomarkers. This kind of diagnosis methods yields precise and early results to the clinicians to start the treatment instead of waiting for biopsies. As an example Ironoxide nanoparticles are recently engineered in such a way it binds specifically well to the tumors and gives a very high resolution and an accurate localization of the lesions with MRI scans. With respect to treatment, therapeutic nanoparticles can be injected directly into the tumor cells which produce heat that destroys cancer cells locally by applying magnetic field or X-rays (Kwatra *et al.* 2013). This local delivery reduces the amount of drug absorbed by the patient and no effects on the healthy cells of the body. These results show that nanocarrier conjugated therapies gives better cure for cancer rather conventional means: Some of the major therapeutic components coupled with nanotechnology for the cancer treatment are summarized in table 1. Similarly the recently developed nanotechnology based imaging techniques for cancer are listed in Table 2. The table illustrates the imaging methods that are under research in the laboratory conditions and in the animal models.

Table 1 : List of the recently developed drugs conjugated with nanoparticles for cancer therapeutics

Nanoparticle	Therapeutic component	Result/ Mechanism	Application	Reference
Albumin nanoparticles	Paclitaxel	Nanocoated chemotherapeutic agent effectively stabilizes microtubules, inducing mitotic arrest.	Different types of cancers	Ruibing *et al.* 2013
Lipid encapsulation	Daunorubicn	Liposome fused drug were found to be effective	Initial treatment for advanced	HIV-associated Kaposi's sarcoma Ruibing *et al.* 2013
Liposome	Cytarabine	Effective inhibition was recorded	Lymphomatous meningitis	Gopalakrishna 2014
PEGylated	Asparaginase	Nanocoated L-asparaginase, controlled leukemic cells.	Acute lymphoblastic leukemia	Ruibing *et al.* 2013
Lipid polymer	Curcumin	The nanoformulation altered vascular inflammation, damaged circulating tumor cells arrest and there by forbidding the metastasis.	For the prevention of metastasis.	Palange, *et al.* 2014
Polymeric micells with DACHPt/m (1,2-diaminocyclohexane) platinum (II)/ metal complex).	Anti tissue factor-Fab'	Maleimide-thiol chemistry aided anti-TF Fab'-DACHPt/ m with increased antigen recognition ability.	Pancreatic tumor	Jooyeon *et al.* 2015
Human serum albumin nanoparticles coated with tumor necrosis factor-related apoptosis-inducing ligand (TRAIL)	Doxorubicin chloride	Nanoformulation exhibited cytotoxic/apoptotic antitumeric property.	Lung cancer	Seong *et al.* 2015

Table 2 : Nanoparticle based imaging methods currently under study in vitro and in animal models.

Nanoparticle/ Formulation	**Method of study**	**Applications**	**Imaging Techniques**	**Reference**
[18F]fluorobenzyl-2-bromoacetamide	*In vivo*	Effective identification of drug release based on nano structures.	PET-CT imaging	Di Mauro *et al.* 2015
Ultra smalliron oxide nanoparticles (NPs) coupled with Arg-Gly-Asp (RGD)-modified dendrimers.	*Invitro & invivo*	The formulation has efficient contrasting property which is dependent on types of tumors.	Magnetic resonance imaging	Yang J *et al.* 2015
Gold nano particles	*Invivo*	Most of the gold nano particles injected intravenously were failed to pass through Blood Brain Barrier (BBB), empowering high resolution	Micro computed tomography tumor imaging	Hainfeld *et al.* 2013
Fluroscent magnetic nano particles	*In vivo* (Human glioblastoma mouse model)	Magnetic assembly of nanoparticles within new blood vessels was noticeable via fluorescence intravital microscopy in real time.	Flurescent microscopy	Fu *et al.* 2012

Treatment and diagnostics of Cardio Vascular Diseases (CVDs)

Cardiovascular diseases (CVDs) are leading cause for the death throughout the globe (WHO, 2015). According to world heart federation report (2011), CVD includes coronary heart disease, cerebrovascular diseases, peripheral arterial disease, rheumatic heart disease, congenital heart disease, deep vein thrombosis and pulmonary embolism. Nanotechnology finds potential applications in diagnosis and therapeutics of cardiovascular diseases. In terms of diagnosis, the nanoscale contrast agents are capable to recognize and characterize early disease stages prior to the progress of coarse disease manifestations. The diagnostic techniques include various contrast agents like fluorescent, superparamagnetic, paramagnetic, radioactive, and electron dense nanoparticles. The applications of various nanoparticles in CVD are shown in the table 3. The imaging property of the nanoparticle is dependent on its size. Each different class of molecules have their application in different imaging techniques. For example Quantum dots nanoparticles are fluorescent exhibiting molecules which have their application in *invivo* cell trafficking, lymphnode detection (Michalet *et al.* 2005). During the course of treatment thenanocarriers offers directed and easy delivery, prolonged effect of the drug and reduced shear effect of the blood flow.For example PEGylated nanoparticles with surface attached [Pyr1]-Apelin-13 polypeptide (anadipokine- regulator of heart dysfunction) into a mouse model of pressure induced heart failure (Vahid *et al.* 2015). Both *in vivo* and *in vitro* studies, it exhibited sustained and extended drug release. This demonstrates that engineered nanocarrier system can provide a delivery platform for treating heart injuries via sustained bioaccessibility of cardioprotective therapeutics. Similarly, nanotechnological applications are being investigated for treatment of atherosclerosis and restenosis, including nanocarriers for drug delivery and devices such as mechanical stents (Wang *et al.* 2013) possessing nanoscale components. A list of nanoparticle anchored therapeutics for CVD is listed in the Table 4.

Table 3 : Various nanotechnology based imaging techniques for CVD

Nanoparticle/ Formulation	Classification	Method of study	Applications	Imaging Techniques	Reference
Quantum dots	Fluorescence	*In vivo*	Recent applications include invivo cell trafficking, tumor target, lymph node detection and diagnostics	Fluorescent based imaging	Michalet X *et al.* 2005
Iron oxide	Super paramagnetic	*In vivo*	Cellular and molecular imaging studies and immune cell trafficking	Magnetic resonance based imaging	Corot C *et. al.* 2006
Gold based nanoparticles	Electron rich	*In vivo*	Intravascular contrasting agents has role in hepatocarcinoma and spleen imaging.	Computed tomography	Kim D *et al.* 2007
Gadolinium chelated with diethylenetria-	Paramagnetic	*In vitro & in vivo*	Improved signal augmentation of the atherosclerotic	Magnetic resonance imaging	Chen W *et. al.* 2008

Contd...

minepentetic acid			wall observed. (24h after the 50μmol Gd/kg injection).		
Dextran with copper-64 isotope	Radioactive	*In vivo*	Organ survival with respect to myeloid cells and heart transplant rejection	PET-CT	Takuya Ueno *et al.* 2013
Semiconductor nanoparticles	photoacoustic	*In vivo*	More efficient than single walled carbon nano tubes and gold nano rods	photoacoustic imaging	Pu *et al.* 2014
Iron oxide nanoparticle-polyethylene glycol complexes with a monoclonal antibody	Superpara magnetic	*In vitro*	No harmful effects in terms of toxicity at cellular and molecular levels. Potential source for both diagnosis and therapy.	Magnetic resonance imaging	Verma *et al.* 2015

Table 4 : Nanotherapeutics for CVD

Nanoparticle/ formulation	Key molecule	Methods of study	Result	Reference
Polyethylene glycol with liposomal nanoparticles	[Pyr1]-apelin-13 polypeptide	*In vivo* & *in vitro* studies	Both the studies indicated constant and elongated drug release. Helpful in avoiding cardiac dysfunction.	Vahid *et al.* 2015
Poly-lactic glycolic acid polymer	Aspirin and Prednisolone anti-inflammatory drugs, co delivery with GFP plasmid	*In vitro*	No toxicity was observed and co uptake of GFP observed. lessened risk for inflammatory myocardial atmosphere	Fargnoli *et al.* 2014
Magnetic mesoporous silica nano particles+ carbon nanotubes	Drug eluting stents	*In vitro* and *In vivo*	Atonement drug release, Hasten endothelialisation and lessened anti inflammatory reactions.	Wang *et al.* 2013

Nanotechnology in diabetics research

Diabetes is a chronic metabolic disorder due to the inability of controlling blood glucose level in a larger population (Shrivastava *et al.* 2013; WHO 2003). The available anti diabetic drugs are having limitations like dosage and adverse effects. For example Sulfonylureas is associated with hypoglycaemia and weight gain (Kunte *et al.* 2007; Patel *et al.* 2008). Eventhough certain phytomedicines are being found with the therapeutic anti-diabetic potentials, bioavailability, exact amount of drug to be administrated and specific target site actions of the biomolecules remains ambiguous.The treatment of diabetes could be transformed by using nanotechnology to create a 'smart delivery' system that regulates glucose levels by effectively replicating the work of the pancreatic cells that produce insulin. Besides the polymer based drug delivery systems provide many advantages as they can carry and deliver the drug to a target site, have the ability to deliver proteins, peptides and genes, increase the therapeutic profit and minimize the side effects of the antidiabetic drugs (Lanza *et al.* 1997). As an example the poly lactide-co-glycolide (PLGA) biocompatible polymers have been used as controlled releasedelivery of diabetic drugs (Wise *et al.* 1979). It has a high drug loading capacity as itpermits a small quantity of the carrier during a single administration.This approach hasbeen used to capsulize a variety of hydrophobic drugs, including plant based molecules like curcumin, coumarin, estradiol, protein and others (Bisht *et al.* 2007; Anand *et al.* 2010; Khuda-Bukhsh *et al.* 2010; Bhattacharyya *et al.* 2011; Hariharan *et al.* 2006; Samadder *et al.* 2013a, b). Nanotechnology also aims to enhance the bioavailability of drugs to target with enhanced efficiency at a least dosage. This is possible because nanoformulated drugs appear to have greater advantages due to their: tiny size, quick entry into target specific cells, biodegradable nature, capability to provide greater bioavailability of the drug, lesser amount of drug requirement. Some of the nanocarriers developed for treating diabetes based on *in vitro* and *in vivo* studies are summarized in Table 5. Nanotechnology also offers potential means to develop non-invasive means of monitoring glucose levels. For instance the typical finger prick test can be painful and cannot be performed when the patient is sleeping, scientists are in the process of developing non-invasive monitoring device that uses nonotechnology to detect acetone in the breath which is a marker that may correlates accurately to blood glucose levels.

Table 5 : Nanocarrier molecules for the treatment of diabetes based on *in vivo* and *in vitro* studies.

Carrier molecule	Therapeutic component	Experimental studies	Effect/ Result	Reference
Poly (lactic-co-glycolic acid); PLGA-NP+ cryoprotectants	Insulin	*In vitro*	Insulin release and stability increased	Fonte *et al.* 2012
Chitosan–decorated selenium nanoparticles (CS-SeNPs)	BAY 55-9837	*In vivo*	increased half life of BAY55-9837	Lei *et al.* 2014
Sterically stabilized micelles (SSM)	Pancreatic polypeptide	*In vivo*	long-lasting half-life	Banerjee and Onyuksel 2013
Poly (lactic-co-glycolic acid)	*Costusspeciosus* extract -NPs	*In vivo*	1. Decreased blood glucose, serum total cholesterol, triglyceride and LDL cholesterol. 2. Alterations in the expression of insulin (I&II) and gluconeogenic genes	Alamoudi *et al.* 2014
Cetylpalmitate based solid lipid nano particles	Insulin	*In vivo*–diabetic rats, (oral administration method)	Insulin protection from partial chemical degradation and promoted intestinal absorption of insulin.	Sarmento *et al.* 2007

Nanotechnology in dental research

Similar to the various potential applications of nanobiotechnology in the treatment of human diseases, nanoparticles are reported to have significant role in oral health. Some of the recent advancements of nanotechnology in dental research are discussed here. A local nanoanaesthesia has been developed that contains a colloidal suspension with millions of anesthesia dental nanorobots, These nanorobots on reaching the pulp via the dentinal tubules close down all sensation in the tooth. When the treatment is done, the nanorobots may be ordered to reestablish all sensations. This method is advantageous as it is fast, reduces apprehension and completely reversible. Nanotechnology finds applications in hypersensitivity cure, tooth repositioning, nanoroboticdentrifice, impression materials, bone replacement materials, nanoencapsulations, dental durability and cosmetics (Neetha *et al.* 2013). Nanotechnology has been explored to apply the beneficial aspects of silver nanoparticles in the field of dentistry (Correa *et al.* 2015; Pal *et al.* 2007). A variety of dental materials used as fillings have their mechanical characteristics enhanced with the introduction of nano fillers of silica or zirconyl salt with particle size of 2-100nm. Hybrid fillers are also added to combine the advantages of nano and microfilled composites. These nano fillers are either in the form of nanomers or nanoclusters impregnated in the composite resin to make it more lustrous, polishable, translucent like tooth and wear resistance. Nano hybrid composites have less polymerization shrinkage and less water sorption as opposed to nano fill composites. Wear and gloss retention and flexural strengthis high with nano fill composites than nano hybrids. Suture needles including nano size stainless steel crystals have been developed. Dental implants which are used to replace teeth are primarily structured with titanium and surface modification is done with controlled surface topography with micro irregularities and impregnation of bioactive materials over the surface to enhance osseointegration. Nano size titanium particles of 20-50 micrometers in the form of a gel emulstion with laser irradiation and activation can be used in periodontal therapy, melanin removal, caries preparation and cutting of hard and soft tissues. Dental impression materials are modified with nanotechnology in siloxane group of impression materials to improvise on the physical properties of flow, hydrophilicity and detail capture of the tissues (Chopra 2007; Alt *et al.* 2004; Samuel and Guggenbichler 2004; Gosheger *et al.* 2004). Several other nanoparticles have been identified for oral and dental problems, some of them are summarized in the table 6.

Table 6: Advancements of nanoresearch in dentistry.

Nano formulation	Purpose	Type of material	Experimental study	Effect/ Result	Reference
Quaternary ammonium dimethacrylate +silver nanoparticles and nanoparticles of amorphous calcium phosphate.	To prevent biofilm accumulation	Composite resin	*Invitro*	Exhibited antimicrobial activity and remineralization. It also reduced the CFU counts, metabolic activity, and lactic acid production	Cheng *et al.* 2012
Titania nanotubes with silver nano particles	Against Biofilm formation	Ti implants	*Invitro* studies	Antibacterial activity up to 30 days. It has promising application in orthopedics and dentistry.	Zhao *et al.* 2011
Embryonic dental follicle tissue + HA-coated dental implant	Potential Artificial (tooth) organ replacement	Biohybrid	Murine tooth loss model	The implant restored physiological functions, as well as bone remodelling, regenerationof severe bone-defect and reaction to noxious stimuli.	Oshima *et al.* 2014
Glyoxylic acid+ Hydroxyapatite nanofibres	Influence of GA on HAP nanofibres dispersion with dental composites and mechanical properties	Dental resins	Characterization Techniques	Enhanced mechanical strength and reinforced dental composites	Chen *et al.* 2013
Amorphous calcium phosphate nanoparticles	Remineralization and Tooth decay inhibition	Composite resin	*Invitro*	Remineralizing property was observed and has caries inhibiton property	Xu, *et al.* 2011

Nanotechnology and gene therapy

Gene therapy involves repairing of the faulty gene usually uses virus as the gene delivery system. Virus,however can themselves cause problems in that they may initiate inflammation. Thus researchers are aiming to use nanoparticles to deliver the genes. It is quite known that, nucleic acids cannot diffuse directly through plasma membrane due to their size and/or their physicochemical properties, for instance, hydrophilicity. By using nanotechnology different strategies for the transfer of nucleic acids, especially genes have been developed (Mannell *et al.* 2012; Iwashita *et al.* 2012; Siu *et al.* 2012; Mulligan *et al.* 1993). In order to transfer the gene with high transfection efficiency, low cellular toxicity and specificity nano particle based chitosan, polyethylenimine, dendrimers and other nano carrier system were developed to deliver genes. The important reasons why thesematerials can be used asthe vectors for gene delivery is that they can interactwith plasmid DNA to form nano-sized complexes, which is the premise to pass through the cellmembrane. With the development of nanomaterials, nano-sized lipids and polymers have shown great potential as non-viral vectors for gene delivery (Jin *et al.* 2014). The nano based gene therapy has more advantage than conventional method. Conventional gene therapy has several limitations such as transfection efficiency, immunogenicity and cytotoxicity. Researchers had foundout a naturally occurring cationic polysaccharide chitosan which has more transfection efficiency with least cytotoxicity. The chitosan nanoparticles have the efficiency to make up the polyelectrolyte complex with DNA and it is highly efficient in non-viral vector based genetherapy (Jayakumar *et al.* 2010) Some of the nanocarriers that are used in gene delivery are shown in table 7.

Table 7 : Nanoparticle based gene delivery Methods

Nanocarrier	Drug formulation	Disease	Target	Outcome	Reference
Nano structured lipid carrier	Doxorubicin or Paclitaxel with Si RNA	Lung cancer	Multiple drug Resistance Protein1 (MRP1) mRNA and B-Cell Lymphoma 2(BCL2) mRNA	Targets only tumor cells and non tumor cells were intact	Taratula *et al.* 2013
Liposome	DSPE-PEG(2000)-COOH-Tf-SL-DOX/ADR (Doxorubicin or Adriamycin)	Cancer	TfR of tumor cells (Human hepato cellular carcinomacell line (HepG2)	Good anti cancer effect with low conc of nanodrug	Li *et al.* 2009, Daniels *et al.* 2012

Conclusion

Early and accurate diagnosis of the diseases is crucial for a timely treatment. Nanotechnology is currently undergoing an explosive development in the field of medicine that offers new promising biodegradable materials and devices with a vast range of translational perspectives in health research. The technique could create better solution and can offer improved features in medical and dental health in terms of accurate diagnosis of disease, effective drug delivery systems and therapy. However before implicating the use of nanotechnology based therapy in human, the side and after effects of these small nanoparticles on human immune responses, if any should be studied. Metabolism, elimination and absorption means of these particles should be well documented in order to achieve safety for human health and environment.

References

Abraham A.M., Kannangai R. and Sridharan G. (2008) Nanotechnology: a new frontier in virus detection in clinical Practice, Indian Journal of Medical Microbiology 26 (4) : 297-301.

Alamoudi E.F., Khalil W.K.B., Ghaly I.S., Hassan N.H.A and Ahmed E.S (2014) Nanoparticles from of Costusspeciosus Extract Improves the Antidiabetic and Antilipidemic Effects Against STZ-induced Diabetes Mellitus in Albino Rats. Int. J. Pharm. Sci. Rev. Res. 29(1): 279-288.

Alt V., Bechert T., Steinrucke P., Wagener M., Seidel P., Dingeldein E., Domann E and Schnettler R. (2004) An in vitro assessment of the antibacterial properties and cytotoxicity of nanoparticulate silver bone cement. Biomaterials. 25(18): 4383–4391.

Anand P., Nair H.B., Sung B., Kunnumakkara A.B., YadavV.R.,Tekmal R.R and Aggarwal B.B. (2010) Design of curcumin-loadedPLGA nanoparticles formulation with enhanced cellularuptake, and increased bioactivity *in vitro* and superiorbioavailability *in vivo*. Biochem. Pharmacol. 79: 330-338.

Andresen T.L., Jensen S.S. and Jorgensen K. (2005) Advanced strategies in liposomalcancer therapy: Problems and prospects of active and tumor specific drug release. Progr. Lipid Res. 44: 68-97.

Banerjee A. and Onyuksel H. (2013) A novel peptide nanomedicine for treatment of pancreatogenic diabetes. Nanomedicine: Nanotechnology, Biology, and Medicine. 9: 722–728.

Bhattacharyya S.S., Paul S., De A., Das D., Samadder A., BoujedainiNandKhuda-Bukhsh A.R. (2011) Poly (lactide-co-glycolide) acidnanoencapsulation of a synthetic coumarin: cytotoxicityand bio-distribution in mice, in cancer cell line andinteraction with calf thymus DNA as target. Toxicol. Appl. Pharmacol. 253: 270-281.

Biana G., Sakamoto J.H., Rita E.S., Alessandro G., Ali B., and Mauro F. (2010) Emerging applications of nanomedicine for therapy and diagnosis of cardiovascular diseases. Trends Pharmacol Sci. 31(5): 199-205.

Bisht S., Feldmann G., Soni S., Ravi R., Karikar C., Maitra A. and Maitra A. (2007) Polymeric nanoparticle-encapsulated curcumin ("nanocurcumin"): a novel strategy for human cancertherapy. J Nanobiotechnology. 5: 3.

Chellat F., Merhi Y, Moreau A and Yahia L.H. (2005) Therapeutic potential of nanoparticulate systems for macrophage targeting. Biomaterials. 26: 7260-7275.

Chen H. and Yada R. (2011) Nanotechnologies in agriculture: New tools for sustainable development. Trends in Food Science & Technology 22: 585-594.

Chen L., Xu C., Wang Y., Shi J., Yu Q. and Li H. (2013) BisGMA/TEGDMA dental nanocomposites containing glyoxylic acid modified high-aspect ratio hydroxyapatite nanofibers with enhanced dispersion, Biomed Mater. 7(4): 045014.

Chen W., Vucic E., Leupold E., Mulder W.J., Cormode D.P., Briley-Saebo K.C., Barazza., Fisher E.A., Dathe M and Fayad Z.A. (2008) Incorporation of an apoE-derived lipopeptide in high-density lipoprotein MRI contrast agents for enhanced imaging of macrophages in atherosclerosis. Contrast Media Mol Imaging. 3 (6): 233–42.

Cheng L, Weir M.D., Xua H.H.K., Antonucci J.M., Kraigsleye A.M., Line N.J., Lin-Gibsone, S. and Zhoub X, (2012) Antibacterial amorphous calcium phosphate nanocomposites with a quaternary ammonium dimethacrylate and silver nanoparticles. Dental materials. 28: 561–572.

Chopra I. (2007) The increasing use of silver-based products asantimicrobial agents: a useful development or a cause forconcern? Journal of Antimicrobial Chemotherapy 59(4): 587–590.

Corot C., Robert P., Idee J.M. and Port M. (2006) Recent advances in iron oxide nanocrystal technology for medical imaging. Adv Drug Deliv Rev. 58(14): 1471–504.

Correa J.M., Mori M., Sanches H.L., da Cruz A.D., Poiate E JrandPoiate I.A. (2015) Silver Nanoparticles in Dental Biomaterials, International Journal of Biomaterials. Article ID 485275, 9.

Daniels T.R., Bernabeu E., Rodriguez J.A., Patel S., Kozman M.,Chiappetta D.A., Holler E., Ljubimova J.Y., Helguera G and Penichet M.L. (2012) The transferrin receptor and the targeted delivery of therapeutic agents against cancer. Biochimica et Biophysica Acta. 1820: 291–317.

DiMauro P.P., Gómez-Vallejo V., Baz Z., Llop J and Borros S (2015) Novel 18F labeling strategy for polyester-based NPs for in vivo PET-CT imaging.BioconjugChem. 26(3): 582-92.

Fargnoli A.S., Mu A., Katz M.G., Williams R.D., Margulies K.B., Weiner D.B., Yang S and Bridges C.R. (2014) Anti-inflammatory loaded poly-lactic glycolic acid nanoparticle formulations to enhance myocardial gene transfer: an *in-vitro* assessment of a drug/gene combination therapeutic approach for direct injection. Journal of Translational Medicine, 12: 171.

Fenniri H. (2006) The Canadian Regenerative Medicine and Nanomedicine Enterprise (CARMENE). International Journal of Nanomedicine. 1(3): 225–227.

Fonte P., Soares S., Costa A., Andrade J.C., Seabra V., Reis S and Bruno S (2012) Effect of cryoprotectants on the porosity and stability of insulin-loaded PLGA nanoparticles after freeze-drying. Biomatter 2(4): 329–339.

FuA., Wilson R.J.,Smith B.R., Joyce M., Chris E., Demir A., Samira G, , Shan X. Wand GambhirS.S. (2012) Fluorescent Magnetic Nanoparticles for Magnetically Enhanced Cancer Imaging and Targeting in Living Subjects, ACS Nano. 6 (8): 6862–6869.

Gopalakrishna P. (2014) Nanomedicines for Cancer Therapy: An Update of FDA Approved and Those under Various Stages of Development. SOJ Pharmacy & Pharmaceutical Sciences. 1(2):13.

Gosheger G., Hardes J., Ahrens H., Streitburger A., Buerger H., Erren M., Gunsel A., Kemper F.H., Winkelmann W and Von EiffC (2004) Silver-coated mega endoprostheses in a rabbit model-An analysis of the infection rate and toxicological side effects. Biomaterials. 25(24): 5547–5556.

Hainfeld J.F., Slatkin D.N., Focella T.M. and Smilowitz H.M. (2006) Gold nanoparticles: A new X-ray contrast agent. Br J Radiol 79: 248-253.

Hainfeld J.F.,SmilowitzH.M.,O'Connor M.J.,Dilmanian, F.A., and Slatkin D.N(2013) Gold nanoparticle imaging and radiotherapy of brain tumors in mice. Nanomedicine: 8(10): 1601–1609.

Hariharan S., Bhardwaj V., Bala I., Sitterberg J., Bakowsky U. and Ravi Kumar M.N. (2006) Design of estradiol loaded PLGA nanoparticulate formulations: a potential oral delivery system for hormone therapy. Pharm Res 23: 184-195.

He H, Li Y., Jia X.R, Du J, Ying X., Lu W.L, Lou J.N. and Wei Y (2011) PEGylated Poly (amidoamine) dendrimer-based dual-targeting carrier for treating brain tumors, Biomaterials 32: 478–487.

Iwashita S., Hiramatsu Y., Otani T., Amano C., Hirai M., Oie K., Yuba E., Kono K., Miyamoto M and Igarashi K (2012) Polyamidoaminedendron-bearing lipid assemblies:Their morphologies and gene transfection ability. J BiomaterAppl.27(4): 445-456.

Jaracz S., Chen, J., Kuznetsova, L.V. and Ojima, I. (2005) Recent advances in tumor targeting anticancer drug conjugates. Bioorg. Med. Chern.13: 5043-5054.

Jayakumar R., Chennazhi K.P., Muzzarelli R.A.A., Tamura H., Nair S.V. and Selvamurugan N (2010) Chitosan conjugated DNA nanoparticles in gene therapy. Carbohydrate Polymers. 79: 1–8.

Jin L., Zeng X., Liu M., Deng Y and He N (2014) Current Progress in Gene Delivery Technology Based on Chemical Methods and Nano-carriers. Theranostics 4(3): 240-255.

JooyeonA., Yutaka M., Naoki Y., Tsukasa C., Xueying L., Ahram K., Ryuta S., Ryo T., Yoshikatsu K., Masahiro Y., NobuhiroN., Yasuhiro M., Horacio C and Kazunori K. (2015) Antibody fragment-conjugated polymeric micelles incorporating platinum drugs for targeted therapy of pancreatic cancer. Biomaterials. 39: 23-30.

Kawadkar J., Meenakshi K. Chauhan and Maheshwari Maharana (2011) Nanobiotechnology: Application of nanotechnology in diagnosis, drug discovery and drug development.Asian Journal of Pharmaceutical and Clinical Research, 4(1): 23-28.

Khalid K.A and Yazeed A.A. (2014) Role and implications of nanodiagnostics in the changing trends of clinical diagnosis. Saudi Journal of Biological Sciences. 21: 109–117.

Khuda-Bukhsh A.R., Bhattacharyya S.S., Paul S. and Boujedaini N. Polymeric nanoparticle encapsulation of a naturally occurring plant scopoletin and its effects on human melanoma cell A375. (2010) Zhong Xi Yi Jie He XueBao. 8(9): 853-862.

Kim D.,Park S., Lee.J.H.,Jeong J.H. and Jon S. (2007) Antibiofouling polymer-coated gold nanoparticles as a contrast agent for in vivo X-ray computed tomography imaging. J Am ChemSoc. 129(24): 7661–5.

Kunte H., Schmidt S., Eliasziw M., del Zoppo G.J., Simard J.M., Masuhr F., Weih M. and Dirnagl U. (2007) Sulfonylureas improve outcome in patients with type 2 diabetes and acute ischemic stroke. Stroke 38: 2526-2530.

Kwatra D., Anand V. and Shrikant A. (2013) Nanoparticles in radiation therapy: a summary of various approaches to enhance radiosensitization in cancer.Translational Cancer Research. 2(4): 330-342.

Lanza R.P., Langer R. and Chick WL (1997) Principles of Tissue Engineering. Austin, TX: Academic Press, pp. 405-427.

Lei R., Ma Y., Zhuang M.,Luo T., Wang Y and Hong A (2014) Chitosan-decorated selenium nanoparticles as protein carriers to improve the *in vivo* half-life of the peptide therapeutic BAY 55-9837 for type 2 diabetes mellitus. International Journal of Nanomedicine. 9: 4819-28.

Li X., Ding L., Xu Y., Wang Y., Ping Q (2009) Targeted delivery of doxorubicin using stealth liposomes modified with transferring., Int. J. Pharm. 373: 116–123.

Mannell H., Pircher J., Fochler F., Stampnik Y., Räthel T., Gleich B., PlankC., Mykhaylyk O., Dahmani C., Wörnle M., Ribeiro A., Pohl U and Krötz F (2012) Site directed vascular gene delivery *in vivo* by ultrasonic destruction of magnetic nanoparticle coated microbubbles. Nanomedicine. 8(8): 1309-1318.

Masahiro Y., Takeshi U., Naoki T., Takayuki I., Kaori N., Norio H., Takeo S and Takahiro O. (2012) Bone integration capability of nanopolymorphic crystalline hydroxyapatite coated on titanium implants. International Journal of Nanomedicine. 7: 859–873.

Michalet X, Pinaud F.F., Bentolila L.A., Tsay J.M., Doose S., Li J.J., Sundaresan G., Wu A.M., Gambhir S.S and Weiss S (2005) Quantum dots for live cells, in vivo imaging, and diagnostics. Science. 307(5709): 538–44.

Mulligan RC. (1993) The basic science of gene therapy. Science 260: 926-932.

Neetha J.S., Swati P and David K (2013) Nanorobots: Future in dentistry, The Saudi Dental Journal. 25: 49–52.

Omid C.F and Robert L. (2009) Impact of Nanotechnology on Drug Delivery. 3(1): 16-20.

Oshima M., Inoue K., Nakajima K., TachikawaT.,Yamazaki H., Isobe T., Sugawara A., Ogawa M., Tanaka C., Saito M., Kasugai S., Takano-Yamamoto T., Inoue T., Tezuka K., Kuboki T., Yamaguchi A and Tsuji T (2014) Functional tooth restoration by next-generation bio-hybrid implant· as a bio-hybrid artificial organ replacement therapy. Scientific Reports 4: 6044.

Pal S., TakY.K and. Song J.M. (2007) Does the antibacterial activity of silver nanoparticles depend on the shape of the nanoparticle?A study of the gram-negative bacterium *Escherichia coli*. Applied and Environmental Microbiology. 73(6): 1712–1720.

Palange A.L., Di Mascolo D., Carallo C., Gnasso A and Decuzzi P (2014) Lipid–polymer nanoparticles encapsulating curcumin for modulating the vascular deposition of breast cancer cells. Nanomedicine. 10: 991-1002.

Patel A., MacMahon S., Chalmers J., Neal B., Billot L., WoodwardM., Marre M., Cooper M., Glasziou P., Grobbee D., Hamet P., Harrap S., Heller S., Liu L., Mancia G., MogensenC.E., Pan C., Poulter N., Rodgers A., Williams B., BompointS., de Galan B.E., Joshi R. and Travert F. (2008) Intensive blood glucose control and vascular outcomes in patients with type 2 diabetes. N Engl J Med 358: 2560-2572.

Pu K., Shuhendler A.J., Jokerst J.V., Mei J., Gambhir S.S., Bao Z. and Rao J. (2014) Semiconducting polymer nanoparticles as photoacoustic molecular imaging probes in living mice, Nature Nanotechnology. 9: 233–239.

Rabin O., Manuel P.J., Grimm J., Wojtkiewicz G and Weissleder R. (2006) An X-ray computed tomography imaging agent based on long-circulating bismuth sulphide nanoparticles. Nat Mater. 5(2): 118-22.

Roby A., Erdogan S. and Torchilin, V.P. (2006) Solubilization of poorly soluble PDTagent, meso-tetraphenylporphin, in plain or immune targeted PEG-PE micelles results in dramatically improved cancer cell killing *in vitro*. Eur. I. Pharm. Biopharm. 62: 235-240.

Rosen J.E., Yoffe S., Meerasa A., Verma M., Gu F.X. (2011) Nanotechnology and Diagnostic Imaging: New Advances in Contrast Agent Technology. J, Nanomedic Nanotechnol. 2: 115.

Ruibing W., Paul S.B. and Wayne M.M. (2013) Nanomedicine in Action: An Overview of Cancer Nanomedicine on the Market and in Clinical Trials. Journal of Nanomaterials. 12 pages.

Rutledge E.B. (2007) Nano Neurology and the Four P's of Central Nervous System Regeneration: Preserve, Permit, Promote, Plasticity. Medical Clinics of North America. 91(5): 937-962.

Sachiko K.N and KeijiN (2013) Biopolymer-Based Nanoparticles for Drug/Gene Delivery and Tissue Engineering. Int. J. Mol. Sci.14: 1629-1654.

Samadder A., Das J., Das S., De A., Saha S.K., Bhattacharyya S.S. andKhuda-Bukhsh A.R (2013a) Poly(lactic-co-glycolic) acid loaded nano-insulin has greater potentials of combating arsenic induced hyperglycemia in mice: some novel findings. Toxicol Appl Pharmacol. 267: 57-73.

Samadder A., Das S., Das J. And Khuda-Bukhsh A.R (2013b) Relative efficacies of insulin and poly (lactic-co-glycolic) acid encapsulated nano-insulin in modulating certain significant biomarkers in arsenic intoxicated L6 cells. Colloids Surf Biointerfaces. 109: 10-19.

Samuel U. and Guggenbichler J.P. (2004) Prevention of catheter relatedinfections: the potential of a new nano-silver impregnated catheter. International Journal of Antimicrobial Agents. 23(1): 575–578.

Sapra P and Allen T.M. (2003) Ligand-targeted liposomal anticancer drugs. Progr. Lipid Res. 42: 439-462.

Sarmento B., Martins S., Ferreira D. And Souto EB (2007) Oral insulin delivery by means of solid lipid nanoparticles. International Journal of Nanomedicine. 2(4): 743–749.

Seong H.C., Hyeong J.B., Ji S.C, LequangT., Insoo K., EunSeong L., BeomSoo S., Kang C.L and Yu S.Y. (2015) Inhalable self-assembled albumin nanoparticles for treating drug-resistant lung cancer. Journal of Controlled Release. 197: 199–207.

Shrivastava S.R.B.L., Shrivastava P.S. and Ramasamy J. (2013) Role of self-care in management of diabetes mellitus.Journal of Diabetes & Metabolic Disorders. 12: 14.

Siu Y.S., Li L., Leung M.F., Lee K.L and Li P. (2012) Polyethylenimine-based amphiphilic-core-shell nanoparticles: Study of gene delivery and intracellular trafficking. Biointerphases. 7(1-4): 16.

Stoimenov P.K., Klinger R.L., Marchin G.L. and Klabunde K.J (2002) Metal oxide nanoparticles as bactericidal agents. Langmuir 18: 6679-6696.

Takuya U., ParthaD., Edmund K., Florian L., MaulikM., Brett M., Yoshiko I., Jose-Luiz F, Christen T, Swirski F.K., Libby P., Weissleder R and Nahrendorf M. (2013) Nanoparticle PET-CT Detects Rejection and Immunomodulation in Cardiac Allografts.Circ Cardiovasc. Imaging. 6: 568-573.

Taratula O., Kuzmov A., Shah M, Garbuzenko O.B.,Minko T (2013) Nanostructured lipid carriers as multifunctional nanomedicine platform for pulmonary co-delivery of anticancer drugs and siRNA. Journal of Controlled Release. 171: 349–357.

Vahid S., Senthilkumar S., Xiaoran H., Morteza M., Andrey V.,Malkovskiy, Zhao M., Inayathullah M., DhananjayW., Zhang X.J, Scott M.,Daniel B., Wu J.C., Ruiz-Lozano P and Rajadas J (2015) [Pyr1]-Apelin-13 delivery via nano-liposomal encapsulation attenuates pressure overload-induced cardiac dysfunction. Biomaterials. 37: 289-298.

Verma V.K., Kamaraju S.R., RavindranathK., Kona L.K., BeeviS.S.,Debnath T.,Usha S P.,Vadapalli R., Syed AA., and Chelluri L.K. (2015) Fluorescent magnetic iron oxide nanoparticles for cardiac precursor cell selection from stromal vascular fraction and optimization for magnetic resonance imaging. Int J Nanomedicine. 10: 711–726.

Wang Y., Zhang W., Zhang J., Sun W., Zhang R and Gu H (2013) Fabrication of a novel polymer-free nanostructured drug-eluting coating for cardiovascular stents, ACS Appl Mater Interfaces. 5(20): 10337-45.

WHO (2003) Global Strategy on diet, physical activity and health.

Wise D.L., Fellman T.D., Sanderson J.E and Wentworth R.L.(1979) Lactide/glycolide acid polymers. In: Geregoriadis G, editor.London: Drug Carriers in Biology and Medicine, Academic Press. pp. 237-270

World Health Organization Report on cardio vascular diseases (2015).

World Heart Federation (2011) Cardiovascular disease: types and symptoms.

Xu H.H., Monreau J.L., Sun L. and Chow L.C. (2011) Nanocomposite containing amorphous calcium phosphate nanoparticles for caries inhibition. Dent Mater. 27:762–769.

Yang J., Luo Y., Xu Y., Li J., Zhang Z., Wang H., Shen M., Shi X and Zhang G. (2015) Conjugation of Iron Oxide Nanoparticles with RGD-Modified Dendrimers for Targeted Tumor MR Imaging, ACS Appl Mater Interfaces. 2015 Feb 26.

Yang J., Luo Y., Xu Y., Li J., Zhang Z., Wang H., Shen M., Shi X. and Zhang G. (2015) Conjugation of Iron Oxide Nanoparticles with RGD-Modified Dendrimers for Targeted Tumor MR Imaging.ACSAppl Mater Interfaces. 7(9): 5420-8.

Yeole MP., Shashikanth N D. and Nilesh S K (2013) Peptide nanomedicine in cancer treatment. Asian Journal of Pharmaceutical and Clinical Research, l6(2): 28-32.

Zhao L., Wang H., Huo K., Cui L., Zhang W., Ni H., Zhang Y., Wu Z and Chu P.K. (2011) Antibacterial nano-structured titania coating incorporated with silver nanoparticles. Biomaterials. 32: 5706-5716.

11

Role of Plant Secondary Metabolites in Cancer Cell Therapy

Jagjit Kaur, Diwakar Aggarwal* and Kuldeep Kumar
Department of Biotechnology, Multani Mal Modi College, Patiala-147001
Punjab, India, **E-mail: diwakar.tu@gmail.com*

Abstract

Cancer is a potent disease causing nearly 6 million deaths per year worldwide. There are a number of cancers such as skin, brain, lung, breast, stomach, colon and blood cancer caused due to number of factors like unbalancing of hormones, smoking *etc*. Cancer could be treated by chemotherapies or through irradiation therapies. But treatment using the natural sources such as plants is an emerging trend. Plants have been known for their medicinal properties since ancient times. A number of records are present which enlist different plants and their medicinal use such as Ayurveda (India), *Materia Medica* (China) and *Ebers Papyrus* (Egypt). Plants produce a number of compounds which can be classified under primary metabolites and secondary metabolites. Plants that are known to possess anti-tumor properties are *Catharanthus roseus, Taxus brevifolia, Betula alba, Curcuma longa, etc*. They produce the secondary metabolites which are responsible for anti-tumor properties: taxenes, vinca alkaloids, vincristine, berbamine, â- lapachone, podophyllotxin, and camptothecins. The secondary metabolites may be produced *in vitro* through callus culture or organ culture. Different bioreactors such as air-lift, bubble column, rotating drum, turbine blade, trickle bed and stirred tank bioreactors are used for the production of secondary metabolites. The choice of the bioreactor is made on the basis of the culture that needs to be cultivated.

Keywords : Cancer, Secondary metabolites, Bioreactors, Callus culture, Organ culture.

Introduction

Plants are an important part of our daily life and are also studied for decades. Plants produce a number of compounds that are therapeutically and biotechnologically active (Quiroga *et al.* 2001). The extracts from plants have been used for the treatment of various diseases in humans and animals. The secondary metabolite content is increased in food and these serve as anticarcinogenic and cardio protective agents. Since ancient times plants have been used for the medicinal purposes in a number of countries such as India, China, and Egypt. The usage of plants for medicinal purposes dates back to 2600 BC and the records are given by the Sumerians and the Akkaidians (Samuelson 1999). The Indian record of medically important plants is given in Susruta and Charaka (1000 BC) (Kappor 1990). *Materia Medica* is the Chinese record of medically important plants accounting for nearly 600 plants (1100BC) (Cragg *et al.* 1997), whereas *Ebers Papyrus* is the Egyptian record which gives the account of 700 drugs (1500 BC). Plants produce a number of products which are classified into primary and secondary metabolites. Primary metabolites such as amino acids, nucleotides, organic acids, etc. are those substances which are used by the plants for their growth and development whereas the secondary metabolites are low molecular compounds do not participate in the growth processes. During photosynthesis plants absorb sun light and produce oxygen and secondary metabolites (Ghasemzadeh *et al.* 2011). Secondary metabolites have complex structure and are isolated from wild and cultivated plants (Marja *et al.* 2004). The secondary metabolites are used as anti-tumor agents, the flavoring agents, dyes, oils, waxes and food additives (Hussain *et al.* 2011). There are different roles played by secondary metabolites in the plants, such as interactions with insects, micro-organisms or other plants (Harborne 2001; Dixon 2001), as protective agents by the production of antifeedants, phytoanticipins (Dixon 2001) and in reproduction. Secondary metabolites also help in enhancing the immunity of humans (Atoui *et al.* 2005). Cancer is a very potent disease caused by hormonal or dietary imbalances and smoking (Ames *et al.* 1995). Cancer causes a number of deaths every year and it ranks the second among the death-causing diseases (Nirmala *et al.* 2011) causing 6 million deaths per year worldwide (Pinar 1998). Cancer could be treated with radiotherapy or chemotherapy but the use of natural products in treatment is an emerging field (Mann 2004). Cancer is the uncontrolled growth of cells which is caused by certain imbalances in the body. It can be controlled by encouraging the strategies which reduces the DNA damage and enhances its repair (Bertan 2001). The most common type of cancer is breast cancer (Koduru *et al.* 2007) and occurs in one woman out of every 31 women in South Africa. Skin cancer is the most common type of cancer among men in United States and causes 37 000 deaths per year and prostate cancer ranking second most common type of cancer (American Cancer Society 1999). *Catharanthus roseus, Taxus brevifolia, Betula alba, Curcuma longa* and many plants produces secondary metabolites which can be used as anti-tumor agents. The discovery of vinca alkaloids, vincristine enlightened the paths towards the use of plant products as the anti-tumor agents nearly around 1950s. After this many other anti-tumor metabolites were identified such as taxanes and camptothecins (Cragg *et al.* 2005). In the earlier stages of growth cycle i.e. lag phase and the log phase carbon is mainly needed for the production of primary metabolites but after that i.e. during the stationary phase carbon is no longer required and plants began to produce secondary metabolites and also the enzymatic activities are the most active during this phase (Payne *et al.* 1991). The secondary metabolites produced are stored in the plant vacuoles which possess the difficulty in their extraction as the cells have to be first broken down to obtain the products. As compared to the microbial cells plant cells have longer life cycle of nearly 0.12 to 0.05 d^{-1} (Dicosma and Misawa 2005). The patients who have become resistant to the traditional therapies or could not survive them due to old age or weakness are recommended for herbal medicines in combination with the chemotherapy and radiotherapy.

Major Secondary Metabolites Used As Anti-Cancer Agents

Vinca Alkaloids

Vinca alkaloids are obtained from periwinkle *Catharanthus roseus* naturally or synthesized chemically (Jordan *et al.* 1992). Vinca alkaloids in conjugation with other chemotherapeutic agents are used for the treatment of breast cancer, lung cancer, testicular cancer (Cragg *et al.* 2005). They block the mitosis process by affecting the micro tubular dynamics leading to the cell death. Vinca alkaloids usually attack on the tubulin protein (Wilson *et al.*, 1974). Mainly used vinca alkaloids are vincristine (VCR), vinblastine (VBL), vinorelbine (VRL). VCR is used in the treatment therapies for acute lymphoblastic leukemia, Hodgkin and non-Hodgkin lymphoma, Wilm's tumor,

Ewing sarcoma, lung cancer and neuroblastoma. VBL is used as chemotherapeutic agent for the treatment of bladder, breast, and brain cancers and also in Kaposi sarcoma. VRL is used in the treatment of ovarian cancer. VCR and VBL differ in their mode of action. For example VCR gives positive results in the treatment of Ridgway osteogenicsarcosoma and CA755 mammary carcinoma but VBL does not give any significant results (Sweeney *et al.* 1974). These two also differ in the side-effects shown by them. VCR causes neuropathy whereas VBL causes marrow suppression. A new form of vinca alkaloids desacetyl vinblastine amide (VDS) is under the phase I clinical trials and also showing both the effects of neuropathy and marrow suppression (Blum *et al.* 1976). A study was performed comparing the effects of VCR, VBL and VDS, it was found that there was hardly any difference in their effect on inhibition of polymerization of tubulin protein and distortion of the microtubules was the common effect of the three drugs (Richards *et al.* 1976).

Camptothecins

Camptothecins (Fig. 1) are water-insoluble alkaloids which are isolated from *Camptotheca acuminate* (Torck *et al.* 1996). These are used in the treatment of solid tumors hematopoietic malignancies (Zunino *et al.* 2004), breast, lung and colorectal cancer (Monroe *et al.* 1996). The mode of action of camptothecins is that they affect the DNA topoisomerase enzyme, which is involved in the replication process. The major side-effects are haemorrhagic cystitis and diarrhoea (Torck *et al.* 1996). The two effective derivatives of camptothecins are topotecan (TPT) and CPT-11. CPT-11 is very effective against lung, cervix, ovary cancer and non-Hodgkin's lymphoma. The side-effects are neutropenia and gastrointestinal effects (Slichenmyer *et al.* 1993). Besides the anti-tumor activity of camptothecins, they also show anti-malarial activity, inhibition of HIV, induction of apoptosis and when exposed to ultra violet light they give out intense blue fluorescence.

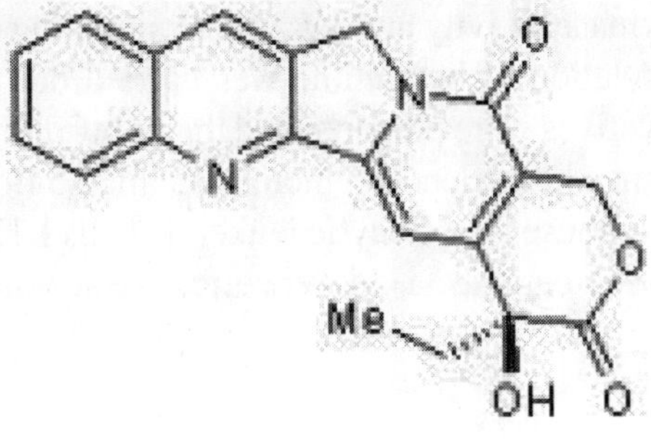

Fig. 1 : Structure of Camptothecin

Taxol

Taxol is an anti-tumor drug isolated from *Taxus brevifolia* (Wani *et al.* 1971) used in the treatment of breast cancer, ovarian cancer and bronchial sarcoma (Pazdur *et al.* 1993). Normally the microtubules get depolymerized by cold or calcium but taxol stabilizes them under these conditions. This drug has some unique characters such as specific binding site and its ability to polymerize the tubulin even when the cofactors and proteins are absent. The drug binds covalently to the â-subunit of tubulin in the presence of ultra-violet rays (Horwitz 1994).

Berbamine

Berbamine (bisbenzylisoquinoline) with molecular weight 753.80 is an anti-tumor agent isolated from *Berberis amurensis* (Du *et al.* 2010) which causes apoptosis of cancer cells by interfaring with the potential of mitochondrial transmembrane and caspase activation (Guan *et al.* 2007). It also shows anti-neoplastic, anti-inflammatory effects besides anti-tumor activity (Zhu and Sui 1986; Xu *et al.* 2004; Kupeli *et al.* 2002). Berbamine inhibits the growth of KM3 cells at G1 stage and also induces apoptosis when given in combination with dexamethasone, doxorubicin and arsenic trioxide. The effects are dependent on the dose level and time of dose induction (Liang *et al.* 2009). In a study conducted to investigate the anti-tumor effects of berbamine in leukopaenic patients it was found that the number of neutrophils and lymphocytes increased but that of bone marrow cells was increased in blood after the time interval of 2 weeks (Li *et al.* 1994). BBMD3 is the most effective anti-tumor agent out of 13 synthetically synthesized berbamine derivatives (BBMDs) which inhibits Jak2 auto phosphorylation kinase activity and Jak2/Stat3 signaling pathway (Nam *et al.* 2012). Glycosylation of berbamine was carried out and a number of glycolysated derivatives were prepared and their anti-tumor activities were studied against human lung adenocarcinoma cell line A549, human leukemia cell line K562 and mouse lymphocytic leukemia cells L1210. It was found that acetyl glycosylberbamine 5a, 5d were effective against the investigated cell lines (Cui *et al.* 2012).

β-Lapachone

β-lapachone (Naphthoquinones) is a pharmacological agent used in the treatment of liver cancer (Woo *et al.* 2006), prostate cancer and promyelocytic leukemia (Planchon *et al.* 1995) and is isolated from the bark of *Tabebuiaavell anedae,* lapacho tree. It causes the apoptosis of human HepG2 cells along with the degradation of poly(ADP-ribose) polymerase protein and activation of caspase-3, caspase-9. Anti-apoptotic Bcl-2 and Bcl-X(L) are down-regulated without affecting the apoptosis inhibitor protein family (Woo *et al.* 2006). This apoptosis is related with the decreased regulation of telomerase reverse transcriptase which further results in decreased telomerase activity (Moon *et al.* 2010). Cytoplasmic two-electron reductase NAD (P) H: quinine oxidoreductase or NQO1, DT-diaphorase and Xip3 causes the activation of the drug which causes cell death while at higher concentrations of drug reactive oxygen species are formed which also facilitates the cell death (Pardee *et al.* 2002). Beta-lapachone is the activator of topoisomerase I which causes chromosomal aberrations and sister-chromatid exchanges (SCE) during G2 phase of cell cycle. These changes are dose-dependent and increase with the increasing dose of the drug (Degrassi *et al.* 1993).

Podophyllotoxin

Podophyllotoxin (podofilox) is isolated from podophyllum plant species and used for the treatment of lung cancer, genital cancer (Fig. 2). Its glycosidic derivatives such as etoposide and teniposide are used in the treatment of cancer and venereal wart (Lu *et al.*, 2011). Podophyllotoxin is isolated from *Podophyllum peltatum,* American Mayapple plant. Necrosis of condylomas is caused in 3-5 days by binding of podophyllotoxin to cellular microtubules and inhibiting mitosis (Krogh 1981). Podophyllotoxin is given twice a day for three days with a gap period of 4-7 days. Podophyllotoxin is used to treat snakebites, skin disorders, coughs, purgatives (Nagar *et al.* 2011). It binds to the tubulin thus preventing the formation of microtubulin and stopping the cell cycle at metaphase (Canel *et al.* 2000). It may cause some side-effects such as itching, sores, peeling of skin, and even fatal poisoning. Due to these side-effects many modifications were carried out in podophyllotoxin. When A-ring was modified it showed a good activity but modification of B-ring reduced the activity. Podophyllotoxin with modified D-ring produced GP-11 whereas that with modified E-ring had no effect on DNA cleavage. When glycosidic portion of the drug is replaced by arylamines then the activity was increased. NK 611, which is under clinical trials, is produced when the sugar ring is modified (Damayanthi *et al.* 1998).

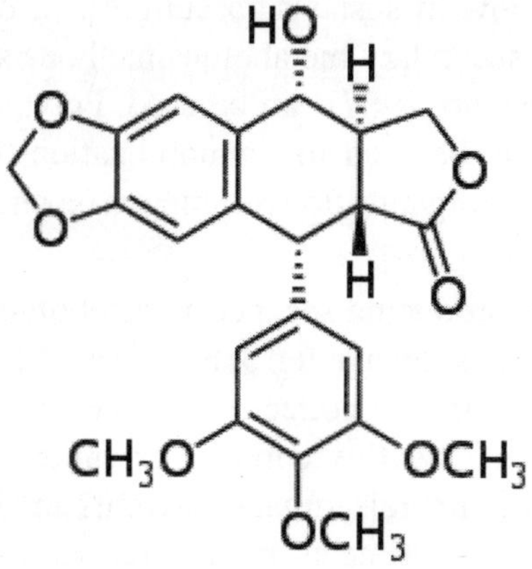

HO H O O O H O CH_3O OCH_3 OCH_3

Fig. 2 : Structure of Podophyllotoxin

Production Methodology

Biotechnological methods

If the biotechnological methods are used for the production of plant secondary metabolites then a number of advantages can be experienced such as: high productivity with the desired qualities, control over the import expenses, and production throughout the year. *Lithospermum erythrorhizon* was the first plant to be produced using *in vitro* methods in 1983 (Fujita and Tabata 1987) and after that some other pharmaceutically important plant products such as berberine, rosmarinic acid, ginseng saponinnes were also produced (Parr, 1989; Yamada and Hashimoto 1990).

Callus and suspension culture

Cultivation of plant cells on solid or semi-solid medium containing all the nutrients gives rise to callus culture, while that the cells growing in liquid medium gives rise to suspension culture. The findings of Zenk *et al.* 1991 that *Morinda citrifolia* could produce 2.5 g of anthraquinones per litre medium used (Zenk *et al.* 1991), discarded the idea that the callus or the suspension cultures could not produce the secondary metabolites (Krikorian and Steward 1969). Suspension cultures are grown on large scale in bioreactors. As compared to micro-organisms plant cells are more sensitive to shear stress and grow slowly so specialized bioreactors are used for their growth (Buitelaar and Tramper 1992). Plant cells are instable and generally the yields are very low in suspension cultures. In order to increase the production of plant secondary metabolites methods such as elicitation or immobilization (Yeoman, 1987) can be used. Polymers such as alginate, carraghenans, etc. can be used for immobilization (Gontier *et al.* 1994) which allows the tissue organization resulting in biochemical differentiation (Brodelius 1984).

The process of producing secondary metabolites through callus and suspension culture is as shown in fig. 3. First of all the plant cells producing the concerned metabolite are collected and the high yielding strains are identified and isolated. After this to induce the callus, the medium with best suited composition of minerals, organic solvents and hormones is selected by incomplete factorial experiments or surface response methods (Box 1954). Through a number of subculturing cycles the calli undergo somaclonal variations (Larkin and Scowcroft 1981) and after a number of generations stable cultures are obtained nearly 2 years (Neto *et al.* 1994). The culture cells are then screened for their growth speed and metabolite concentrations both intracellular and extracellular. The best cells are then cultured in the bioreactors. Scaling up of cultivation from flask level to bioreactors presents a number of difficulties such as: sedimentation and death of cells due to oxygen mass transfer limitations (Hulst 1985), nutrients (Jones and Veliky 1981) and inhomogeneous culture system, shear sensitiveness of plant cells which increases with cell size (Steward *et al.* 1999). To overcome these problems air-lift or bubble reactors may be used (Jolicoeur *et al.* 1992). The cells are then subjected to batch, fed-batch, perfusion fermentations so that they become capable of producing the secondary metabolites. The metabolites may be intracellular and extracellular. When the metabolites are intracellular then the cells are given treatment with pH (Mukundan *et al.* 1998) or chemicals (Brodelius 1988).

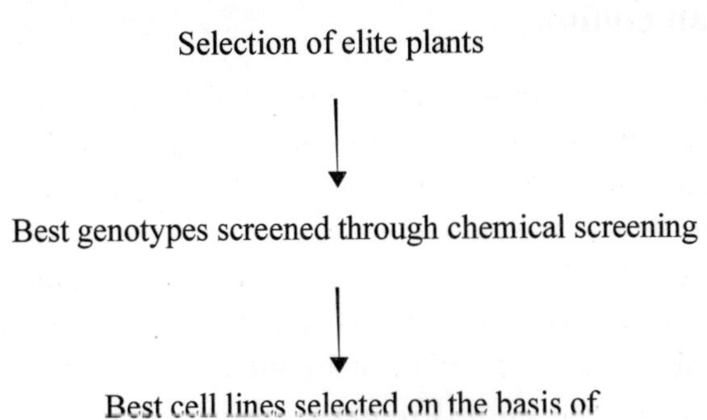

1. Genetic stability
2. Callus initiation
3. Subculture cycles

Cell suspensions studied through elicitation, genetic modifications, immobilization and biosynthetic pathways

Permeabilization, Growth, Process Engineering and Productivity studies of the bioreactor

Fig. 3 : Production of Plant Secondary Metabolites from Plant Cells Through Callus Culturing

Elicitors

When the microbial, physical and chemical factors are induced in plants to enhance their secondary metabolite production, this is known as elicitation and these factors are known as elicitors (Karuppusamy 2009; Kiong *et al.* 2005). It was found through GC and GC-MS analysis that in elicited tissues there is large amount of umbelliferone. In *Rubia akane* cultures, chitosan, a biotic elicitor, enhanced the production of anthraquinone (Jin *et al.* 1999). The production of ginseng saponin by hairy roots of *Panax ginseng* is enhanced by several abiotic elicitors such as selenium and $NiSO_4$ (Jeong and Park 2007). Hairy growth is inhibited by tannic acid and sodium chloride but the production of ginseng saponin is enhanced.

Plant organ culture

Root or shoot cultures can also be used for the production of plant products (Parr 1989) and this is achieved by creating hormonal imbalance. The root cultures also grow slowly; moreover, their manipulation is bit difficult rendering the large scale cultivation hard to be achieved (Charlwood *et al.* 1988). In order to obtain good results transformed cultures can be used which can be obtained by any method like callus or suspension culture (Flores *et al.* 1987; Rhodes *et al.* 1990) or by infection with the soil bacterium *Agrobacterrium rhizogenes* (Chilton *et al.* 1982). The transformed cells obtained by infection with *Agrobacterrium rhizogenes* are called hairy roots and are genetically and biochemically stable. Actively growing hairy roots can produce secondary metabolites which allow their continuous supply (Holmes *et al.* 1997). The production of secondary metabolites can be increased by elicitation (Rijhwani *et al.* 1998) or by changing the medium composition (Tone *et al.* 1997). Plant cells can also be cultivated in the presence of light which shows different biosynthetic capacities (Bhadra *et al.* 1998) but this can reduce the growth rates (Bourgaud *et al.* 1999). Organ cultures are less sensitive to shear stress with few exceptions such as in *Catharanthus roseus* and therefore it is usually grown in air-sparged bioreactors (Nuutila *et al.* 1994) (Figure 4). Apart from growth in air-sparged bioreactors, some other techniques such as polymer matrix (Taya *et al.* 1989), immobilization (Holmes *et al.* 1997), growth in stirred bioreactors (Rhodes *et al.* 1986) can be used to reduce the shear stress. The products produced are usually intracellular (Bourgaud *et al.* 1999). The major problem encountered during organ culturing is the inhomogeneous character of cells. To overcome this submerged cultures (Wyslouzil *et al.* 2000), drip-tube techniques (Holmes *et al.* 1997), two-phase systems (Buitelaar *et al.* 1991; Bassetti *et al.* 1997) can be used. It facilitates the easy recovery of the products in the medium and also prevents their inhibition (Holmes *et al.* 1997; Corry *et al.* 1993).

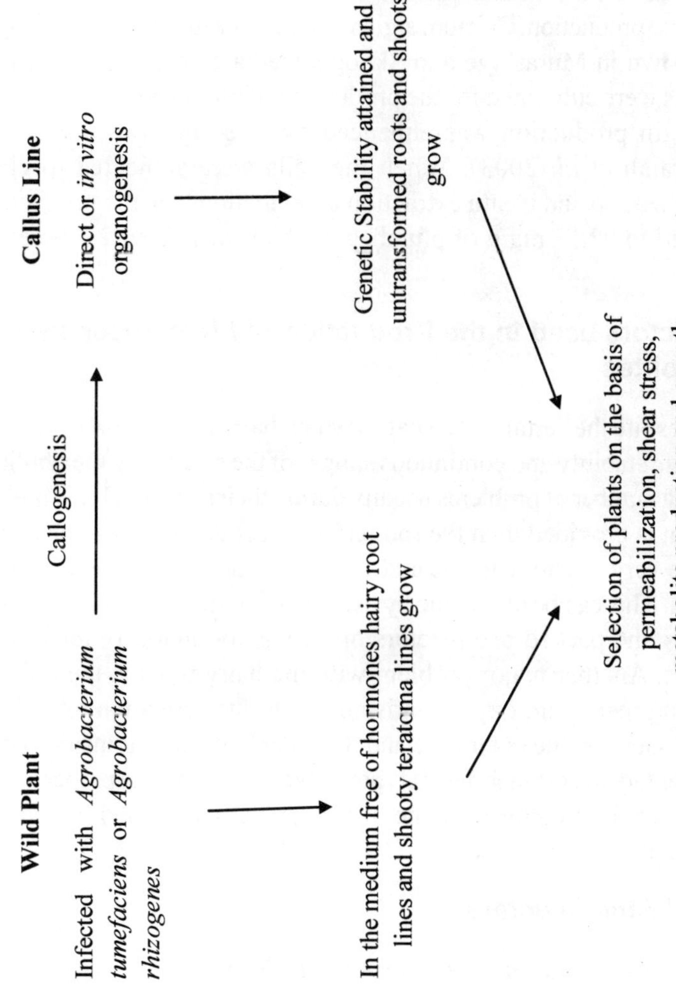

Fig. 4 : Production of Plant Secondary Metabolites through Organ Culture

Immobilization

Increase in the number of high value products synthesis has been possible due to advances in immobilization techniques. A study was conducted to demonstrate the effect of immobilization on the secondary metabolite production. Calcium alginate immobilized cells of *Plumbago rosea* were grown in Murashige and Skoog's media containing 10 mM $CaCl_2$. The cells were cultivated for the production of plumbagin. It was found that plumbagin production was enhanced by three folds to that of control (Komaraiah *et al.* 2003). When the cells were subjected to chitosan, immobilization and in situ extraction combination then the production was increased to 92.13 mg/g of plumbagin which was about 2.5 times that of control.

Bioreactors used in the Production of Plant Secondary Metabolites

Despite the certain advantages which hairy root cultures provide such as genetic stability and continuous supply of the secondary metabolites they possess a number of problems usually during their scale-up. If the mechanical agitation is provided then the root cultures get damaged because the plant cells are very sensitive to shear stress due to their large size and inflexible cell wall. This can be overcome by using batch fermentations for high valued products and packed bed reactors operating continuously for low-valued products. Another major problem with the hairy roots is the difficulty in supplying them with oxygen, arising due to the formation of root ball by young tissues on the outer core and the older ones locked inside. Therefore it is necessary to design different types of bioreactors for the proper cultivation of the hairy root cultures. Some of the specially designed bioreactors are listed below:

Stirred Tank Reactors

The traditional reactors were not suitable for hairy root cultures because the rotation of impellors caused injury to the cultures (Taya *et al.* 1989; Hilton *et al.* 1990), but recently the reactors with steel cage and having large impellors and baffles have been developed which are rotated at low speed. This facilitates the proper cultivation of hairy root cultures.

Airlift or Submerged Bioreactors

Hairy roots are grown well in these reactors as these reactors lack the impellors. The oxygen is supplied through glass grids in which humid air is passed at the rates of 0.05-0.4 vol of air/vol of liquid/min (Taya *et al.*, 1989; Tescionen *et al.* 1997).

Bubble Column Reactor

The reactor is provided with multiple spargers which increases the transfer rates of oxygen and biomass. To attain a better growth of the hairy roots the bubbling rate should be increased as the hairy roots grow (Buitelaar *et al.* 1991).

Trickle Bed Reactor

The medium is supplied from the top of the reactor along with the humid air. This creates a mist and hence trickle bed reactor is also known as mist reactor. The medium percolates through Whatman filter papers with biomass over them. The spent medium is recirculated after it has been collected at the bottom in a reservoir (Dilorio *et al.* 1992; Whitney 1992).

Turbine Blade Reactor

These reactors also provide better cultivation of hairy roots as in these reactors the agitation space and cultivation space are well separated from each other by a stainless steel mesh. A special type of impellor with eight blades is used and air is introduced from the bottom (Kondo *et al.* 1989).

Rotating Drum Bioreactor

It was first used for the cultivation of hairy roots from carrot (Kondo *et al.* 1989). It consists of drum-like container which is fixed to rollers and is rotated at the speed of 2-6 rpm. These rollers provide it support and also help in rotation. The drums are lined inside with polyurethane foam sheet so that the hairy roots adhere to the surface and grow well without detaching from it.

A study was conducted using *Tagete spatula* hairy roots growth and thiophene production to compare between the air-lift bioreactors, bubble column bioreactors and stirred tank reactors. It was found that cultivation in bubble column bioreactors gave the best results (Buitelaar *et al.* 1991). Air-lift reactors were used for successful cultivation of *Lithospermum erythrorhizon* hairy roots and production of shikonin from them (Shimomura *et al.* 1991), *Panax ginseng* and production of hernandulin from hairy roots of *Lippiadulcis* (Sauerwein *et al.* 1991).

Conclusion

Human life is threatened by a number of diseases out of which cancer ranks the second. Nearly 6 million deaths are caused by cancer every year worldwide. It could be treated by chemotherapy and irradiation therapy but the treatments may cause some side-effects. Therefore, natural sources such as plants have been now days used for the treatment of cancer. *Catharanthus roseus, Taxus brevifolia, Betula alba, Curcuma longa*, **etc.** are known to possess anti-tumor properties. These may be attributed to the production of secondary metabolites which can help in getting rid of cancer. Plants produce the phytochemicals which can be classified as primary metabolites and secondary metabolites. Primary metabolites are the compounds which are used in the growth of the plant whereas secondary metabolites are not used in the growth metabolism. Taxol, vinca alkaloids, vincristine, camptothecins, berbamine and podophyllotoxins are some of the secondary metabolites responsible for the anti-cancer properties. The secondary metabolites can be produced through callus culturing or organ culturing. In callus culturing the plants are selected on the basis of the secondary metabolites produced by them and grown through callus culturing. The best genotypes are selected and then grown in the bioreactor. In organ culturing the plants are infected with *Agrobacterium tumefaciens* or *Agrobacterium rhizogenes* which leads to the formation of hairy roots. Depending on the type of the culture to be grown the bioreactor is chosen such as for the growth of hairy roots submerged or airlift bioreactors are used as with the impellors the hairy root culture may be distorted. Therefore, great care must be taken when the choice of the bioreactor is to be made for the growth of plant cultures to produce secondary metabolites.

References

Ames B.N. and Gold L.S. (1995) The causes and prevention of cancer procedure. Natl. Academic Sciences USA. 92: 5258-5265.

Atoui A.K., Mansouri A., Boskou G. and Kefalas P. (2005) Tea and herbal infusions: their antioxidant activity and phenolic profile. Food Chem. 89: 27-36.

Bassetti L., Buitelaar R.M. and Tramper J. (1997) Hairy roots of *Tagetes* in two-phase system, in: P.M. Dora (ed.), Hairy Roots, Culture and Applications, Harwood Academic pp. 227-233.

Bhadra R., Morgan J.A. and Shanks J.V. (1998) Transient studies of light-adapted cultures of hairy roots of *Catharanthus roseus:* growth and indole alkaloid accumulation. Biotechnology and Bioengineering 60: 670-678.

Blum A.H. and Dawson D.I.A. (1976) Vindesine-Phase I. Study of a Vica alkaloid. Proc Am Assoc Cancer Res 17: 108.

Bourgaud F., Bouque V. and Guckert A. (1999) Production of flavonoids by *Psoralea* hairy root cultures. Plant Cell Tissue and Organ Culture 56: 97-104.

Brodelius P. (1984) Immobilization of cultured plant cells and protoplasts, in: IK Vasil (Editor), Cell Culture and Somatic Cell Genetic of Plants, Academic Press pp. 535-546.

Brodelius P. (1988) Permeabilization of plant cells for the release of intracellularly stored products: viability studies, Applied Microbiology and Biotechnology 27: 561-566.

Bryan W.L. (1974) Biochemical and pharmacological properties of microtubules. Advan Cell Mol Bio 3: 21-73.

Buitelaar R.M. and Tramper J. (1992) Strategies to improve the production of secondary metabolites with plant cell cultures: a literature review. J Biotechnol 23: 111-141.

Buitelaar R.M., Lanfenhoff A.A.M., Heidstra R. and Trampers J. (1991) Growth and thiophene production by hairy root cultures of *Tagates patula* in various two-liquid-phase bioreactors. Enzyme Microbiology Technology 13: 487-494.

Canel C., Moraes R.M., Dayan F.E. and Ferreira D. (2000) Molecules of interest: podophyllotoxin. Phytochemistry 54: 115-120.

Charlwood B.V., Charlwood K.A. and Molina-Torres J. (1988) Properties of plant cell cultures. Organized cultures. In: PSE International Symposium of Secondary Producers from Plant Tissue Culture. Kings College, London.

Chilton M.D., Drummond M.H., Merlo D.J., Sciaky D., Montoya A.L., Gordon M.P. and Nester E.W. (1977) Stable incorporation of plasmid DNA into higher plant cells: the molecular basis of crown gall tumourigenesis. Cell 11: 263-271.

Chilton M.D., Tepfer D.A., Petit A., Casse-Delbart F. and Tempe J. (1982) *Agrobacterium* inserts T-DNA into the genome of the host plant root cells. Nature 295: 432.

Corry J.P., Reed W.L. and Curtis W.R. (1993) Enhanced recovery of Solavetivone from *Agrobacterium* transformed root cultures of *Hyoscymus muticus* using integrated product extraction. Biotechnology and Bioengineering 42: 503-508.

Cragg G.M., Newmann D.J. and Snader K.M. (2005) Plants as a source of anti-cancer agents. Journal of Ethno Pharmacology 100: 72-79.

Cui Y., Xu M., Mao J., Ouyang J., Xu R. and Yu Y. (2012) Synthesis of berbamine acetyl glycosides and evaluation of anti-tumour activity. Eur J. Med Chem 54: 867-872.

Damayanthi Y. and Lown J.W. (1998) Podophyllotoxins: current status and recent developments. Curr Med Chem 5: 205-252.

Degrassi F., De Salvia R. and Berghella L. (1993) The production of chromosomal alterations by beta-lapachone, an activator of topoisomerase I. Mutat Res. 288: 263-267.

Dicosmo F. and Misawa M. (2005) Plant cell and tissue culture: alternatives for metabolite production. Biotechnology Advances 13: 425-453.

Dilorio A.A., Cheetham R.D. and Weathers P.J. (1992) Growth of transformed roots in a nutrient mist bioreactor: Reactor performance and evaluation. Appl Microbiol Biotechnol 37: 457-462.

Dixon R.A. (2001) Natural products and plant disease resistance. Nature 411: 843-847.

Du H.P., Shen J.K., Yang M., Wang Y.Q., Yaun X.Q., Ma Q.L. and Jin J. (2010) 4-Chlorobenzoyl berbamine induces apoptosis and G2/M cell cycle arrest through the P13K/Akt and NF-kappa B signal pathway in lymphoma cells. Oncol Rep 23: 709-716.

Fett Neto A.G., Zhang W.Y. and Cosmo F.D. (1994) Kinetics of taxol production, growth and nutrient uptake in cell suspensions of *Taxus cuspidate.* Biotechnology and Bioengineering 44: 205-210.

Flores H.E., Hoy M.W. and Pickard J.J. (1987) Secondary metabolites from root cultures. Trends Biotechnol 5: 64-68.

Fujita Y. and Tabata M. (1987) Secondary metabolites from plant cells. Pharmaceutical applications and progress in commercial production. In: C.E. Green, D.A. Somers, W.P. Hackett, D.D. Deesboer (Editors) Plant Biology New York, pp. 169-185.

Ghasemzadeh A. and Ghasemzadeh N. (2011) Flavonoids and phenolic acids: role and biochemical activity in plants and human. Nat Prod Rep 5: 6697-6703.

Gontier E., Sangwan B.S. and Barbotin J.N. (1994) Effects of calcium, alginate and calcium-alginate immobilization on growth and tropane alkaloid levels in a stable suspension cell line of *Daturainnoxia* Mill, Plant Cell Report 13: 533-536.

Harborne J.B. (2001) Twenty-five years of chemical ecology. Nat Prod Rep 18: 361-379.

Hilton M.G. and Rhodes M.J.C. (1990) Growth and hyoscyamine production of "hairy root" cultures of *Datura stramonium*in a modified stirred tank reactor. Appl Microbiol Biotechnol 33: 132-138.

Holmes P., Li S.L., Green K.D., Ford-Lloyd B.V. and Thomas J. (1997) Drip- tube technology for continuous culture of hairy roots with integrated alkaloid extraction, in: P.M. Doran (Editor), Hairy Roots, Culture and Applications, Harwood Academic pp. 201-208.

Horwitz S.B. (1994) Taxol (paclitaxel): mechanisms of action. Ann Oncol 6: 3-6.

Hulst A.C., Tramper J., Brodelius P., Eijkenboom L.J.C. and Lyben K.C.A. (1985) Immobilized plant cells: respiration and oxygen transfer. Journal of Chemical Technology and Biotechnology 35: 198-204.

Hussain M.S., Fareed S., Ansari S, Rahman M.A., Ahmad I.Z. and Saeed M. (2011) Current approaches toward production of secondary plant metabolites. J Pharm Bioall Sci 4: 10-20.

Jeong G.A. and Park D.H. (2007) Enhanced secondary metabolite biosynthesis by elicitation in transformed plant root system. Appl Biochem Biotechnol 130: 436-446.

Jin J.H., Shin J.H., Kim J.H., Chung I.S. and Lee H.J. (1999) Effect of chitosan elicitation and media components on the production of anthraquinone colorants in madder (*Rubiaakane* Nakai) cell culture. Biotechnol Bioprocess Eng 4: 300-304.

Jolicoeur M,, Chavarie C., Carreau P.J. and Archambault J. (1992) Development of a helicoidal-ribbon impellor bioreactor for high density plant cell suspension culture.Biotechnology and Bioengineering. 39: 511-521.

Jones A. and Veliky K.C.A. (1981) Effects of medium constituents on the viability of immobilized plant cells. Canadian Journal of Botany 39: 511-521.

Jordan M.A., Thrower D. and Wilson L. (1992) Effects of vinblastine, podophylotoxin and nocodazole on mitotic spindles.Implication for the role of microtubule dynamics in mitosis. J Cell Sci 102: 401.

Kappor L.D. (1990) Handbook of ayurvedic medicinal plants. Boca Raton, Florida, CRC Press pp. 416-417.

Karuppusamy S. (2009) A review on trends in production of secondary metabolites from higher plants by *in vitro* tissue, organ and cell cultures. J. Med Plants Res 3: 1222-1239.

Kiong A.L., Mahmood M., Fodzillan N.M. and Daud S.K. (2005) Effects of precursor supplementation on the production of triterpenes by *Centella asiatica* callus culture. Pak J Biol Sci 8: 1160-1169.

Komaraiah P., Ramakrishna S.V., Reddanna P. and Kavikishore P.B. (2004) Enhanced production of plumbagin in immobilized cells of *Plumba gorosea* by elicitation and in situ adsorption. J. Biotechnol 10: 181-187.

Kondo O., Honda H., Taya M. and Kobayashi T. (1989) Comparison of growth properties of carrot hairy root in various bioreactors. Appl Microbiol Biotechnol 32: 291-294.

Krikorin A.D. and Steward F.C. (1969) Biochemical differentiation: the biosynthetic potentialities of growing and quiescent tissue, in: FC Steward (Editors), Plant Physiology. A Treatise, Academic, Press pp. 227-326.

Kupeli E., Kosar M., Yasilada E., Husnu K. and Baser C. (2002) A comparative study on anti-inflammatory, antinociceptive and antipyretic effects of isoquinoline alkaloids from the roots of Turkish *Berberis* species. Life Sci 72: 645-657.

Larkin P.J. and Scowcroft W.R. (1981) Somaclonal variation- A novel source of variability from cell cultures for plant improvement. Theoretical and Applied Genetics 60: 197-214.

Li S.Y., Jei W., Seow W.K. and Thong Y.H. (1994) Effect of berbamine on blood and bone marrow stem cells of cyclophosphamide-treated mice. Int J Immuno pharmacol 16: 245-249.

Liang Y., Xu R.Z., Zhang L. and Zhao X.Y. (2009) Berbamine, a novel nuclear factor kappa B inhibitor, inhibits growth and induces apoptosis in human myeloma cells. Acta Pharmacol Sin 30: 1659-1665.

Lu W., Fu C. and Zhao Y. (2011) Review on biosynthesis of podophyllotoxin. Zhongguo Zhong Yao ZaZhi 36: 1109-1114.

Majra K., Caldentey O. and Inze D. (2004) Plant cell factories in the post-genomic era:new ways to produce designer secondary metabolites. Trends in Plant Science. 9: 1360-1385.

Moon D.O., Kang C.H., Kim M.O., Jeon Y.J., Lee J.D., Choi Y.H. and Kim G.Y. (2010) Beta-lapachone (LAPA) decreases cell viability and telomerase activity in leukaemia cells: suppression of telomerase activity by LAPA. J Med Food 13: 481-488.

Mukundan U., Bhide V., Singh G. and Curtis J. (1998) pH-mediated release of betalains from transformed root cultures of *Beta vulgaris.* Applied Microbiology and Biotechnology 50: 241-245.

Nagar N., Jat R.K., Saharan R.K., Verma S., Sharma D. and Bansal K. (2011) Podophyllotoxin and their glycosidic derivatives. Pharmacophore 2: 124-134.

Nam S., Xie J., Perkins A., Ma Y., Yang F., Wu J., Wang Y., Xu R.Z., Huang W., Horne D.A. and Jove R. (2012) Novel synthetic derivatives of the natural product berbamine inhibit Jak2/Stat3 signaling and induce apoptosis of human melanoma cells. Mol Oncol 6: 484-493.

Nirmala M.J., Samundeeswari A. and Sankar P.D. (2011) Natural plant resources in anti-cancer therapy-a review. Research in Plant Biology. 1: 1-14.

Nuutila A.M., Toivonen L. and Kauppinen V. (1994) Bioreactor studies on hairy root cultures of *Catharanthus roseus:* comparison of three bioreactor types. Biotechnology Techniques 8: 61-66.

Pardee A.B., Li Y.Z. and Li C.J. (2002) Cancer therapy with beta-lapachone. Curr Cancer Drug Targets 2: 227-242.

Parr A.J. (1989) The production of secondary metabolites by plant cell cultures. J Biotechnol 10: 1-26.

Payne G.F., Bringi V., Prince C. and Shuler M.L. (1991) Immobilized plant cells. In: Payne G.F., Bringi V., Prince C., Shuler M.L. (Eds.), Plant Cell and Tissue Culture in Liquid Systems, Hanser, pp. 179-223.

Pazdur R., Kudelka A.P., Kavanagh J.J., Cohen P.R. and Raber M.N. (1993) The toxoids: paclitaxel (Taxol) and docetaxel (Taxotere). Cancer Treat Rev 19: 351-386.

Pinar R. (1998) To investigate knowledge of Turkish people about cancer. Cancer agenda 2: 66-73.

Planchon S.M., Wuerzberger S., Frydman B., Witiak D.T., Hutson P., Church D.R., Wilding G. and Boothman D.A. (1995) Beta-lapachone-mediated apoptosis in human promyelocytic leukaemia (HL-60) and human prostate cancer cells: a p53-independent response. Cancer Res 55: 3706-3711.

Quiroga E.N., Sampietro A.R. and Vattuone M.A. (2001) Screening antifungal activities of selected medicinal plants. J Ethnopharmacol 74: 89-96.

Rhodes M.J.C., Hilton M., Parr A.J., Hamill J.D. and Robins R.J. (1986) Nicotine production by 'hairy roots' cultures of *Nicotiana rustica:* fermentation and product recovery. Biotechnology Letters 8: 415-420.

Rhodes M.J.C., Robins R.J., Hamill J.D., Parr A.J., Hilton M.G. and Walton N.J. (1990) Properties of transformed root cultures. In: BV Charlwood and MJC Rhodes (Editors). Secondary products from Plant Tissue Culture. Clarendon Press, Oxford pp. 201-225.

Richard H., Himes H., Kersey R.N., Bettinger I.H. and Samson F.E. (1976) Action of the Vinca alkaloids vincristine, vinblastine and desacetyl vinblastine amide on microtubules *in vitro*. Cancer Res 36: 3798-3802.

Rijhwani S.K. and Shanks J. (1998) Effects of elicitor dosage and exposure time on biosynthesis of indole alkaloids by *Catharanthus roseus* hairy root cultures, Biotechnology Progress 14: 442-449.

Samuelson G. (1999) Drugs of natural origin: a textbook of Pharmacognosy. 4th edition, Stockholm, Swedish Pharmaceutical Press, Sweden.

Sauerwein M., Yamazaki T. and Shimomura K. (1991) Hernandulcin in hairy root cultures of *Lippiadulcis.* Plant Cell Reports 9: 579-581.

Shimomura K., Sudo H., Saga H. and Kamada H. (1991) Shikonin production and secretion by hairy root cultures of *Lithospermumery throrhizon.* Plant Cell Reports 10: 282-285.

Slichenmyer W.J., Rowinsky E.K., Donehower R.C. and Kaufmann S.H. (1993) The current status of camptothecin analogues as anti-tumour agents. J Natl Cancer Inst 85: 271-291.

Steward N., Martin R., Engasser J.M. and Goergen J.L. (1999) Determination of growth and lysis kinetics in plant cell suspension cultures from the measurement of esterase release. Biotechnology and Bioengineering 66: 114-121.

Sweeney M.J., Cullinan G.J., Poore G.A. and Gerzon K. (1974) Experimental anti-tumour activity of vinblastine amides. Proc Am Assoc Cancer Res 15: 37.

Taya M., Yoyama A., Kondo O., Kobayashi T. and Matusui V. (1989) Growth characteristics of plant hairy roots and their cultures in bioreactors. Journal of Chemical Engineering Japan 22: 84-89.

Tescione L.D., Ramakrishan D. and Curtis W.R. (1997) The role of liquid mixing and gas-phase dispersion in a submerged, sparged root reactor. Enzyme Microbiol Tech 20: 207-213.

Tone S., Taya M. and Kino-oka M. (1997) Alteration of metabolite formation and morphological properties of hairy roots by environmental stimuli, in: PM Dora (Editor), Hairy Roots, Culture and Applications, Harwood Academic pp. 65-72.

Torck M. and Pinkas M. (1996) Camptothecins and derivatives: a new class of anti-tumour agents. J Pharm Belg 51: 200-207.

Von Krogh G. (1981) Podophyllotoxin for condylomata acuminate eradication.Clinical and experimental comparative studies on Podophyllumligans, colchicine and 5-fluorouracil. Acta Dermatovenereol Suppl.98: 1-48.

Wall M.E. and Wani M.C. (1996) Camptothecin and taxol: from discovery to clinic. J Ethno Pharmacology 51: 239-254.

Wang G.Y., Zhang J.W., Lu Q.H., Xu R.Z. and Dong Q.H. (2007) Berbamine induces apoptosis in human hepatoma cell line SMMC7721 by loss in mitochondrial transmembrane potential and caspase activation. Zhejiang Univ Sci B. 8: 248-255.

Whitney P.J. (1992) Novel bioreactors for the growth of roots transformed by *Agrobacterium rhizogenes.* Enzyme Microbiology Technology 14: 13-17.

Woo H.J., Park K.Y., Rhu C.H., Lee W.H., Choi B.T., Kim G.Y., Park Y.M. and Choi Y. H. (2006) Beta-lapachone, a quinone isolated from Tabebuiaavellanedae, induces apoptosis in HepG2 hepatoma cell line through induction of Bax and activation of caspase. J Med Food 9: 161-168.

Wyslouzil B.E., Waterbury R.G. and Weathers P.J. (2000) The growth of single roots of *Artemisia annua* in Nutrient mist reactors. Biotechnology Bioengineering 70:143-150.

Xu C.Q., Dong D.L., Du Z.M., Chen D.W., Gong D.M. and Yang B.F. (2004) Comparison of the anti-arrhythmic effects of matrine and berbamine with amiodarone and RP58866. Yao Xue Xue Bao 39: 691-694.

Yamada Y. and Hashimoto T. (1990) Possibilities for improving yields of secondary metabolites in plant cell cultures. In Nijkam H.J.J., van der Plas L.H.W. and van Aartrijk J. (Eds). Progress in Plant Cellular and Molecular Biology. Kluwer Academic, Dordrecht. pp. 547-554.

Yeoman M.M. (1987) Techniques, characteristics, properties and commercial potential of immobilized plant cells. Cell Culture and Somatic Cell Genetics of Plants 4: 197-215.

Zenk M.H. (1991) Chasing the enzyme of secondary metabolism: plant cultures as a pot of gold. Phytochemistry 30: 3861-3863.

Zunino F. and Pratesi G. (2004) Camptothecins in clinical development. Expert Opin Investig Drugs 13: 269-284.

12

RNA Interference: An Efficient Tool for the Study of Plant Functional Genomics

Kulbhushan Chaudhary[1], Anirudha Chattopadhyay[2] and Dharmendra Pratap[3*]

[1]Advanced Centre for Plant Virology, Indian Agricultural Research Institute, New Delhi-110012, India

[2]Department of Plant Pathology, C.P. College of Agriculture, S. D. Agricultural University, S. K. Nagar, Gujarat

[3]Department of Horticulture, Sikkim University, 6th Mile, Samdur, Tadong, Gangtok, Sikkim-737102, India

E-mail: pratapbiotech@gmail.com

Abstract

As we have entered in the post genomic era where we have successfully completed the full genome sequencing of many important crop species and now the future challenge is to ascertain the functions of all genes present in plant genome. There are various approaches that were tried in the past viz. to generate gene knockouts through insertional mutagenesis, which make the mutant phenotype and is suggestive of its functionality. The tool has its own limitations such as gene redundancy, lethal knock-outs, non tagged mutant and the inability to target the specific gene. RNA interference (RNAi) which results in gene silencing through cleavage of mRNAs and obstruction of protein synthesis made it an attractive tool for plant functional genomics. RNAi is now been widely used in plant biotechnology, both as a useful tool for discovering or validating gene function as well as an alternating way of crop improvement using small RNA.

Introduction

The relatively recent discovery of RNAi has now embarked in an exciting time where we are finding the description of many molecular mechanism that just a few years back was unimagined. The huge repository of genome sequences made it possible. Genome sequencing projects have identified large numbers of plant genes for which no role has yet been defined. Therefore, more tools are required to validate the function of these unknown genes and to address this, number of reverse genetics based techniques have been developed such as Insertional mutagenesis, transposons tagging and RNA interference (RNAi).

The discovery of RNAi was very accidental, unfolded by a surprising observation made in petunia by Napoli and co-workers. Further, Fire and Mello in 1998 first discover the phenomenon of RNA silencing or RNAi in the nematode worm *Caenorhabditis elegans* as a response to double stranded RNA (dsRNA), which restricted in a sequences specific gene silencing (Fire *et al.*, 1998). Andrew Fire and Craig Mello won the Nobel Prize for medicine to their discoveries of gene silencing (Daneholt 2006). Although, the first report of RNA Silencing was appeared in 1928 by Wingard, where he showed that tobacco plants infected with *Tobacco ring spot virus* were recovered after some time and become somehow resistant to secondary infection (Wingard 1928). At that time the recovery was a mystery and there was no obvious way to explain the specificity of the resistance to secondary infection. RNA interference (RNAi) is a double stranded RNA (dsRNA) mediated gene silencing phenomenon that is conserved among various organisms including animal and plants. Co-suppression, post-transcriptional gene silencing and virus induced gene silencing (VIGS) all describe homology dependent gene silencing phenomenon in plant that involve what is more broadly known as RNAi. Over the previous decade, small RNAs (sRNAs) have emerged as ubiquitous and versatile repressors (occasionally activators as well) of gene expression and have fundamentally altered our understanding of the biological world. The true story began in 1990, when two groups published the experimental result, in which over expression of a gene coding for a chalcone synthase to produce deep purple petunia flowers gave white flowers instead (Napoli *et al.* 1990; Smith *et al.* 1990; van der Krol *et al.* 1990). At this time, the phenomenon named as "cosuppression" but was ineffective to claim the rationality. The molecular mechanism was first discovered in the worm *Caenorhabditis elegans* by exploring RNA interference (Guo and Kemphues 1995; Fire *et al.* 1998; Montgomery *et al.* 1998; Elbashir *et al.* 2002). RNAi is also referred as post transcriptional gene silencing (PTGS) in plants (Hamilton and Baulcombe 1990), and quelling in fungi (Cogoni *et al.* 1996) and algae

(Wu-Scharf *et al.* 2000). The RNAi pathway is thought to act as an immune system against invading nucleic acids coming from viruses, transposons or transgenes (Plasterk 2002). The silencing is triggered by the presence of long double-stranded (ds) RNA molecule in the cell. These can be synthetic RNAs, replicating viruses or even the result of transcription of nuclear genes. These small interfering RNAs (siRNAs) are generated via processing of longer dsRNA precursors by an RNaseIII-like enzyme termed Dicer (Bernstein *et al.* 2001). Despite the overall conservation of RNA- mediated silencing mechanism, this key enzyme has diversified structurally and numerically in different taxa (Schauer *et al.* 2002). These siRNAs are then incorporated in a RNA induced silencing complex (RISC) that will recognize, bind and inducc clcavagc of pcrfcctly match complementary mRNAs (Hamilton and Baulcombe 1999; Hammond *et al.* 2000; Hannon 2002; Zammore *et al.* 2002). The core component of the RISC complex is a member of Argonaute protein family which has RNA-binding ability (Hammond *et al.* 2000). This fundamental discovery makes the artificial silencing of virtually any gene possible with artificially engineered siRNAs, without the knowledge of complete gene sequence, a technology now routinely and widely used in functional genomics

Components of RNAi Machinery

Gene silencing process involves various components, some serves as inhibitors and others as effectors, amplifiers and transmitters. Some of the important factors are discussed below. The overall mechanism is illustrated in schematic diagram (Fig. 1).

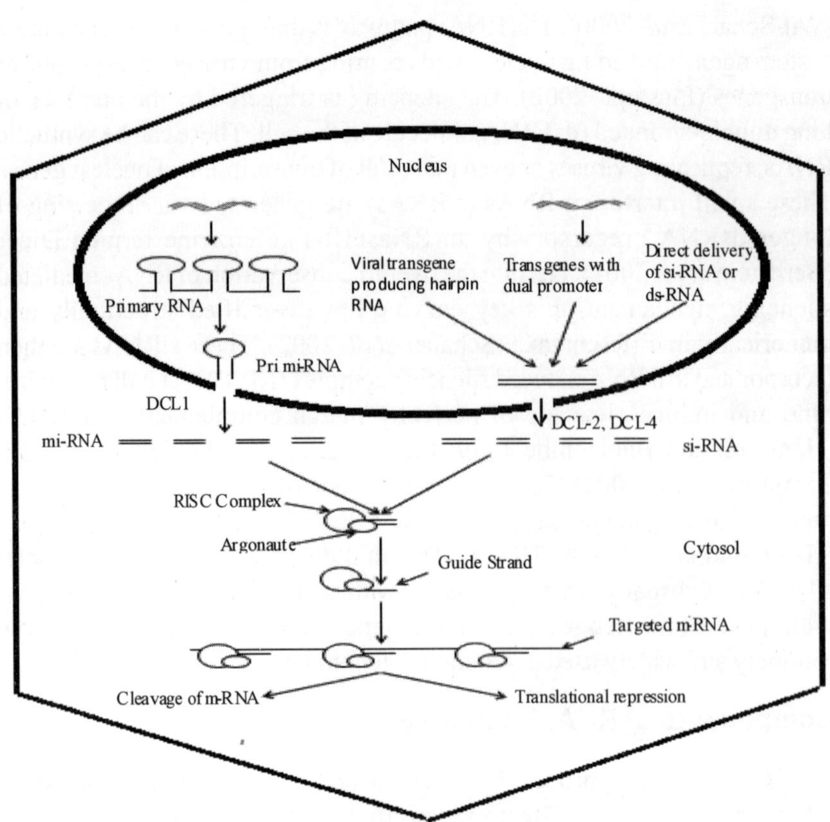

Fig. 1 : Schematic diagram of RNAi mechanism.

Inducer: That induces the game

A dsRNA produced either by a transgene introduced into an organism or by direct injection of dsRNA (in case of animal system) homologous to target mRNA actually induces the onset of whole process. Such a RNA molecule is called an inducer (Palaqui and Balzergue 1999).

DICER: The initiator to "Dice" the dsRNA.

The double-stranded nature of miRNA and siRNA precursors was apparent, and RNase III enzymes had long been characterized as a dsRNA-specific nucleases therefore, enzymes with RNase III domains were promptly recognized as primary candidates in the search for miRNA/siRNA biogenesis factors (Meister and Tuschl 2004; Tomari and Zamore 2005). One class of large RNase III enzymes is characterized by several domains in a specific order, amino-to-carboxyl terminus: a DEAD/H ATPase domain, a DUF283 domain, a PAZ domain, two tandem RNase III domains, and a dsRNA-binding domain (dsRBD). In *C. elegans* and *D. melanogaster* the orthologs of these proteins were shown to be required for RNAi and the members of this protein family came to be known as DICER enzymes. Some organisms including mammals and nematodes have only a single DICER that does double duty in the biogenesis of both miRNAs and siRNAs, whereas other organisms divide the labor among multiple DICER proteins. In *Arabidopsis*, four Dicer-like (DCL) proteins are known, each having a distinct function in different small RNA pathways (Schauer *et al.* 2002; Xie *et al.* 2004; Dunoyer *et al.* 2005; Gasciolli *et al.* 2005; Xie *et al.*2005b). DCL1 produces miRNAs, DCL2 produces siRNAs involved in the silencing of at least some viral sequences (Xie *et al.* 2004) and DCL3 produces siRNAs involved in DNA methylation and heterochromatin formation (Xie *et al*. 2004). Recent work suggests that DCL4 produces siRNAs triggered by inverted-repeat transgenes in plants (Dunoyer *et al.* 2005) and is also associated with the tasiRNA pathway (Gasciolli *et al.* 2005; Xie *et al.* 2005b; Yoshikawa *et al.* 2005).

Argonaute family: At the centre of RNA Induced Silencing Complex (RISC)

The Argonaute superfamily can be divided into three separate subgroups: the Piwi clade that binds piRNAs, the Ago clade that associates with miRNAs and siRNAs, and a third clade that has only been described thus far in nematodes (Yigit *et al.* 2006). All gene-regulatory phenomena involving ~20–30 nt RNAs are thought to require one or more Argonaute proteins, and these proteins are the central, defining components of the various forms of RNA Induced Silencing Complex (RISC). There are five, eight and twenty seven paralogs exist in *Drosophila*, humans and *C. elegans* respectively. In *Arabidopsis*, AGO1 is associated with miRNAs, *trans*-acting siRNAs and transgene-derived siRNAs but not with virus-derived siRNAs and siRNAs involved in chromatin silencing (Vaucheret *et al.* 2004; Kidner and Martienssen 2005). Mutants of AGO1 were shown to be hypersensitive to virus infections. Argonaute proteins are defined by the presence of four domains: the PAZ domain (shared with DICER enzymes), the PIWI domain that is unique to the Argonaute superfamily, and the N and Mid domains. The overall protein structure is bilobed, with one lobe consisting of the PAZ domain and the other lobe consisting of the PIWI domain flanked by N-terminal (N) and middle (Mid) domains. The Argonaute PAZ domain has RNA 30 terminus binding activity, and the co-crystal structures reveal that this function is used in guide strand binding. The other end of the guide strand engages a 52 phosphate binding pocket in the mid domain, and the remainder of the guide tracks along a positively charged surface to which each of the domains contributes. The protein-DNA contacts are dominated by sugar-phosphate backbone interactions, as expected for a protein that can accommodate a wide range of guide sequences. Guide strand nucleotides 2–6, which are especially important for target recognition, are stacked with their Watson- Crick faces exposed and available for base pairing. A critical breakthrough was the demonstration that the PIWI domain adopts an RNase H-like fold that in some cases can catalyze guide strand-dependent endonucleolytic cleavage of a base paired target. This initial cut represents the critical first step in a subset of small RNA silencing events that proceed through RNA destabilization (Parker *et al.* 2004; Song *et al.* 2004).

RNA-induced silencing complex (RISC): The effector to "Slice" the mRNA.

The siRNAs were originally observed during transgene and virus induced silencing in plants, consistent with a natural role in genome defense (Mello and Conte 2004). The canonical inducer of RNAi is long, linear, perfectly base paired dsRNA, introduced directly into the cytoplasm or taken up from the environment (Mello and Conte 2004). These dsRNAs are processed by Dicer into the double stranded siRNAs which enter into a RISC assembly pathway that involves duplex unwinding, culminating in the stable association of only one of the two strands with the Ago effector protein (Meister and Tuschl 2004; Tomari and Zamore 2005). The strand selection does not require or involve the presence of a cognate mRNA target but the strand selection is dictated by the relative thermodynamic stabilities of the two duplex ends: whichever strands has its 52 terminus at the less stably base-paired end will be favored as the guide strand. The siRNAs with equal base-pairing stabilities at their ends will incorporate either strand into RISC with approximately equal frequency (Tomari and Zamore 2005). The guide strand (single stranded siRNA) loads into AGO proteins and directs target recognition by Watson-Crick base pairing and perfectly complementary targets are then degraded. Target RNA degradation is induced by the PIWI domain of AGO protein and this "slicer" activity is very precise: the phosphodiester linkage the target nucleotides that are base paired to siRNA residues 10 and 11 (counting from the 52 end) is cleaved to generate products with 52 -monophosphate and 32 -hydroxy termini. Once this initial cut is made, cellular exonuclease attacks the fragments to complete the degradation process. The target dissociates from the siRNA after cleavage, freeing RISC to cleave additional targets. Mismatches at or near the centre of the siRNA/target duplex suppress endonucleolytic cleavage, moreover, some siRNA-programmed Ago proteins lack endonuclease activity even with perfectly paired targets (Tomari and Zamore 2005). Nevertheless, targets that are partially mismatched or are recognized by endonuclease-inactive siRISCs can still be silenced at posttranscriptional level. In such cases, silencing can involve translational repression or exonucleolytic cleavage in a manner similar to miRNA silencing. The other strand of the original small RNA duplex (the passenger strand) is discarded. In many species, the siRNA populations can be further amplified by the action of RNA-dependent RNA polymerases (RDRs) thus, greatly amplifying the response. The RDR1 and RDR6 are required in the siRNA pathway that silences viruses and transgenes (Dalmay *et al.* 2000; Xie *et al.* 2001). RDR change single-stranded (ss) RNA into dsRNA, with or without a siRNA as a primer and as a result, a single RNA or primary siRNA molecule can induce synthesis of secondary siRNAs (if the target mRNA is present), therefore, strengthening and perpetuating the response (Baulcombe 2004). This secondary pool of siRNAs can highly sustain the response and in some organisms, such as plants and nematodes, where they can lead to systemic silencing that spread throughout the organism. Virus dsRNA sequences are recognized by the RNA silencing machinery, which produces siRNAs that will silent viral genes and prevent the accumulation of the pathogen (Dunoyer and Voinnet 2005). Virus defense via siRNA is likely to be an ancient mechanism and therefore, viruses have evolved various ways to bypass this barrier (Baulcombe 2004; Dunoyer and Voinnet 2005; Simon-Mateo and Garcia 2006). The mechanism best known involves the production of a protein (suppressor protein) by the virus that will block the silencing pathway of the host (Voinnet *et al.* 1999; Mallory *et al.* 2002; Ye *et al.*, 2003; Chapman *et al.* 2004; Dunoyer *et al.* 2004). Another striking feature of siRNA-mediated silencing in plants and in some animals is its systemic nature: the effect of silencing can extend beyond the site of initiation and spread through the organism (Voinnet 2005). Recent work suggests that DCL4 is responsible for the production of the 21-nt long siRNAs involved in the cell-to-cell silencing signal (Dunoyer *et al.* 2005). The movement of siRNAs or miRNAs could be important for the regulation of endogenous genes. Many siRNAs and miRNAs were detected in the phloem sap of pumpkin, where a protein has been characterized that binds specifically to small RNAs (Yoo *et al.* 2004). Recently, it is shown that siRNA can also induce heterochromatin formation (Lippman and Martienssen 2004).

RNA-dependent RNA polymerase (RdRp)

RNA-dependent RNA polymerase (was first discovered in RNA viruses (Blumenthal and Carmichael 1979) which was responsible for transcription and replication of the viral genome. RdRp activity has also been detected in plants. In 1993, RdRp was isolated and purified from tomato leaves (Schiebel *et al.* 1993) and characterize its catalytic properties. RdRp amplifies the siRNA produced by the action of Dicer. This siRNA then systemically moves from one cell to another cell, so if has systemic inheritance from one part of a plant to another part of the same plant.

Molecular Mechanism of RNAi

The mechanism by which RNAi is thought to work is very complicated. The process by which RNAi occurs in different organisms varies slightly. This is usually because different proteins are used in the process of RNAi. Basically, the molecular mechanism of RNAi is multi-step process. When the dsRNA entered into the cell, it is cleaved by the enzyme dicer. The dicer cut the dsRNA into 21-25 nucleotide segments called small interfering RNA (siRNA). The siRNA can be form a ribonuclease protein complex (RISC). This complex includes "slicer" –an argonaute protein with an RNase H like domain called PIWI. RISC first mediate the unwinding of the siRNA duplex, then binds to the target mRNA in a sequence specific manner. This binding mediates the target mRNA cleavage by slicer. This prevent translation from occurring, silencing the expression of the gene from which the mRNA was transcribed.

miRNAs

miRNAs are fascinating group of noncoding small RNAs and have recently came out as a key element of genetic control in many organisms. miRNAs are ~20-24 nucleotides endogenous single stranded RNAs that repress post-transcriptional gene expression either through translational inhibition or endonucleolytic cleavage (slicing) within miRNA-mRNA base pairing (Ambros *et al.* 2003; Bartel 2004; Voinnet 2009). A majority of the miRNA genes exist as independent transcriptional units and these miRNA genes resemble protein coding genes in that they may contain intron and they are transcribed by RNA polymerase II. Because, these transcripts are generated by pol II and therefore possess 52 cap and 32 ploy-A tail hallmarks (Kim 2005). From this transcript (pri-miRNA), arise the 70-100 nt long pre-miRNA, which are subsequently cleaved into mature miRNAs. Since, the first discovery of miRNAs in plant (*Arabidopsis)* in 2002 (Reinhart *et al.* 2002), more than 100 plant miRNA genes have identified, located mainly in the intergenic regions throughout the genome and most of among them are found to bc conscrved in many different species of plant kingdom. Although miRNAs mediate diverse aspects of development and physiology in both plants and animals, there are substantial differences between them. First, plant pre-miRNAs have larger and more variable stem-loop structures. Second, mature plant miRNAs pair target sites with near perfect complementarity. Third, in animals, miRNAs usually recognize several target sequences in the 32 UTR of mRNAs and cause translation inhibition, while plant miRNAs often recognize a single target site in the coding region and guide the mRNA to cleavage (Bartel 2004). Still, plant and animal miRNAs are processed from precursors containing imperfect stem-loops and generally regulate longer target mRNAs in both systems.

Plant miRNA biogenesis

Most plant miRNAs are typically derived by transcription of a large (~1 kb) pri-miRNA from a segment of genome which is independent and distinct from protein coding regions. This step is RNA pol II dependent and may include splicing. From this transcript (pri-miRNA) arise the 70-100 nt long pre-miRNA. Mature miRNAs are processed from the imperfect duplex region of such pre-miRNA transcripts. The region of the precursor that pairs to the miRNA is called miRNA*. Experimental evidence indicates that the structure, rather the sequence of the pre-miRNA, directs their correct processing. In animals, the pri-miRNA is processed in the nucleus into hairpin pre-miRNA by the double-strand specific ribonuclease III (RNase III) enzyme, DROSHA. The pre miRNA is transported to the cytoplasm *via* EXPORTIN-5 (EXP-5), a transportin-5 dependent mechanism (Lund *et al.* 2004). It is then digested by a second, double-strand specific RNase III enzyme, called DICER. The resulting 19-24 nt miRNA is bound by a complex that is similar to the RNA-induced Silencing Complex (RISC) that participates in repressing the target transcript resulting in a reduced expression of the corresponding gene. The plant machinery is different in sense that both the pri- and pre-miRNA are processed in the nucleus by the dsRNA specific RNAse III enzymes, DICER LIKE 1 (DCL 1) in conjunction with HYPONASTIC LEAVES 1 (HYL 1) and Serrate to release the mature miRNA (Tang *et al.* 2003). Dicer generates small double stranded products with 2 nucleotide long 3' overhangs. One strand represents the miRNA, while the other strand is called miRNA*. The HUA ENHANCER 1 (HEN 1) methylates the terminal sugar residues of miRNA to prevent 32 -end uridylation and enhance the stability of miRNA. DCL1, HYL1, and SE interact with each other and are co-localized in the nuclear bodies, termed Dicing bodies or D-bodies in which some pri-miRNAs have been found, suggesting that D-bodies serve as a center for active miRNA processing (Fang and Spector 2007), The miRNA: miRNA* duplex is then transported to the cytoplasm by EXP-5 homolog, HASTY (Bollman *et al.* 2003) where it is incorporated into RISC and guides it to repress target mRNA(s). In the double strand, one end is normally thermodynamically less stable, since it is either mismatched or contains less GC (but more AU) pairs relative to the other end. The strand with lower thermodynamic stability at its 5' end (5' instability) is preferentially incorporated into RISC and the other is short lived (Khvorova *et al.* 2003; Schwarz *et al.* 2003). PAZ domain of AGO family may allow interaction with Dicer and other proteins and is also believed to help align miRNA with its target. The proposed model of the biogenesis of miR319 is given to understand importance of stem-loop structure in miRNA biogenesis. The structure of PIWI domain is related to RNaseH and is believed to be involved in target cleavages. Earlier, it was believed that plant miRNA lead to cleavage of target while in animals translational repression is the norm. However, now we know that wide spread translational repression also occur in plants (Broderson *et al.* 2008). MiRNA biogenesis is under feedback regulation such that the DCL1 and AGO1 genes are themselves regulated by miRNAs. DCL1 is itself a target of miRNA 162, which leads to the cleavage of the DCL1 mRNA (Xie *et al.* 2003), while AGO1 targeted miR168. Also, over the years few non-canonical miRNA biogenesis pathway have also been discovered where miRNA were processed from introns (termed as mirtrons) (Ruby *et al.* 2007) or viral tRNA –like sequences. Recently, in a phenomenon sometimes referred to as RNA activation, miRNA are shown to up regulate translation as well, adding to further complexity (Vasudevan *et al.* 2007).

Difference between siRNA and miRNA

MiRNAs and siRNAs appeared to be distinguished in two primary ways. First, miRNAs were viewed as endogenous and purposefully expressed products of an organism' own genome, whereas siRNAs were thought to be primarily exogenous in origin, derived directly from the virus, transposons, or transgene trigger. Second, miRNAs appeared to be processed from stem-loop precursors with incomplete double-stranded character, whereas siRNAs were found to be excised from long, fully complementary dsRNA (Tomari and Zamore 2005). Despite these differences, the size similarities and sequence-specific inhibitory functions of miRNAs and siRNAs suggest relatedness in their biogenesis and mechanism. Both classes of small RNAs depend upon the same two families of proteins: DICER enzymes to excise them from their precursors, and Ago proteins to support their silencing effector functions (Tomari and Zamore 2005).

Vector used in RNAi based functional genomics

Currently, several vectors are used for performing RNAi for the use of functional genomics. The pHELLSGATE, high throughput gene silencing vector and high throughput tobacco rattle virus (RTV) based virus-induced gene silencing (VIGS) vector are binary vectors developed by Invitrogen are used for expression of GUS and GFP proteins. Two vectors developed by Waterhouse and coworkers at CSIRO in Australlia allow for relatively state forward incorporation of targeting sequences in the sence and antisence direction with an intron sequence between them (Wesely -*et al.,* 2001). The first pHANNIBAL vector, whose T-DNA (the portion of the plasmid transfer to the plant genome *via Agrobacterium* mediated transformation) includes a selectable marker gene and a strong promoter upstream of a pair of multiple cloning sites flanking an intron. This structure allows one to clone sence and antisence copies of a gene of interest, separated by the intron. A derivative of the pHANNIBAL vector, pHELLSGATE2, was engineered to facilitate high trroughput cloning of targeting sequences. The pHANNIBAL vector was modified by replacing the poly linker with *aatB* site specific recombination sequences, allowing introduction of targeting sequences of interest using the gateway technology from Invitrogen. pHELLSGATE4 allows the production of an hpRNA construct in a single step from a single polymerase chain reaction product, while pHELLSGATE8 requires a two step process *via* an intermediate vector. pHELLSGATE8 is identical to pHELLSGATE2 but contains the more efficient *aatP* recombination sites. These vectors are based on gateway recombination-based technology, which replace the conventional cloning strategy. It is based on the phage lambda system of recombination. It enables segments of DNA to be transferred between different vectors while orientation and reading frame are maintained. It can also be used for transfer of PCR products.

It saves valuable times, because once the DNA has been cloned into a gateway vector, it can be used as many genome functional analysis system as is required. So by this process, the method of plant functional genomics has been made much easier, while the process has also been made faster (Invitrogen 2004). Another set of RNAi vectors are chrome DB vectors, are based on pCAMBIA plasmids, developed by the Centre for Application of Molecular Biology to International Agriculture (CAMBIA), which contain two origin of replication one for replication in *Agrobacterium tumefaciens* and another for replication in *Escherichia coli* (http://www.cambia.org/).

Delivery of RNAi in plants

Plant cell cell wall prevents the easy delivery of siRNA into the cell. Thus, RNAi is often achieved by expressing hairpin RNAs that fold back to create double stranded region that acts as substrates for the dicer-like enzymes (Small 2007). Such hairpin RNAs are potent inducer of RNAi and gives rise to copious siRNAs derived from the double stranded region. In an hpRNA-producing vector, that target gene is cloned as an inverted repeat into an intron sequence and is derived by strong promoter such as 35S CaMV for dicots or the maize ubiquitin promoter for monocots. When an intron used as the spacer, the efficiency become very high, almost 100% of transgenic plants show gene silencing (Smith *et al.* 2000). dsRNA can be delivered to plants in several ways: microprojectile bombardment with dsRNA or intron containing hairpin RNA (ihpRNA). Expression vector, infiltration of plant tissue with an *Agrobacterium* strain carrying a T-DNA expressing an hpRNA target gene; VIGS, in which the target sequence is integrated into viral sequences which are used to infect the plant or expressed by *Agrobacterium*-introduced transgene. Each technique has advantages and disadvantages with respect to how persistent their effect is and the range of plants to which they can be applied.

Virus-induced gene silencing is another approach often used to analyze gene function in plants (Waterhouse and Helliwell 2003). VIGS has the advantage that it negates the need to introduce a trasnsgene into the target plant genome, and thus permit RNAi to be carried out in plant species that are recalcitrant to genetic transformation (Robertson ,2004; Burch-Smith 2004). Additionally, there are alternative methods that can be utilized to induce gene silencing. Artificial miRNA (Ossowski *et al.* 2008) and zinc finger nucleases mediated-genome modification (Shukla *et al.* 2009) all have been developed for modifying gene expression in plants.

RNAi as a tool for functional genomics

Functional genomics is an important aspect of genomics, determining the function of all genes in the plant genome is very challenging. For this purpose, RNAi has been widely used as a compared to other insertional mutagenesis method. Although RNAi is not a knockout but a known down technology, its high efficiency and case of application make it applicable to genome wide analysis of gene functions. So far, RNAi technology has been applied in various plant species (Table 1) for revealing gene function. For example, Travella and coworker in 2006 introduced dsRNA expressing construct of two genes, phyloene desalturase (PDS) and ethylene insensitive 2 (EIN 2) into hexaploid wheat. Transformed plants showed phenotypic changes that were stably inheritable over at least two generations. They should that RNAi mediated gene silencing of both genes resulted in reduction of transcript by up to 93% for PDS and 99% for EIN2, the endogenous target mRNA levels of homologous genes were decreased in RNAi transgenic lines, which displayed a strong correlation between decreased mRNA levels and increased severity of phenotypes. The results showed the potentiality of dsRNA in being a useful and efficient tool for functional genomics.

Table 1 : Use of RNAi technology in various plant species for revealing gene function.

Plant species	Targeted genes	References
Arabidopsis thaliana	AGAMOUS, CLAVATA3, APETALA1, PERIANTHIA	Chaung *et al.* 2000
Cotton	Myb transcription factor	Machado *et al.* 2009
Soybean	Phytoene desaturase	Zhang *et al.* 2010
Tomato	Phyoene desaturase, CTR1	Liu *et al.* 2002
Wheat	Phytoene desaturase, EIN2	Travella *et al.* 2006

Advantages and limitations of RNAi as a functional genomics tool

RNAi in plants has numerous advantages over other reverse genetics approaches used in functional genomics. The first and primary advantage is that RNAi is a sequence-specific target a gene. RNAi can also be used to reduce the expression of several related genes in parallel by targeting conserved region of the gene, facilitating the study of redundant gene functions. In addition to this, RNAi can be controlled in a tissue specific or time dependent manner. There are also some limitations for RNAi when used in plant functional genomics. Firstly, knowledge of target gene sequence is mandatory. Delivery methods for dsRNA are also a limiting step in a number of species for RNAi based approach.

Conclusion

RNAi is an efficient knockdown technology and thought to be useful for genetic improvement even in plants with low transformation efficiencies. Since its discovery, the field of RNAi progressed at an amazing rate and within a short period of time it has become the technology of choice as a tool for plants functional genomics. As per as future prospects of this technology are concerned, it looks promising and seemed to be advanced at a very rapid pace. Finally it could be stated that it just the starting of unfolding many unknown genes functioning and the exciting journey of plant science will move ahead.

References

Ambros V., Bartel B., Bartel D.P., Burge C.B., Carrington J.C., Chen X., Dreyfuss, G., Eddy S.R., Griffiths-Jones S., Marshall M., Matzke M., Ruvkun G., Tuschl T.A. (2003) Uniform system for microRNA annotation. RNA 9: 277–279.

Bartel D.P. (2004) MicroRNAs: genomics, biogenesis, mechanism, and function. Cell 116: 281–297.

Baulcombe D. (2004) RNA silencing in plants. Nature 431: 356–363.

Bluementhal T. and Carmichael G.G. (1979) RNA replication: Function and structure of Qbeta-replicase. Ann. Rev. Biochem. 48: 525-548.

Bollman K.M., Aukerman M.J., Park M. Y., Hunter C., Berardini T.Z. and Poethig R.S. (2003) HASTY, the Arabidopsis ortholog of exportin 5/MSN5, regulates phase change and morphogenesis. Development 130: 1493–1504.

Brodersen, P., Sakvarelidze-Achard L., Bruun-Rasmussen M., Dunoyer P., Yamamoto, Y.Y., Sieburth L. and Voinnet O. (2008) Widespread translational inhibition by plant miRNAs and siRNAs. Science 320: 1185–1190.

Chapman E. J., Prokhnevsky A. I., Gopinath K., Dolja V. V., Carrington J. C. (2004) Viral RNA silencing suppressors inhibit the microRNA pathway at an intermediate step. Genes and Development 18: 1179–1186.

Chauang C.F. and Meyerowitz E.M. (2000) Specific and heritable genetic interference by double-stranded RNA in Arabidopsis thaliana. PNAS 97: 4985-4990.

Cogoni C., Irelan J. T., Schumacher. M., Schmidhauser T. J., Selker, E. U. and Macino G. (1996) Transgene silencing of the al-1 gene in vegetative cells of Neurospora is mediated by a cytoplasmic effector and does not depend on DNA–DNA interactions or DNA methylation. EMBO Journal 15: 3153–3163.

Dalmay T., Hamilton A., Rudd S., Angell S., Baulcombe D. C. (2000) An RNA-dependent RNA polymerase gene in *Arabidopsis* is required for posttranscriptional gene silencing mediated by a transgene but not by a virus. Cell 101: 543–553.

Dunoyer P., Lecellier C. H., Parizotto E. A., Himber C. and Voinnet O. (2004) Probing the microRNA and small interfering RNA pathways with virus-encoded suppressors of RNA silencing. Plant Cell 16: 1235–1250.

Dunoyer P. and Voinnet O. (2005) The complex interplay between plant viruses and host RNA-silencing pathways. Current Opinions in Plant Biology 8: 415–423.

Elbashir S. M., Harborth J., Weber K. and Tuschl T. (2002) Analysis of gene function in somatic mammalian cells using small interfering RNAs. Methods 26: 199–213.

Fang Y. and Spector D. L. (2007) Identification of nuclear dicing bodies containing proteins for microRNA biogenesis in living Arabidopsis plants. Curr. Biol. 17: 818–823.

Fire A., Xu S., Montgomery M. K., Kostas S. A. and Driver S. E. (1998) Potent and specific genetic interference by double-stranded RNA in *C elegans*. Nature 391: 806-811.

Gasciolli V., Mallory A. C., Bartel D. P. and Vaucheret H. (2005) Partially redundant functions of *Arabidopsis* DICER-like enzymes and a role forDCL4 in producing trans-acting siRNAs. Current Biology 15: 1494–1500.

Guo S. and Kemphues K. J. (1995) par-1, a gene required for establishing polarity in C. elegans embryos, encodes a putative Series/Thr kinase that is asymmetrically distributed. Cell 81: 611-620.

Hamilton A.J. and Baulcombe D.C. (1999) A species of small antisense RNA in posttranscriptional gene silencing in plants. Science 286: 950–952.

Hammond S.M., Bernestein E., Beach D. and Hannon G.J. (2000) An RNA directed nuclease mediates posttranscriptional gene silencing in Drosophila cells. Nature 404: 293-296.

Hannon G.J. (2002) RNA interferance. Nature 418: 244-251.

Invitrogen (2004) Gateway TM Technology. Quest. 1(2): 32-33.

Khvorova A., Reynolds A. and Jayasena S. D. (2003) FunctionalsiRNAs and miRNAs exhibit strand bias. Cell 115: 209–216.

Kidner C.A. and Martienssen R.A. (2005) The developmental role of microRNA in plants. Current Opinions in Plant Biology 8: 38–44.

Kim V. N. (2005) MicroRNA biogenesis: coordinated cropping and dicing. Nat.Rev. Mol. Cell Biol. 6: 376–385.

Lippman Z. and Martienssen R. (2004) The role of RNA interference in heterochromatic silencing. Nature 431: 364–370.

Liu Y., Schiff M. and Dinesh-Kumar S.P. (2002) Virus-induced gene silencing in tomato. Plant J. 31: 777-786.

Lund E., Guttinger S., Calado A., Dahlberg J.E. and Kutay U. (2004) Nuclear export of microRNA precursors. Science 303: 95–98.

Machado A., Wu Y., Yang Y. and Denis E.S. (2009) The MYB transcription factor GhMYB25 regulates early fibre and trichome development. Plant J 59: 52-62.

Mallory A. C., Reinhart B. J., Jones-Rhoades M. W., Tang, G., Zamore P. D., Barton M. K. and Bartel D. P. (2004) MicroRNA control of PHABULOSA in leaf development: Importance of pairing to the microRNA 59 region. EMBO J. 23: 3356–3364.

Meister G. and Tuschl T. (2004) Mechanisms of gene silencing by double stranded RNA. Nature 431: 343–349.

Mello C. C. and Conte D. (2004) Revealing the world of RNA interference. Nature 431: 338–342.

Montgomery M. K., Xu S. and Fire A. (1998) RNA as a target of double-stranded RNA-mediated genetic interference in Caenorhabditis elegans. Proceedings of the National Academy of Sciences, USA 95: 15502–15507.

Palauqui J.C. and Balzergue S. (1999) Activation of systemic acquired silencing by localized introduction of DNA. Curr. Biol. 9: 59-66.

Parker J. S., Roe S. M. and Barford D. (2004) Crystal structure of a PIWIprotein suggests mechanisms for siRNA recognition and slicer activity. EMBO J. 23: 4727–4737.

Plasterk R. (2002) RNA silencing: the genome's immune system. Science 296: 1263–1265.

Reinhart B. J., Weinstein E. G., Rhoades M. W., Bartel B. and Bartel D. P. (2002) MicroRNAs in plants. Genes and Development 16: 1616–1626.

Ruby J. G., Jan C. H. and Bartel D. P. (2007) Intronic microRNA precursors that bypass Drosha processing. Nature 448: 83–86.

Schauer S. E., Jacobsen S. E., Meinke D. W. and Ray A. (2002) DICER-LIKE1: blind men and elephants in *Arabidopsis* development. Trends in Plant Science 7: 487–491.

Schiebel W., Haer B., Marinkovic S., Kianner A. and Sanger H.L. (1993) RNA-directed RNA polymerase from tomato leaves, purification and physical properties. J. Biol. Chem. 268: 11851-11857.

Schwab R., Ossowski S., Riester M., Warthmann N. and Weigel D. (2006) Highly specific gene silencing by artificial microRNAs in *Arabidopsis*. The Plant Cell Online 18: 1121.

Schwarz D. S., Du, Hutvágner G. T., Xu Z., Aronin N. and Zamore P. D. (2003) Asymmetry in the assembly of the RNAi enzyme complex. Cell 115: 199–208.

Simon-Mateo C. and Garcia J. A. (2006) MicroRNA-guided processing impairs Plum pox virus replication, but the virus readily evolves to escape this silencing mechanism. Journal of Virology 80: 2429–2436.

Small I. (2007) RNAi for revealing and engineering plant gene functions. Curr. Opin. Biotechnol. 18: 148-153.

Smith N., Singh S., Wang M.B., *et al.* (2000) Total silencing by intron-spliced hairpin RNAs. Nature 407: 319–320.

Song J. J., Smith S. K., Hannon G. and Joshua-Tor L. (2004) Crystal structure of Argonaute and its implications for RISC slicer activity. Science 305: 1434–1437.

Tang G., Reinhart B.J., Bartel D.P. and Zamore P. D. (3003) A biochemical framework for RNA silencing in plants. Genes and Development 17: 49–63.

Tomari Y. and Zamore P. D. (2005) Perspective: machines for RNAi. Genes Dev. 19: 517–529.

Travella S., Klimm T.E. and Keller B. (2006) RNA interference based gene silencing as an efficient tool for functional genomics in hexaploid bread wheat. Plant Physiol. 142: 7-20.

VanderKrol A.R., Mur L. A., Beld M., Mol J.N. and Stuitje A.R. (1990) Flavonoid genes in petunia: addition of a limited number of gene copies may lead to a suppression of gene expression. Plant Cell 2: 291–299.

Vasudevan S., Tong Y. and Steitz J.A. (2007) Switching from repression to activation: microRNAs can up-regulate translation. Science 318: 1931–1934.

Vaucheret H., Vazquez F., Crété P. and Bartel D. P. (2004) The action of ARGONAUTE1 in the miRNA pathway and its regulation by the miRNA pathway are crucial for plant development. Genes and Development 18: 1187–1197.

Voinnet O. (2005) Non-cell autonomous RNA silencing. FEBS Lett. 579: 5858-5871.

Voinnet O. (2009) Origin, biogenesis, and activity of plant microRNAs. Cell 136: 669–687.

Voinnet O. (2004) Shaping small RNAs in plants by gene duplication. Nature Genetics 36: 1245–1246.

Voinnet O., Pinto Y. M. and Baulcombe D. C. (1999) Suppression of gene silencing: a general strategy used by diverse DNA and RNA viruses of plants. Proceedings of the National Academy of Sciences, USA 96: 14147–14152.

Waterhouse P. and Helliwell C. (2003) Exploring plant genomes by RNA-induced gene silencing. Nature Genet. 4: 29–38.

Wingard S.A. (1928) Host and symptoms of ring spot, a virus disease of plants. J. Agric. Res. 37:127-153.

Wu-Scharf D., Jeong B., Zhang C. and Cerutti H. (2000) Transgene and transposon silencing in *Chlamydomonas reinhardtii* by a DEAH-box RNA helicase. Science 290: 1159–1162.

Xie Z., Allen E., Fahlgren N., Calamar A., Givan S.A. and Carrington J.C. (2005a) Expression of *Arabidopsis* MIRNA genes. Plant Physiology 138: 2145–2154.

Xie Z., Allen E., Wilken A. and Carrington J.C. (2005b) DICER-LIKE 4 functions in trans-acting small interfering RNA biogenesis and vegetative phase change in *Arabidopsis* thaliana. Proceedings of the National Academy of Sciences 102: 12984–12989.

Xie Z., Fan B., Chen C. and Chen Z. (2001) An important role of an inducible RNA-dependent RNA polymerase in plant antiviral defense. Proceedings of the National Academy of Sciences, USA 98: 6516–6521.

Xie Z., Johansen L.K., Gustafson A.M., Kasschau K.D., Lellis A.D., Zilberman D., Jacobsen S.E. and Carrington J.C. (2004) Genetic and functional diversification of small RNA pathways in plants. PloS Biology 2: E104.

Xie Z., Kasschau K.D. and Carrington J.C. (2003) Negative feedback regulation of *Dicer-Like1* in *Arabidopsis* by microRNA-guided mRNA degradation. Current Biology 13: 784–789.

Ye K., Malinina L. and Patel D. J. (2003) Recognition of small interfering RNA by a viral suppressor of RNA silencing. Nature 426: 874–878.

Yigit E., Batista P.J., Bei Y., Pang K.M., Chen C.C., Tolia N.H., Joshua-Tor L., Mitani S., Simard M.J. and Mello C.C. (2006) Analysis of the C. elegans Argonaute family reveals that distinct Argonautes act sequentially during RNAi. Cell 127; 747–757.

Yoo B.C., Kragler F., Varkonyi-Gasic E., Haywood V., Archer-Evans S., Lee Y. M., Lough T. J. and Lucas W. J. (2004) A systemic small RNA signaling system in plants. Plant Cell 16: 1979–2000.

Yoshikawa M., Peragine A., Park M.Y. and Poethig R.S. (2005) A pathway for the biogenesis of *trans*-acting siRNAs in *Arabidopsis*. Genes and Development 19: 2164–2175.

Zamore P.D., Tuschl T., Sharp P.A. and Bartel D.P. (2000) RNAi: double-stranded RNA directs the ATP dependent cleavage of mRNA at 21 to 23 nucleotide intervals. Cell 101: 25–33.

Zhang C., Bradshaw J.D., Whitham S.A. and Hill J.H. (2010) The development of an efficient an multipurpose bean pod mottle virus viral vector set for foreign gene expression and RNA silencing. Plant Physiol 153: 52-65.

13

Meta Analysis: An Advanced Statistical Tool for Biological Research

Arundhati Bag[1]* and Niladri Bag[2]
[1]Department of Medical Biotechnology, Sikkim Manipal University, 5th Mile Tadong, Gangtok- 737102, India
[2]Department of Horticulture, Sikkim University, 6th Mile, Tadong, Gangtok – 737102, India
**E-mail:* *arundhatis5@rediffmail.com*

Abstract

Meta analysis quantitatively summarizes the findings of a large collection of individual experiments. It is a powerful statistical tool since it integrates a vast literature and thus includes large sample size. It leads to a stronger conclusion than traditional literary review as studies are analyzed applying statistical methods. Meta- analysis is increasingly being applied in different fields of biological sciences. It has been emerged as an independent subfield of modern health science researches as in this field very few studies achieve statistical power in isolation. Meta- analysis has also become an integral part of plant science research, *e.g.*, agricultural research, ethnobotanical research, tea and health science research *etc.* Meta-analysis has few limitations although rigorously conducted meta- analyses are considered more valuable than any single study contributing to the analysis.

Introduction

Single isolated experiments are rarely sufficient for drawing a definitive conclusion on a particular research topic in biological sciences either in the field of plant science research, agricultural research or health research. Usually there are hundreds of studies available on a single topic those follow different experimental methodologies, samples, statistical analyses, and their conclusions are often contradictory, non- significant. These disparities are commonly explained by small sample size, different sample organisms etc. Finally no acceptable conclusions can be drawn and confusions are created making all the research efforts fruitless. Thus the vast literature is required to be organized so that a message can be extracted that would be useful for other researchers in the field, policy makers, administrators and teachers. Review articles fulfill this purpose; they provide a survey and examination of existing literature on a particular topic. It benefits readers by providing a summary of published articles, a conclusion, limitations of the current knowledge and future directions of research. However, traditional reviews have many limitations. An expert in the field usually selects studies in haphazard way as no systematic approach has been defined for obtaining primary data or to integrate findings. Research findings are described in vague and imprecise narrative summaries and little detail of methodologies are provided (Kulik and Kulik 1989). Further, these reviews are often subjected to reviewer's bias as studies are included on the basis of reviewer's own impressionistic view about the quality of the studies (Wolf 1986). Traditional reviews are mostly narrative and data are not synthesized quantitatively across literature (Thacker 1988).

According to Glass *et al.* (1981) review should be more technical and statistical rather than being merely narrative one and the findings of the multiple studies should be treated as a complex data set for statistical analyses. Meta- analysis is a quantitative method to combine data and an alternative method of analyzing existing literature. It carries out statistical analyses of quantitative summaries of individual experiments (Kulik and Kulik 1989).

Meta-analyses are a subset of systematic reviews (Haidich 2010). Systematic reviews, as the name implies, typically involve a detailed and comprehensive plan and search strategy derived a priori, with the goal of reducing bias by identifying, appraising, and synthesizing all relevant studies on a particular topic (Uman 2011). A protocol for the review, with the detailed description of the methods to be used, is prepared in advance and is to be maintained unchanged until the study is finished. Detailed description of the whole process allows assessment of quality of the reviews and validity of their conclusions (Lunet 2012). Systematic reviews bring together a number of separately conducted studies sometimes with contradictory results, and synthesize their results. Following a systematic review, meta-analysis is carried out where data from individual studies are pooled quantitatively and reanalyzed using established statistical methods to synthesize the data into a single quantitative estimate.

Definition, Advantages and Limitations of Meta- analysis

Glass (1976) first defined meta-analysis in the social science literature; he defined meta- analysis as the statistical analysis of a large collection of analysis results from individual studies for the purpose of integrating the findings. Meta- analysis permits quantitative reviews and synthesis of research literature (Wolf 1986). It focuses on size of treatment effects, not just statistical significance (Kulik and Kulik 1989). It is an application of statistical tools to summary statistics, not raw data. Its observations are means, standard deviation, and results from statistical tests. It includes a large number of studies. An analysis of raw scores is done in primary or secondary analysis. According to Glass (1976) three types of analyses are important, primary, secondary and meta- analysis. Primary analysis is the original analysis of data in a research study and secondary is the re- analysis of data for the purpose of answering the original research with better statistical technique, or answering new questions with new data. Meta-analysis offers a mechanism for estimating the magnitude of effect in terms of a statistically significant effect size or pooled odds ratio (Thacker 1988) as it combines the samples of individual studies by effectively increasing the sample size and thereby improving the statistical power and precision of the estimates. In addition, the combination of data from several studies potentially increases statistical power and enable the reader to assess more completely the impact of a procedure or variable (Pillemer and Light 1980). It is expected that this systematic approach would decrease investigator bias.

In recent years the term "meta- analysis' has been used in broader sense (Kulik and Kulik 1989), for example, Rosenthal (1984) uses the term to describe almost any attempt to combine or compare statistical results from two or more studies. However, many researchers show dissatisfaction in its name as it seems grander it need to be. Yet another group of scientists consider 'synthesis' as a better word than analysis.

Despite its several advantages meta- analysis is not above criticism. While it includes maximum literature, it does not differentiate between poorly designed studies and good ones, and incorporates all of them. Further, it covers multiple results derived from the same studies giving false impression of large sample size. Meta- analysis may be biased as it depends on published results, and publication favours significant findings while non- significant findings are rarely published. In addition, comparing studies employing different measuring techniques, definitions of variables and subjects ('mixing apples and bananas' problem) may lead to erroneous conclusion. In one study of psychotherapy, for example, Glass and his colleagues (Smith *et al.* 1980) mixed results from different types of outcome measures. Meta-analysis also ignores interaction effects and is more focused on the main effect (Wolf 1986). Finally, it emphasizes on unwarranted sense of scientific validity, rather than true understanding that may result from quantification (Thacker 1990). In biology variables like treatment outcome, disease risk etc. are consequences of gene- gene interactions, gene- environment interactions and cross talk between signal transduction pathways. Thus a phenomenon requires deeper understanding of the causative mechanisms. Therefore, statistical validation of a single main effect could be too simplistic to draw a conclusion.

Process of conducting Meta- analysis

Meta-analysis is applicable to collections of research that are empirical, produce quantitative results, rather than qualitative findings, examine the same constructs and relationships and have findings that can be configured in a comparable statistical form (Lipsey and Wilson 2001). Meta-analysis combines effect sizes from different studies to an overall measurement (Kock 2009). In statistics, an effect size is a quantitative measure of the strength of a phenomenon. It is independent of sample size. Effect-size indices usually gauge the association between two variables (Johnson *et al.* 2007). Any standardized index can be an "effect size" *e.g.*, standardized mean difference (*d*), correlation coefficient (*r*), odds- ratio (OR) as long as it is comparable across studies, and represents the magnitude and direction of the relationship of interest (Lipsey and Wilson 2001). The convention is to use *r* as the effect size, if most of the studies that are integrated report correlations between two quantitative variables. If most of the studies report ANOVAs, t-tests, and chi-squares for comparisons between two groups (e.g., experimental vs. control), *d* is typically selected. If both variables are dichotomous then odds ratio is selected (Johnson *et al.* 2007). Different meta-analyses use different effect size indices. Odds ratios are particularly used in medicine and public health related meta- analyses.

There are various statistical programs available to calculate effects sizes for meta-analyses, such as the Review Manager (RevMan) program endorsed by the Cochrane Collaboration (Uman 2011). Effect sizes are calculated along with a 95% confidence interval (CI) range, and presented in both quantitative format and graphical representation (e.g., forest plots). Forest plots visually present each trial usually as a horizontal diamond shape with the middle representing the effect size. A horizontal line represents 95% confidence interval for each of them. The diamonds are presented on a graph with a centre solid vertical line representing the zero mark. At the bottom of the graph is an overall estimate of effect size or diamond representing all of the individual studies pooled together (Uman 2011). The centre of the diamond represents the combined effect, and the horizontal tips represent the 95% CI. Further, heterogeneity between studies is calculated to see whether the individual studies are similar enough to compare. Meta-analysis should be conducted only when the individual studies are homogeneous regarding their methodological characteristics (Lunet 2012). However, it is still possible to pool studies when significant heterogeneity exists, although these results should be interpreted with caution and reasons for the heterogeneity should be explored (Uman 2011).

Applications of Meta- analysis

The method of meta- analysis can be applied within all the biological sciences, including health sciences, agricultural sciences and plant science research.

Health Sciences

Health care profession is dependent on the best available evidence. From the late 1970s application of meta- analyses in health science has been increased sharply (Haidich 2010). It has been emerged as an independent subfield within the statistical sciences, especially in medicine, where isolated clinical trial usually has too few observations to achieve statistical validity (Alston *et al.* 2000). Moreover, it has been found that meta-analyses are the most frequently cited form of clinical research- review in medical literature (Patsopoulos *et al.* 2005). It is used to assess the clinical effectiveness of healthcare interventions and very commonly to determine genetic association with a disease (cancer, cardiovascular disease) risk etc. Meta-analysis is being increasingly used in assessing chemotherapy outcome, performance of diagnostic tests, patient education interventions in clinical medicine, spinal manipulation, the effects of exercise on serum lipid levels, duodenal ulcer therapy, to quantify the impact on survival and safety of a wide range of surgical and anesthetic innovations etc. (Thacker 1990).

Meta-analysis in health research is a two stage process. The first stage involves calculation of a measure of a variable with its 95% confidence intervals (CI) for each individual study. The summary statistics that are usually used to measure include odds ratios (OR), relative risks (RR), and mean differences. Odds ratio or RR are used if the outcome is binary or dichotomous (risk vs. no risk, live vs. dead, effect *vs.* resistance etc.) whereas mean difference is calculated for the continuous outcome (blood glucose measurements *etc.*). In the second stage, an overall measure is calculated as a weighted average of the individual summary statistics (Akobeng 2005).

An odds ratio demonstrates association between an exposure and an outcome. The OR represents the odds that an outcome will occur given a particular exposure, compared to the odds of the outcome occurring in the absence of that exposure (Szumilas 2010). They are most commonly used in case-control studies, however, they can also be used in cross-sectional and cohort study designs (Szumilas 2010). An OR is calculated by the following formula:

$$\mathrm{OR} = \frac{\text{exposed cases (n) / unexposed cases (n)}}{\text{Exposed non cases (n) / unexposed non cases (n)}}$$

An OR = 1 indicates that the exposure does not affect odds of outcome, OR>1 means that exposure is associated with higher odds of outcome and an OR<1 means that the exposure is associated with lower odds of outcome. A relative risk (RR) is the ratio of the risk of an event in two groups.

The 95% confidence interval (CI) is used to estimate the precision of the OR. A large CI indicates a low level of precision and vice versa. The 95% CI is calculated to get an idea of statistical significance if it does not overlap the null value (OR=1 for a measure). Other intervals such as 90% or 99% are also sometimes used. The 95% CI value is calculated using the following formula:

95% Cl = e^ [In (OR) – 1.96√(1/a + 1/b + 1/c + 1/d)] to e^ [In (OR) + 1.96√(1/a+1/b+1/c+1]

Here we incorporate few examples of published meta- analyses to illustrate the utility of this statistics in cancer research.

Example 1

Xu *et al.* (2014) calculated a meta- analysis to determine genetic association of Glutathione S-transferase omega (*GSTO*) polymorphisms with cancer risk. These polymorphisms are often associated with disease susceptibility. Glutathione S- transferases (GSTs) constitute a multigene family of phase II metabolizing enzymes, which catalyze conjugation of tripeptide glutathione (GSH) to a wide variety of compounds including those with electrophilic functional groups and render them more water soluble (Hayes and Pulford 1995). Thus they play a major role in phase II cellular detoxification, and protect cellular components from reactive electrophiles including environmental carcinogens. Genetic polymorphisms that decrease or abolish their detoxifying ability of carcinogens are thus expected to play important role in cancer development. Human *GST* genes are polymorphic, either due to presence of single nucleotide polymorphisms (SNPs) or due to deletions. There are eight distinct classes designated as: Alpha, Kappa, Mu, Omega, Pi, Sigma, Theta, and Zeta (Xu *et al.* 2014).

Several experiments have been conducted on genetic polymorphisms of GSTs and susceptibility to various cancers. GSTO has two members, *GSTO1* and *GSTO2*. Two human GST polymorphisms, e.g., *hGSTO1**A140D and *hGSTO2**N142D have been extensively studied for an associated cancer risk. But results of these studies have been largely inconsistent. While some studies demonstrated increased risk of *GSTO1* A140D for breast cancer, hepatocellular carcinoma, bile duct carcinoma, urothelial cancer, acute lymphoblastic leukemia and non-small cell lung cancer, some studies did not find significant association with thyroid, colorectal as well as breast cancer risk. Therefore Xu *et al.* (2014) conducted a meta-analysis on the pre- existing literature.

Study Selection

Only case- control studies were included in this analysis. Selection of articles was a step-by-step process that was carried out carefully. After deleting duplicate articles initial 1843 articles were collected from the databases. After screening by abstracts, 420 articles were found for GSTO. Among them 379 articles were excluded (282 were not for polymorphism and 97 were not for cancer). From remaining 41 articles only full articles on case- control design and ample sample size were chosen (n= 20). Among the selected final 20 articles, some studied both *GSTO1* and *GSTO2* polymorphisms, thus each one of them was regarded as two separate studies. Finally 13 studies for *GSTO1* and 13 for *GSTO* were included in the meta-analysis. 13 eligible studies involving 3540 cancer cases and 3879 controls were pooled for analysis of *GSTO1* A140D.

Interpreting result

No significant association of *GSTO1* polymorphism with cancer risk was observed form the meta- analysis. By pooling 13 eligible studies with 3399 cases and 4135 controls, a significantly increased risk of cancer susceptibility was observed for *GSTO2* polymorphism homozygotes (GG vs. AA: OR = 1.20, 95%CI: 1.02). Subgroup analyses revealed significant increased cancer risk for Caucasians. Typically in meta-analysis on population based case- control studies heterogeneity between studies are tested. Although no significant heterogeneity was detected for *GSTO1* polymorphism, for *GSTO2* heterogeneity was detected (not absolute), which however, was not due to ethnicity, cancer types, sample size and sources of control. Further, the authors carried out sensitivity analyses of their meta-analysis by repeatedly deleting the single studies each time from pooled analysis and found the results were robust and convincing. In addition, Begg's funnel plot and Egger's test were performed to assess the publication bias of literatures. No publication bias for *GSTO1* or *GSTO2* polymorphism was observed throughout the study.

In summary, while individual studies demonstrated contradictory findings, from the above meta- analysis it was possible to conclude that *GSTO2* may significantly increase cancer risk in Caucasians and is associated with elevated risk of breast cancer. Important limitation of this analysis was that it included studies written in English only which may be a source of publication bias.

Example 2

In another meta- analysis Liu *et al.* (2012) studied association between Manganese superoxide dismutase (MnSOD) target sequence polymorphism Val9Ala (rs4880) with breast cancer risk. MnSOD constitutes first line of defense in the cell against oxidative stress. The Val16Ala polymorphism leads to change in MnSOD precursor protein that leads to an impaired transport of MnSOD to mitochondria and influences cellular ROS level. There is increasing evidence demonstrating the association between Val16Ala polymorphism with several types of malignancies, including breast, lung, bladder, colorectal and prostate cancer (Bag and Bag 2008; Liu *et al.* 2012).

A number of studies have been carried out for the risk of *MnSOD* Val16Ala polymorphism for breast cancer. However, the results are largely inconclusive. One of the causes that led to inconclusive results is that small sample size included in each study. Thus the authors performed a meta-analysis from 14 published case-control studies including 17842 subjects of which 8,102 were breast cancer cases and 9,740 were controls.

Study Selection

Study inclusion criteria were like previous example, *i.e.*, it included full text articles written in English only and case-control or cohort studies that quantitatively evaluated association between breast cancer risk and Val16Ala polymorphism, with sufficient sample size.

Interpreting results

No significant association was observed between the polymorphism (either heterozygous Val/Ala or homozygous variant Ala/Ala) and breast cancer risk (Val/Ala vs. Val/Val: OR, 1.04; 95% CI, 0.93-1.17; Ala/Ala vs. Val/Val: OR, 1.12; 95% CI, 0.95-1.33).

However, subgroup analysis revealed significantly increased risk of the variant allele for breast cancer in premenopausal women and for certain lifestyle habits, *e.g.*, tobacco smoking, using oral contraceptives, higher body mass index. To analyze the sensitivity of the association between *MnSOD* gene polymorphism and breast cancer risk, they identified 4 studies that did not follow the Hardy- Weinberg equilibrium. However, there was no change in the final result either they were included or excluded from the meta-analysis. No publication bias was revealed from Egger's test.

Above two examples show that meta- analysis can bring out a definite conclusion on a particular research topic while individual studies demonstrate contradictory results. Further, it can draw more confirmatory conclusion in comparison to traditional review. For example, in a conventional review Park *et al.* (2009) observed that polymorphism of DNA repair gene X-ray repair cross-complementing group 1, *XRCC1* R399Q, was associated with prostate cancer risk in 5 of 7 selected studies. They concluded impressionistically that *XRCC1* R399Q allele is associated with increased prostate cancer risk for carriers alone or when the variant allele is combined with other DNA repair polymorphisms or low antioxidant diet. In a quantitative review or meta- analysis (Yan *et al.* 2015) the conclusion was more confirmatory: there is a significant association between the *XRCC1* R399Q polymorphism with prostate cancer risk.

Meta-analysis is no longer a novelty in health science rather numerous meta-analyses have been conducted for the same medical topic by different researchers. For example, different meta- analyses are now available in literature for the same *XRCC1* R399Q polymorphism and cancer risk, that too, with different conclusions. As stated earlier, in meta- analysis by Yan *et al.* (2015), significant association of *XRCC1* R399Q polymorphism and prostate cancer risk was found. However, in another meta- analysis (Wei *et al.* 2011), no overall association between this polymorphism with prostate cancer risk was documented. Therefore, there is recent a trend to combine the results of different meta-analyses, known as a meta-epidemiological study, to assess the risk of bias (Haidich 2010).

Plant Science

A growing interest has been observed for meta- analysis in different fields of plant science research as well. Following are some examples which show how plant scientists embrace this statistical concept in their data analysis:

Agriculture

In agriculture meta-analyses are widely being used in the syntheses of studies measuring the value of a natural resource and the effect of farm size on measures of crop yield risk (Alston, 2000). An example of application of meta- analysis in the agriculture can be cited from the recent work of Corbeel *et al.* (2014) in the field of conservation- agriculture. Conservation agriculture involves reduced or no-tillage, permanent soil cover and crop rotations to enhance soil fertility and crop yields. Conservation agriculture practices are increasingly promoted on smallholder farms in sub-Saharan Africa as a means to overcome continuing poor-profitability and soil degradation (Corbeel *et al.* 2014). In their meta-analysis mean differences were applied. Yield differences between treatment and control were used. The results of their meta-analysis revealed that reduced or no-tillage without mulch and/or crop rotation leads to depressed crop yields compared to conventional farmers' practices based on tillage. However, crop grain yields were higher overall in no-tillage treatments in comparison to conventional tillage-based practices, when mulch was applied whether or not in combination with crop rotations.

In another example of meta- analysis in conservation agriculture Rusinamhodzi *et al.* (2011) assessed the effect of long-term tillage and residue retention on maize grain yield under contrasting soil textures, nitrogen input and climate. In their meta- analysis they also used the mean difference in yield between the treatment and control. The result shows an increase in maize yield over time with conservation agriculture practices that include rotation and high input use in low rain fall areas. They observed a strong relationship between maize grain yield and annual rainfall.

In another study to get an overview of technical efficiency in Nigerian agriculture a meta- analysis was conducted on a total of 64 existing studies on technical efficiency in Nigerian agriculture (Kolawle 2009). Empirical findings revealed that mean technical efficiency in Nigerian agriculture increased significantly over the years.

Earthworms are generally thought to be essential to sustainable agroecosystems. Although it is well known that earthworms positively influence plant growth, studies on the effect of earthworm on crop yields have not been quantitatively synthesized. Recently Groenigen *et al.* (2014) used meta-analysis and demonstrated that on average earthworm presence in agroecosystems leads to a 25% increase in crop yield and a 23% increase in aboveground biomass. This was first meta-analysis to summarize the effect of earth- worms on plant production across the globe. They conducted it from 58 studies that were published between 1910 between 2013. Studies included three main global staple crops (maize, rice and wheat), pastures, as well as many other food crops and were conducted on all continents except Antarctica. For each observation within every study they collected the means of the control treatment (that is, without earthworm presence) and the experimental treatment (that is, with earthworm presence), as well as their standard deviation (SD) and replicate numbers (n). Their result also revealed that increased nitrogen availability is a dominant pathway by which earthworms stimulate plant growth.

Ethnobotany

In the field of ethnobotany one hypothesis is often being tested: plants that are used more frequently by local populations are available in greater numbers in the environment. According to Albuquerque and Medeiros (2012) the combination of data from different studies can answer how the abundance of a given resource influences its local importance for human populations.

There exists a wealth of published ethnobotanical and pharmacological studies on different medicinal plants for different diseases, however, this information has to be reviewed systematically and studied quantitatively by meta- analysis for determining their role in medical interventions. One example is *Hypericum perforatum*, commonly known as St. John's wort (SJW). This is an herbaceous aromatic perennial plant native to Europe, Asia, Africa. The primary pharmacological activity attributed to SJW is antidepressant. It is currently prescribed by physicians as a treatment for depression, and the World Health Organization's "Monographs on Selected Medicinal Plants" classifies SJW for the symptomatic treatment of mild to moderate depression. These therapeutic uses and recommendations are based on systematic reviews and meta-analyses of the randomized controlled clinical trials that support the use of SJW for the symptomatic treatment of mild to moderate depression (McClatchey *et al.* 2009).

Tea and its impact on health

Tea is the most consumed drink in the world after water, well ahead of coffee, beer, wine and carbonated soft drinks (Rietveld and Wiseman 2003). Black tea is the most common form of world's tea consumption and other forms are green tea, oolong, white tea etc. A considerable number of studies suggest that consumption of green and black tea beverages may bring positive health effects. This protective effect is attributed to the presence of high levels of antioxidants in both black and green tea that can protect cells and tissues from oxidative damage by scavenging free radicals and thus reducing cellular oxidative stress.

Oxidative damage to DNA, lipids and proteins may contribute to the development of different diseases including cardiovascular disease, cancer and neurodegenerative diseases. Human intervention studies with green and black tea demonstrates a significant increase in plasma antioxidant capacity in humans 1 h after consumption of moderate amounts of tea, *i.e.*, 1–6 cups/d (Rietveld and Wiseman, 2003). Health-promoting effects of tea are mainly attributed to catechins, which belong to a family of compounds known as flavonoid-like polyphenols or flavanols (Peng *et al.* 2014) and other antioxidants.

Effect of tea compounds on cardiovascular disease (CVD) has been widely studied. However, studies investigating the effects of green tea on CVD have generated inconsistent results. Peters *et al.* (2001) conducted meta- analysis on the effect of tea by including 10 cohort studies and seven case-control studies and calculated relative risk (RR). They found that incidence rate of myocardial infarction decreased by 11% with an increase in tea consumption of 3 cups per day (fixed-effects relative risk estimate = 0.89, 95% confidence interval: 0.79, 1.01) although a publication bias was evidenced. One cup measures 237 ml. In another study Peng *et al.* (2014) carried out a meta- analysis to quantitatively evaluate the effects of green tea on blood pressure (BP) control. They included 13 studies comprising a total of 1,367 subjects in the analysis. The overall outcome of the meta-analysis suggested that green tea consumption significantly decrease systolic blood pressure (SBP) level by -1.98 mmHg. Green tea also showed a significant lowering effect on diastolic blood pressure in treatment group in comparison to the control group. In another meta- analysis black tea, like green tea was found to have blood pressure lowering effect (Greyling *et al.* 2014). In this meta- analysis pooled mean effect of regular tea ingestion was calculated from 11 randomized controlled intervention studies including 378 subjects. The results revealed that daily consumption of 4-5 cups of black tea for ≥ 1 week is associated with a statistically significant reduction in SBP and DBP of 2 and 1 mmHg, respectively.

Other than CVD, meta- analysis has been explored for the effect of tea on different types of cancer risk, *e.g.*, gall bladder cancer, esophageal cancer, pancreatic cancer, renal carcinoma etc. For example, a meta-analysis including six studies (4 case-control and 2 cohort studies) involving a total of 753 patients and 115,349 controls showed that tea consumption may reduce the risk of gallbladder cancer in women although not in men (Zhu *et al.*, 2015). It is the first meta- analysis on the effect of tea consumption on gall bladder cancer. In this study OR with 95% CI was used as a common measure of the association between tea consumption and the risk of gallbladder cancer across studies. The overall combined odds ratio for tea consumption and gallbladder cancer was 0.67 with 95% confidence interval 0.40-1.12.

Conclusion

Vast literature can be efficiently summarized by quantitative review following meta- analysis (Wolf 1986). Meta- analysis has wide application in almost every field of biological research, in health sciences and plant sciences. Its strength lies in the facts that this approach addresses the limitations of sample size in individual studies, can include diverse populations, leads to stronger conclusion as studies can be analyzed in statistical methods than in impressionistic literary review, can bring effects into sharper focus particularly when the results of all the studies are not consistent (Haidich 2010; Wolf 1986). It is assumed that heterogeneity damages a meta- analysis, however, some believes that the identification of outliers in meta- analysis can lead to increased understanding and new hypothesis (Wolf 1986). Despite its importance in modern scientific literature meta- analysis has its own limitations like any other statistical tool. Meta- analysis oversimplifies the results of a research area by focusing on overall effects and ignoring the interaction effects. A meta-analysis was carried out by us on total of 7,366 cancer cases and 9,102 controls from 13 published case-control studies, which showed no overall association of MnSOD Val16Ala polymorphism for cancer risk following meta- analysis (Bag and Bag 2008). However, considering the fact that MnSOD constitutes a first-line defense against oxidative stress in cell, it may not be right to conclude on the basis of meta-analysis alone that this polymorphism has no role in cancer development. It is important to remember that gene- gene, gene- environment interactions should also be considered in a statistical analysis be it meta- analysis or other statistical approach that aims to determine association between genetic polymorphism and complex diseases. Therefore, meta- analysis should be regarded as a better tool to quantify information across the studies but not as an ultimate method of analysis beyond which nothing is required to be known.

References

Akobeng A.K. (2005) Understanding systematic reviews and meta-analysis. Archives of Disease in Childhood 90: 845–848.

Albuquerque U.P. and Medeiros P.M. De (2012) Systematic reviews and meta-analysis applied to ethnobiological research. Ethnobiology and Conservation 1:6.

Alston J.M., Chan-Kang C., Marra M.C., Pardey P.G. and Wyatt T.J. (2000) A meta-analysis of rates of return to agricultural R&D. Ex Pede Herculem? International Food Policy Research Institute. www.ifpri.org/sites/default/files/publications/rr113.pdf.

Bag A. and Bag N. (2008) Target sequence polymorphism of human Manganese Superoxide Dismutase Gene and its association with cancer risk: A review. Cancer Epidemiology, Biomarkers & Prevention 17: 3298-3305.

Corbeel M., Sakyi R.K., Kühne R.F. and Whitbread A. (2014) Meta-analysis of crop responses to conservation agriculture in sub-Saharan Africa. CCAFS Report No. 12. Copenhagen: CGIAR Research Program on Climate Change, Agriculture and Food Security (CCAFS).

Glass G.V. (1976) Primary, secondary, and meta-analysis of research. Educational Researcher 5: 3-8.

Glass G.V., McGaw B. and Smith M.L. (1981) Meta-analysis in social research. Beverly Hills, CA: Sage.

Greyling A., Ras R.T., Zock P.L., Lorenz M., Hopman M.T., Thijssen D.H.J. and Draijer R. (2014) The effect of black tea on blood pressure: a systematic review with meta-analysis of randomized controlled trials. PLoS ONE 9: e103247. doi:10.1371/journal.pone.0103247.

Haidich A.B. (2010) Meta-analysis in medical research. Hippokratia 14 (Suppl 1): 29-37.

Hayes J.D. and Pulford D.J. (1995) The glutathione S-transferase supergene family: regulation of GST and the contribution of the isoenzymes to cancer chemoprotection and drug resistance. Critical Reviews in Biochemistry and Molecular Biology 30: 445-600.

Johnson B.T., Scott-Sheldon L.A.J., Snyder L.B., Noar S.M. and Huedo-Medina T.B. (2007) Contemporary approaches to meta-analysis in communication research. http://www.sagepub.com/upm-data/23659_Chapter11.pdf.

Kock A. (2009) A guideline to meta- analysis. https://www.tim.tuberlin.de/fileadmin/fg101/TIM_ Working_Paper_Series/Volume_2/TIM_WPS_Kock_2009.pdf.

Kolaowle O. (2009) A meta- analysis of technical efficiency in Nigerian agriculture. http://ageconsearch.umn.edu/bitstream/50327/2/238.pdf.

Kulik J.A. and Kulik C.L.C. (1989) The concept of meta-analysis. International Journal of Educational Research 13: 227-340.

Lipsey M. and Wilson D.B. (2001) Practical meta- analysis. London, Sage.

Liu G., Sun G., Wang Y., Wang D., Hu W. and Zhang J. (2012) Association between manganese superoxide dismutase gene polymorphism and breast cancer risk: A meta-analysis of 17,842 subjects. Molecular Medicine Reports 6: 797-804.

Lunet N. (2012) The use of systematic review and meta-analysis in modern epidemiology. *In* Epidemiology - Current Perspectives on Research and Practice. Nuno Lunet (Ed.), Intech.

McClatchey W.C., Mahady G.B., Bennett B.C., Shiels L. and Savo V. (2009) Ethnobotany as a pharmacological research tool and recent developments in CNS-active natural products from ethnobotanical sources. Pharmacology & Therapeutics 123: 239–254.

Park J., Huang Y. and Seller T.A. (2009) Single nucleotide polymorphisms in DNA repair genes and prostate cancer risk. Methods in Molecular Biology 471: 361–385.

Patsopoulos N.A., Analatos A.A. and Ioannidis J.P. (2005) Relative citation impact of various study designs in the health sciences. JAMA 293: 2362-2366.

Peng X., Zhou R., Wang B., Yu X., Yang X., Liu K. and Mi M. (2014) Effect of green tea consumption on blood pressure: A meta-analysis of 13 randomized controlled trials. Scientific Reports 4: 6251; DOI:10.1038/srep06251.

Peters U., Poole C. and Arab L. (2001) Does tea affect cardiovascular disease? A meta-analysis. American Journal of Epidemiology 154: 495–503.

Pillemer D.B. and Light R.J. (1980) Synthesizing extremes: How to use research evidence from many studies. Harvard Educational Review 50: 176-195.

Rietveld A. and Wiseman S. (2003) Antioxidant effects of tea: evidence from human clinical trials. The Journal of Nutrition 133: 3285S-3292S.

Rosenthal R. (1984) Meta-analytic procedures for social research. Beverly Hills, CA: Sage.

Rusinamhodzi L., Corbeels M., van Wijk M.T., Rufino M.C., Nyamangara J. and Giller K.E. (2011) A meta-analysis of long-term effects of conservation agriculture on maize grain yield under rain-fed conditions. Agronomy for Sustainable Development DOI 10.1007/s13593-011-0040-2.

Smith M.L., Glass G.V. and Miller T.I. (1980) The benefits of psychotherapy. Baltimore, MD: Johns Hopkins University Press.

Szumilas M. (2010) Explaining odds ratios. Journal of the Canadian Academy of Child and Adolescent Psychiatry 19: 227-229.

Thacker S.B. (1990) Meta-analysis: A quantitative approach to research integration. *In* Modern methods of clinical investigation. Annetine C. Gelijns (Ed.), National Academy Press, Washington DC.

Uman L.S. (2011) Systematic reviews and meta-analyses. Journal of the Canadian Academy of Child and Adolescent Psychiatry 20: 57-59.

van Groenigen J.W., Lubbers I.M., Vos H.M.J., Brown G.G., De Deyn G.B. and van Groenigen K.J. (2014) Earthworms increase plant production: a meta-analysis. Scientific Reports 4: 6365; DOI:10.1038/srep06365.

Wei B., Zhou Y., Xu Z., Ruan J., Zhu M., Jin K., Zhou D., Hu Q., Wang Q., Wang. Z and Yan. Z (2011) XRCC1 Arg399Gln and Arg194Trp polymorphisms in prostate cancer risk: a meta-analysis. Prostate Cancer and Prostatic Diseases 14: 225-231.

Wolf F.M. (1986) Meta- analysis: quantitative methods for research synthesis. SAGE Publication, USA.

Xu Y.T., Wang J., Yin R., Qiu M.T., Xu L., Wang J. and Xu L. (2014) Genetic polymorphisms in Glutathione S-transferase Omega (GSTO) and cancer risk: a meta-analysis of 20 studies. Scientific Reports 4: 6578; DOI:10.1038/srep06578.

Yan J., Wang X., Tao H., Deng Z., Yang W. and Lin F. (2015) Meta-analysis of the relationship between XRCC1-Arg399Gln and Arg280His polymorphisms and the risk of prostate cancer. Scientific Reports 5: 9905; DOI:10.1038/srep09905.

Zhu G., Hua J., Wang Z., She F. and Chen Y. (2015) Tea consumption and risk of gallbladder cancer: A meta-analysis of epidemiological studies. Molecular and Clinical Oncology 3: 613-618.

Abbreviations

μm	Micrometer
μM/μmol	Micromolar
2,4- D	2,4- Dichlorophenoxyacetic acid
2-ip	6-γ,γ-Dimethylallylamino purine
ACC	1-aminocyclopropane-1-carboxylate
ACP	Acyl-Carrier Protein
Amsl	Above means sea level
ANOVA	Analysis of variance
BAP/ BA	6-benzyle amino purine
CCD	Central Composite Design
CFC	Continuous flow configuration
Chl	Chlorophyll
cm	Centimetre
CT	Computed Tomography
DNA	Deoxyribo Nucleic Acid
ED	Entner–Doudoroff pathway
EU	European Union
FT-IR spectroscopy	Fourier Transform Infrared Spectroscopy
G	Gram
GA_3	Gibberellic acid
GC-MS	Gas chromatography / mass spectrometry
GD	Gresshoff and Doy
HAA	3-(3-hydroxyalkanoyloxy) alcanoic acid
HPLC	High Performance Liquid Chromatography

IAA	Indole -3- acetic acid
IBA	Indole -3- butyric acid
ISI	Indian Standards Institution
ISR	Induced systemic resistance
KDPG	2-keto-3-deoxy-6-phosphogluconate
LSD	List significant difference
$m^{-2}s^{-1}$	Per meter square per second
mg	Milligram
MP	Mother plant
MRI	Magnetic Resonance Imaging
MS	Murashigue and Skoog 1962, medium
MS	Mass Spectrometry
NAA	Naphthaleneacetic acid
NMR	Nuclear magnetic resonance
PET	Positron Emission Tomography
PGPR	Plant Grwoth Promoting Rhizobacteria
PGR	Plant Growth Regulator
PSB	Phosphate solubilizing bacteria
PTGS	Post Transcriptional Gene Silencing
RISC	RNA-induced silencing complex
siRNA	Small-interfering RNA
TC	Tissue culture
TDZ	Thidiazuron
THMPD	Traditional Herbal Medicinal Products Directive
VIGS	Virus-Induced Gene Silencing
W/V	Weight by volume
WA	Water Agar medium
WP	Woody Plant

Index

C

D

E

L

M

N

O

P

Q

R

S

T

U